Applied Mathematics for Scientists and Engineers

Mathematicians, physicists, engineers, biologists, and other scientists who study related fields frequently use differential equations, linear algebra, calculus of variations, and integral equations. The purpose of *Applied Mathematics for Scientists and Engineers* is to provide a concise and well-organized study of the theoretical foundations for the development of mathematics and problem-solving methods. A wide range of solution strategies are shown for real-world challenges. The author's main objective is to provide as many examples as possible to help make the theory reflected in the theorems more understandable. The book's five chapters can be used to create a one-semester course as well as for self-study. The only prerequisites are a basic understanding of calculus and differential equations.

The five main topics include:

- Ordinary differential equations
- Partial differential equations
- Matrices and systems of linear equations
- Calculus of variations
- Integral equations

The author strikes a balance between rigor and presentation of very challenging content in a simple format by adopting approachable notations and using numerous examples to clarify complex themes. Exercises are included at the end of each section. They range from simple computations to more challenging problems.

Textbooks in Mathematics
Series editors:
Al Boggess, Kenneth H. Rosen

Abstract Algebra
A First Course, Second Edition
Stephen Lovett

Multiplicative Differential Calculus
Svetlin Georgiev, Khaled Zennir

Applied Differential Equations
The Primary Course
Vladimir A. Dobrushkin

Introduction to Computational Mathematics: An Outline
William C. Bauldry

Mathematical Modeling the Life Sciences
Numerical Recipes in Python and MATLAB™
N. G. Cogan

Classical Analysis
An Approach through Problems
Hongwei Chen

Classical Vector Algebra
Vladimir Lepetic

Introduction to Number Theory
Mark Hunacek

Probability and Statistics for Engineering and the Sciences with Modeling using R
William P. Fox and Rodney X. Sturdivant

Computational Optimization: Success in Practice
Vladislav Bukshtynov

Computational Linear Algebra: with Applications and MATLAB® Computations
Robert E. White

Linear Algebra With Machine Learning and Data
Crista Arangala

Discrete Mathematics with Coding
Hugo D. Junghenn

Applied Mathematics for Scientists and Engineers
Youssef N. Raffoul

https://www.routledge.com/Textbooks-in-Mathematics/book-series/CANDHTEX-BOOMTH

Applied Mathematics for Scientists and Engineers

Youssef N. Raffoul

CRC Press
Taylor & Francis Group
Boca Raton London New York

CRC Press is an imprint of the
Taylor & Francis Group, an **informa** business

A CHAPMAN & HALL BOOK

First edition published 2024
by CRC Press
6000 Broken Sound Parkway NW, Suite 300, Boca Raton, FL 33487-2742

and by CRC Press
4 Park Square, Milton Park, Abingdon, Oxon, OX14 4RN

CRC Press is an imprint of Taylor & Francis Group, LLC

ISBN: 978-1-032-58257-3 (hbk)
ISBN: 978-1-032-58394-5 (pbk)
ISBN: 978-1-003-44988-1 (ebk)

DOI: 10.1201/9781003449881

Typeset in Nimbus Roman font
by KnowledgeWorks Global Ltd.

Publisher's note: This book has been prepared from camera-ready copy provided by the authors.

Dedication
To my beautiful and adorable granddaughter
Aurora Jane Palmore

Contents

Preface

The author is very excited to share his book with you, and hopes you will find it beneficial in broadening your education and advancing your career. The main objective of this book is to give the reader a thorough understanding of the basic ideas and techniques of applied mathematics as they are employed in various engineering fields. Topics such as differential equations, linear algebra, the calculus of variations, and integral equations are fundamental to scientists, physicists, and engineers. The book emphasizes both the theory and its applications. It incorporates engineering applications throughout, and in line with that idea, derivations of the mathematical models of numerous physical systems are presented to familiarize the reader with the foundational ideas of applied mathematics and its applications to real-world problems.

For the last twenty-four years, the author has been teaching a graduate course in applied mathematics for graduate students majoring in mathematics, physics, and engineering at the University of Dayton. The course covered various topics in differential equations, linear algebra, calculus of variations, and integral equations. As a result, the author's lecture notes eventually became the basis for this book.

The book is self-contained, and no knowledge beyond an undergraduate course on ordinary differential equations is required. A couple of sections of Chapters 4 and 5 require knowledge of Fourier series. To make up for this deficiency, an appendix was added. The book should serve as a one-semester graduate textbook exploring the theory and applications of topics in applied mathematics. Educators have the flexibility to design their own three-chapter, one-semester course from Chapters 2–5. The first chapter is intended as a review on the subject of ordinary differential equations, and we refer to particular sections of it in later chapters when dealing with calculus of variations and integral equations. The author made every effort to create a balance between rigor and presenting the most difficult subject in an elementary language while writing the book in order to make it accessible to a wide variety of readers. The author's main objective was to provide as many examples as possible to help make the theory reflected in the theorems more understandable. The purpose of the book is to provide a concise and well-organized study of the theoretical foundations for the development of mathematics and problem-solving methods. This book's text is organized in a way that is both very readable and mathematically sound. A wide range of solution strategies are shown for a number of real-world challenges.

The author's presentational manner and style have a big impact on how this book develops mathematically and pedagogically. Some of the concepts from the extensive and well-established literature on many applied mathematics topics found their way into this book. Whenever possible, the author tried to deal with concepts in a more conversational way, copiously illustrated by 165 completely worked-out examples. Where appropriate, concepts and theories are depicted in 83 figures.

Exercises are a crucial component of the course's learning tool and are included at the end of each section. They range from simple computations to the solution of more challenging ones. Before starting the exercises, students must read the mathematics in the pertinent section. The book is divided into five chapters and an appendix.

Chapter 1 is a review of ordinary differential equations and is not intended to be formally covered by the instructor. It is recommended that students become acquainted with it before proceeding to the following chapters. The main reason for including Chapter 1 is that by the time students take a graduate course in applied mathematics, they have already forgotten most techniques for solving ordinary differential equations. In addition, it will save class time by not formally reviewing such topics but rather asking the students to read them beforehand. The chapter covers first-order and higher-order differential equations. It also includes a section on the Cauchy-Euler equation, which plays a significant role in Chapters 4 and 5.

The second chapter is devoted to the study of partial differential equations, with the majority of the content aimed toward graduate students pursuing engineering degrees. The chapter begins with linear equations with constant and variable coefficients and then moves on to quasi-linear equations. Burger's equation occupies an important role in the chapter, as do second-order partial differential equations and homogeneous and nonhomogeneous wave equations.

The third chapter discusses matrices and systems of linear equations. Gauss elimination, matrix algebra, vector spaces, and eigenvalues and eigenvectors are all covered. The chapter concludes with an examination of inner product spaces, diagonalization, quadratic forms, and functions of symmetric matrices.

Chapter 4 delves deeply into fundamental themes in the calculus of variations in a functional analytic environment. The calculus of variations is concerned with the optimization of functionals over a set of competing objects. We begin by deriving the Euler-Lagrange necessary condition and generalizing the concept to functionals with higher derivatives or with multiple variables. We provide a nice discussion on the theory behind sufficient conditions. Some of the topics are generalized to isoperimetric problems and functionals with constraints. Toward the end of the chapter, we closely examine the connection between the Sturm-Liouville problem and the calculus of variations. We end the chapter with the Rayleigh-Ritz method and the development of Euler-Lagrange to allow variational computation of multiple integrals.

Chapter 5 is solely devoted to the study of Fredholm and Volterra integral equations. The chapter begins by introducing integral equations and the connections between them and ordinary differential equations. The development of Green's function occupies an important role in the chapter. It is used to classify kernels, which in turn leads us to the appropriate approach for finding solutions. This includes integral equations with symmetric kernels or degenerate kernels. Toward the end of the chapter, we develop iterative methods and the Neumann series. We briefly discuss ways of approximating non-degenerate kernels and the use of the Laplace transform in solving integral equations of convolution types. Since not all integral equations can be reduced to differential equations, one should expect odd behavior from solutions. For such reasons, we devote the last section of the chapter to the qualitative analysis of solutions using fixed point theory and the Liapunov direct method.

Appendix A covers the basic topics of Fourier series. We briefly discuss Fourier series expansion, including sine and cosine, and the corresponding relations to periodic odd extension and periodic even extension. We provide applications to the heat problem in a finite slab by utilizing the concept of separation of variables. We transform the Laplacian equation in different dimensions to polar, cylindrical, and spherical coordinates. We end this appendix by studying the Laplacian equation in circular domains, such as the annulus. Materials in this section will be useful in several places in the book, especially Chapters 2, 4, and 5.

The author owes a debt of gratitude to Drs. Sam Brensinger and George Todd for reading the first and third chapters, respectively, and for their insightful remarks and recommendations. I'd like to express my gratitude to the hundreds of graduate students at the University of Dayton who helped the author polish and refine the lecture notes so that a significant portion of them made it into this book over the course of the last 22 years.

This book would not exist without the encouragement and support of my wife, my children Hannah, Paul, Joseph, and Daniel, and my brother Melhem.

Youssef N. Raffoul
University of Dayton
Dayton, Ohio
July, 2023

Author

Youssef N. Raffoul is Professor and Graduate Program Director at the University of Dayton. After receiving his PhD in mathematics from Southern Illinois University, he joined the faculty at Tougaloo College in Mississippi, serving as Department Chair. Prof. Raffoul has published 160 articles in prestigious journals in the areas of functional differential, difference equations, and dynamical systems on time scales. He is twice recipient of the University of Dayton College of Arts and Sciences' Award for Outstanding Scholarship as well as recipient of the University of Dayton Alumni Award in Scholarship. He was honored by the Lebanese government with the Career in Science Award. The Archbishop of Lebanon awarded him the Lifetime Achievement Award. Most notably, he is the recipient of the Order of Merit, Silver Medal with Distinction, presented to him by President General Michel Aoun.

1

Ordinary Differential Equations

In this chapter, we briefly go over elementary topics from ordinary differential equations that we will need in later chapters. The chapter provides the foundations to assist students in learning not only how to read and understand differential equations but also how to read technical material in more advanced-setting texts as they progress through their studies. We discuss basic topics in first-order differential equations, including separable and exact equations and the variation of parameters formula. We provide applications for infections. At the end of the chapter, we study higher-order differential equations and some of their theoretical aspects. The chapter is not intended to be taught as a part of a graduate course but rather as a reference for the students for later chapters.

1.1 Preliminaries

Let I be an interval of the real numbers $\mathbb{R}$ and consider the function $f : I \to \mathbb{R}$. For $x_0 \in I$, the *derivative* of f at x_0 is

$$f'(x_0) = \lim_{h \to 0} \frac{f(x_0 + h) - f(x_0)}{h} \tag{1.1}$$

provided the limit exists. When the limit exists, we say that f is *differentiable* at x_0. The term $f'(x_0)$ is the instantaneous rate of change of the function f at x_0. If x_0 is one of the endpoints of the interval I, then the above definition of the derivative becomes a one-sided derivative. If $f'(x_0)$ exists at every point $x_0 \in I$, then we say f is differentiable on I and write $f'(x)$. The derivative of a function f is again a function f'; its domain, which is a subset of the domain of f, is the set of all points x_0 for which f is differentiable. Other notations for the derivative are $D_x f, df/dx$, and dy/dx, where $y = f(x)$. The function f' may in turn have a derivative, denoted by f'', which is defined at all points where f' is differentiable. f'' is called the *second derivative* of f. For higher-order derivatives, we use the notations

$$f'''(x), \ f^{(4)}(x), \dots, f^{(n)}(x), \ \text{ or } \ \frac{d^n f}{dx^n} \ \text{ for } \ n = 1, 2, 3, \dots.$$

DOI: 10.1201/9781003449881-1

Example 1.1 For $x \in \mathbb{R}$, we set $f(x) = x|x|$. Then we have that $f(x) = x^2$, for $x > 0$ and $f(x) = -x^2$, for $x < 0$. Next, we compute $f'(x_0)$. For $x_0 > 0$, we may choose $|h|$ small enough so that $x_0 + h > 0$. Then by (1.1), we see that

$$
\begin{aligned}
f'(x_0) &= \lim_{h \to 0} \frac{f(x_0 + h) - f(x_0)}{h} = \lim_{h \to 0} \frac{(x_0 + h)^2 - x_0^2}{h} \\
&= \lim_{h \to 0} \frac{2x_0 h + h^2}{h} = 2x_0.
\end{aligned}
$$

On the other hand, if $x_0 < 0$, we may choose $|h|$ small enough so that $x_0 + h < 0$, and so from (1.1), we obtain

$$
\begin{aligned}
f'(x_0) &= \lim_{h \to 0} \frac{f(x_0 + h) - f(x_0)}{h} = \lim_{h \to 0} \frac{-(x_0 + h)^2 + x_0^2}{h} \\
&= \lim_{h \to 0} \frac{-2x_0 h - h^2}{h} = -2x_0.
\end{aligned}
$$

Finally,

$$
f'(0) = \lim_{h \to 0} \frac{f(0 + h) - f(0)}{h} = \lim_{h \to 0} \frac{h|h|}{h} = 0.
$$

In conclusion,

$$
f'(x) = 2|x| \quad \text{for all} \ x \in \mathbb{R}.
$$

$\square$

A *differential equation* is an equation involving a function and derivatives of this function. Differential equations are divided into two classes: ordinary and partial. *Ordinary differential equations* contain only functions of a single variable, called the independent variable, and derivatives with respect to that variable. *Partial differential equations* contain a function of two or more variables and some partial derivatives of this function.

The *order of a differential equation* is defined by the highest derivative present in the equation. An *nth-order ordinary differential equation* is a functional relation of the form

$$
F\left(x, y, \frac{dy}{dx}, \frac{d^2y}{dx^2}, \frac{d^3y}{dx^3}, \ldots, \frac{d^ny}{dx^n}\right) = 0, \ x \in \mathbb{R}, \tag{1.2}
$$

between the independent variable x and the dependent variable y, and its derivatives

$$
\frac{dy}{dx}, \frac{d^2y}{dx^2}, \frac{d^3y}{dx^3}, \ldots, \frac{d^ny}{dx^n}.
$$

We shall always assume that (1.2) can be solved for $y^{(n)}$ and put in the form

$$
y^{(n)} = f\left(x, y, y', \ldots, y^{(n-1)}\right). \tag{1.3}
$$

Loosely speaking, by a solution of (1.2) on an interval I, we mean a function $y(x) = \varphi(x)$ such that

$$
F\left(x, \varphi(x), \varphi'(x), \ldots, \varphi^{(n)}(x)\right)
$$

is defined for all $x \in I$ and

$$F\left(x, \varphi(x), \varphi'(x), \ldots, \varphi^{(n)}(x)\right) = 0$$

for all $x \in I$. If we require, for some initial time $x_0 \in \mathbb{R}$, a solution $y(x)$ to satisfy the initial conditions

$$y(x_0) = a_0, \quad y'(x_0) = a_1, \quad \ldots \quad , y^{n-1}(x_0) = a_{n-1}, \qquad (1.4)$$

for constants a_i, $i = 0, 1, 2, ..., n - 1$, then (1.3) along with (1.4) is called an initial value problem (IVP).

Before we state the next theorem, we define partial derivatives.

Given a function of several variables $f(x, y)$, the *partial derivative* of f with respect to x is the rate of change of f as x varies, keeping y constant, and it is given by

$$\frac{\partial f}{\partial x} = \lim_{h \to 0} \frac{f(x+h, y) - f(x, y)}{h}.$$

Similarly, the *partial derivative* of f with respect to y is the rate of change of f as y varies, keeping x constant, and it is given by

$$\frac{\partial f}{\partial y} = \lim_{h \to 0} \frac{f(x, y+h) - f(x, y)}{h}.$$

More often we write f_x, f_y, to denote $\frac{\partial f}{\partial x}$ and $\frac{\partial f}{\partial y}$, respectively.

For the (IVP) (1.3) and (1.4), the following existence and uniqueness result is true. For more discussion on the topic and on the proof of the next theorem, we refer to [3], [4], or [5].

Theorem 1.1 *Consider the (IVP) defined by (1.3) and (1.4), where f is continuous on the $(n+1)$-dimensional rectangle D of the form*

$$D = \{(x, y_0, y_1, \ldots, y_{n-1}) : b_{k-1} < y_{k-1} < d_{k-1} \text{ and } b_n < x < d_n, \ k = 1, 2, \ldots, n\}.$$

If the initial conditions are chosen so that the point $(x_0, a_0, a_1, \ldots, a_{n-1})$ is in D, then the (IVP) has at least one solution satisfying the initial conditions. If, in addition, f has continuous partial derivatives

$$\frac{\partial f}{\partial y}, \frac{\partial^2 f}{\partial y'}, \ldots, \frac{\partial^{n-1} f}{\partial y^{n-1}},$$

in D, then the solution is unique.

Suppose f satisfies the hypothesis of Theorem 1.1 in D. Then a general solution of the (IVP) in D is given by the formula

$$y = \varphi(x, c_1, c_2, \ldots, c_n)$$

if y solves (1.3) and if for any initial condition $(x_0, a_0, a_1, \ldots, a_{n-1})$ in D one can choose values $c_1, c_2, \ldots, c_n$ so that the solution y satisfies these initial conditions. We have the following corollary concerning existence and uniqueness of solutions of first-order initial value problems, which is an immediate consequence of Theorem 1.1.

Corollary 1 *Let $D \subset \mathbb{R} \times \mathbb{R}$, and denote the set of all real continuous functions on D by $C(D, \mathbb{R})$. Let $f \in C(D, \mathbb{R})$ and suppose $\frac{\partial f}{\partial x}$ is continuous on D. Then for any $(t_0, x_0) \in D$, the (IVP)*

$$y' = f(x, y), \quad y(x_0) = y_0,$$

has a unique solution on an interval containing x_0 in its domain.

Example 1.2 As an example, consider

$$y'(x) = xy^{1/2},$$

then

$$f(x, y) = xy^{1/2} \quad \text{and} \quad \frac{\partial f}{\partial y} = \frac{x}{2y^{1/2}}$$

are continuous in the upper half-plane defined by $y > 0$. We conclude from Corollary 1 that for any point (x_0, y_0), $y_0 > 0$, there is some interval around x_0 on which the given differential equation has a unique solution.   □

Example 1.3 Consider

$$y'(x) = y^2, \quad y(0) = 1.$$

Here, we have that

$$f(x, y) = y^2 \quad \text{and} \quad \frac{\partial f}{\partial y} = 2y$$

are continuous everywhere in the plane and in particular on the rectangle

$$D = \{(x, y) : -2 < x < 2, \ 0 < y < 2\}.$$

Since the initial point $(0, 1)$ lies inside the rectangle, Corollary 1 guarantees a unique solution of the (IVP).   □

In the next example, we illustrate the existence of more than one solution.

Example 1.4 Consider the differential equation

$$y'(x) = \frac{3}{2}y^{1/3}, \quad y(0) = 0, \ x \in \mathbb{R}.$$

Here, we have

$$f(x, y) = \frac{3}{2}y^{1/3} \quad \text{and} \quad \frac{\partial f}{\partial y} = \frac{1}{2}y^{-2/3}.$$

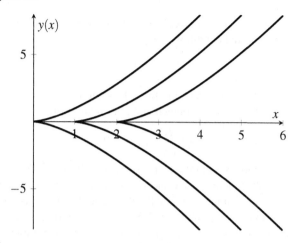

FIGURE 1.1
This example displays three solutions at $x_1 = 0, 1, 2$.

Clearly $f(x,y)$ is continuous everywhere , but the partial derivative is discontinuous when $y = 0$, and hence at the point $(0,0)$. This explains the existence of many solutions, as we display below. It is clear that $y(x) = 0$ is a solution. Hence, we may consider a solution $y_1(x) = 0$ and let

$$y_2(x) = \begin{cases} 0, & \text{for } x \leq 0 \\ x^{3/2}, & \text{for } x > 0 \end{cases}$$

which is also a solution that is continuous and differentiable. Likewise, for $x_1 > 0$ we have

$$y_3(x) = \begin{cases} 0, & \text{for } x \leq x_1 \\ (x-x_1)^{3/2}, & \text{for } x > x_1 \end{cases}.$$

Continuing in this way, we see that the differential equation has infinitely many solutions. Similarly, if y is a solution then $-y$ is also a solution (see Fig. 1.1). ☐

1.2   Separable Equations

Consider the first-order differential equation

$$y' = f(x,y) \tag{1.5}$$

where f and $\frac{\partial f}{\partial y}$ are continuous for all (x,y) in a rectangle D of the form

$$D = \{(x,y) : b_1 < x < b_2 \text{ and } d_1 < y < d_2\}.$$

In some cases, we may want to solve (1.5) and, at different times, we may want to solve it subject to the initial condition

$$y(x_0) = y_0, \tag{1.6}$$

where the point (x_0, y_0) is specified and in D. In some instances, the first-order differential equation (1.5) can be rearranged and put in the form,

$$g(y)\frac{dy}{dx} + h(x) = 0 \tag{1.7}$$

where $h, g : \mathbb{R} \to \mathbb{R}$ and are continuous on some subsets of $\mathbb{R}$. Here, the functions $h(x)$ and $g(y)$ are only functions of x and y, respectively. When (1.5) can be put in the form (1.7), we say it *separates* or that the differential equation is *separable*. Now we discuss the method of solution when (1.5) is separable. In this case, we consider (1.7) and assume that the functions $H(x)$ and $G(y)$ are the definite integrals of $h(x)$ and $g(y)$, respectively. Then (1.7) can be written in the form

$$H'(x)dx = -G'(y)dy.$$

By integrating the left-side with respect to x and the right-side with respect to y, we get the general solution

$$H(x) + G(y) = c, \tag{1.8}$$

where c is an arbitrary constant. If (1.8) permits us to solve for y in terms x and c, then we obtain a general solution of (1.7) of the form $y = \varphi(x, c)$. If we cannot solve for y in terms of x and c, then (1.8) represents an *implicit* solution of (1.7).

The technique used to obtain (1.8) is simple and easy to understand and apply. However, if we try to be precise and justify the procedure, some care would have to be implemented. To see this, we take the initial point (x_0, y_0) in D and suppose that

$$h(x_0) \neq 0 \text{ and } g(y_0) \neq 0.$$

Let $y = \varphi(x)$ be a solution of (1.7) on an interval $I = \{x : |x - x_0| < d\}$, which satisfies the initial condition $\varphi(x_0) = y_0$. Then for all x in I we see that

$$h(x) = -g(\varphi(x))\frac{d\varphi(x)}{dx}.$$

Integrating both sides from x_0 to x, we arrive at

$$\int_{x_0}^{x} h(s)ds = -\int_{x_0}^{x} g(\varphi(s))\varphi'(s)ds.$$

Since $\varphi'(x_0) = -\frac{h(x_0)}{g(\varphi(x_0))} = -\frac{h(x_0)}{g(y_0)} \neq 0$, then in the neighborhood of x_0 the solution $\varphi(x)$ is either increasing or decreasing. In either case, we are able to use the change of variable $u = \varphi(x)$ in the right-hand integral and obtain

$$\int_{x_0}^{x} h(s)ds = -\int_{\varphi(x_0)}^{\varphi(x)} g(u)du = -\int_{y_0}^{y} g(u)du, \tag{1.9}$$

which is equivalent to (1.8).

Example 1.5 Solve $y' = 5/2 - y/2$, $y(0) = 2$. The differential equation can be written as

$$\frac{1}{5-y}\frac{dy}{dx} = \frac{1}{2}.$$

Using (1.9), we obtain

$$\int_0^x \frac{1}{2}ds = \int_2^y \frac{1}{5-u}du.$$

For finite x, we must have $u < 5$; otherwise, the integral on the right diverges. Therefore, $5 - u > 0$ and integration yields

$$\frac{x}{2} = -\ln(5-y) + \ln(3),$$

or

$$\frac{x}{2} = \ln\left(\frac{3}{5-y}\right).$$

Taking the exponential of both sides, we arrive at the solution

$$y(x) = 5 - 3e^{-\frac{x}{2}}.$$

$\square$

Example 1.6 (Logistic Equation) The logistic equation is a simple model of population dynamics. Suppose, we have a population of size y with initial population size of $y_0 > 0$ that has a birth rate αy and a death rate βy. With this model, we obtain

$$\frac{dy}{dt} = (\alpha - \beta)y, \text{ which has the solution}$$

$$y = y_0 e^{(\alpha-\beta)t}.$$

Our population increases or decreases exponentially depending on whether the birth rate exceeds death rate or vice versa. However, in reality, there is fighting for limited resources. The probability of some piece of food (resource) being found is proportional to y. The probability of the same piece of food being found by two individuals is proportional to y^2. If food is scarce, they fight (to the death), so the death rate due to fighting (competing) is γy^2 for some γ. So

$$\frac{dy}{dt} = (\alpha - \beta)y - \gamma y^2$$

or

$$\frac{dy}{dt} = ry\left(1 - \frac{y}{d}\right),$$

where $r = \alpha - \beta$ and $d = r/\gamma$. This is the differential logistic equation. Note that it is separable and can be solved explicitly. $\square$

Example 1.7 The law of mass action is a useful concept that describes the behavior of a system that consists of many interacting parts, such as molecules, that react with each other, or viruses that are passed along from a population of infected individuals to non-immune ones. The law of mass action was derived first for chemical systems but subsequently found wide use in epidemiology and ecology. To describe the law of mass action, we assume m substances $s_1, s_2, \ldots, s_m$ together form a product with concentration p. Then the law of mass action states that $\frac{dp}{dt}$ is proportional to the product of the m concentrations $s_i, i = 1, \ldots, m$. That is,

$$\frac{dp}{dt} = k s_1 s_2 \ldots s_m.$$

Suppose we have a homogeneous population of fixed size, divided into two groups. Those who have the disease are called infective, and those who do not have the disease are called susceptible. Let $S = S(t)$ be the susceptible portion of the population and $I = I(t)$ be the infective portion. Then by assumption, we may normalize the population and have $S + I = 1$. We further assume that the dynamics of this epidemic satisfy the law of mass action. Hence, for some positive constant λ we have the nonlinear differential equation

$$I'(t) = \lambda S I \tag{1.10}$$

Let $I(0) = I_0$, $0 < I(0) < 1$ be a given initial condition. It follows that by substituting $S = 1 - I$ into (1.10),

$$I'(t) = \lambda I(1 - I), \quad I(0) = I_0. \tag{1.11}$$

If we can solve (1.11) for $I(t)$, then $S(t)$ can be found from the relation $I + S = 1$. We separate the variables in (1.11) and obtain

$$\frac{dI}{I(1 - I)} = \lambda \, dt.$$

Using partial fractions on the left side of the equation and then integrating both sides yields

$$\ln(|I|) - \ln(|1 - I|) = \lambda t + c,$$

or for some positive constant c_1 we have

$$I(t) = \frac{c_1 e^{\lambda t}}{1 + c_1 e^{\lambda t}}.$$

Applying $I(0) = I_0$ gives the solution

$$I(t) = \frac{I_0 e^{\lambda t}}{1 - I_0 + I_0 e^{\lambda t}}. \tag{1.12}$$

Now for $0 < I(0) < 1$, the solution given by (1.12) is increasing with time as expected. Moreover, using L'Hospital's rule, we have

$$\lim_{t \to \infty} I(t) = \lim_{t \to \infty} \frac{I_0 e^{\lambda t}}{1 - I_0 + I_0 e^{\lambda t}} = 1.$$

Hence, the infection will grow, and everyone in the population will get infected eventually. $\qquad \square$

1.2.1 Exercises

In Exercises 1.1, verify that the given expression is a solution of the equation. Where appropriate, c_1, c_2 denote constants.

Exercise 1.1 *(a)* $y'' + y' - 2y = 0$; $y = c_1 e^x + c_2 e^{-2x}$.

(b) $y' = 25 + y^2$; $y = 5\tan(5x)$.

(c) $y' = \sqrt{\dfrac{y}{x}}$; $y = (\sqrt{x} + c_1)^2$, $x > 0, c_1 > 0$.

(d) $3x^2 y\,dx + (x^3 y + 2y)dy = 0$; $x^3 y + y^2 = c_1$.

(e) $\dfrac{dy}{dx} = y(2 - 3y)$; $y = \dfrac{2c_1 e^{2x}}{1 + 3c_1 e^{2x}}$.

(f) $x^2 y'' - xy' + 2y = 0$; $y = x\cos(\ln x)$, $x > 0$.

In each of Exercises 1.2–1.7, decide whether the existence and uniqueness theorem or corollary of this section does or does not guarantee the existence of a solution of each of the initial value problems. In the case a solution exists, determine whether uniqueness is guaranteed or not and determine the region of existence and uniqueness.

Exercise 1.2
$$y' = 3x^2 y, \ \ y(0) = 3.$$

Exercise 1.3
$$y' = y^2, \ \ y(1) = 5.$$

Exercise 1.4
$$y' = \sqrt{x - y}, \ \ y(2) = 1$$

Exercise 1.5
$$y' = \ln(1 + y^2), \ \ y(2) = 2.$$

Exercise 1.6
$$y' = x\ln(y), \ \ y(1) = 0$$

Exercise 1.7
$$y' = \sqrt{1 - y^2}, \ \ y(0) = 0.$$

Exercise 1.8 *Show that the (IVP)*
$$y' = -\sqrt{4 - y^2}, \ \ y(0) = 2$$

has the two solutions $y_1(x) = 2$ *and* $y_2(x) = 2\cos(x)$ *on the interval* $[0, 2\pi]$. *Why doesn't this contradict Corollary 1?*

In 1.9–1.16, solve the given differential equation by separation of variables.

Exercise 1.9
$$y' + xy = y, \ \ y(1) = 3.$$

Exercise 1.10

$$(y^2 - 1)\frac{dy}{dx} = xe^x, \ y(0) = 5.$$

Exercise 1.11

$$\frac{dy}{dx} - 1 = \sqrt{x - y}.$$

Exercise 1.12

$$x\frac{dy}{dx} = y^2 - y, \ y(0) = 1.$$

Exercise 1.13

$$y'\tan(x) = y, \ y(\pi/2) = \pi/2.$$

Exercise 1.14

$$\frac{dy}{dx} + 2 = \sin(2x + y + 1).$$

Exercise 1.15

$$y' = 2x^3y^2 + 3x^2y^2, \ y(0) = 2.$$

Exercise 1.16

$$y' = \frac{x^3y - y}{y^4 - y^3 + 1}, \ y(0) = 1.$$

1.3   Exact Differential Equations

In this section, we look at a special type of differential equations called *exact differential equations*. Given a function $f(x,y)$, the *total derivative* of f is given by

$$df = f_x dx + f_y dy.$$

Definition 1.1 (Exact equation) $Q(x,y)\frac{dy}{dx} + P(x,y) = 0$ *is an* exact equation *if and only if the differential form* $Q(x,y)\,dy + P(x,y)\,dx$ *is exact, i.e. there exists a function* $f(x,y)$ *for which*

$$df = Q(x,y)\,dy + P(x,y)\,dx.$$

If $P(x,y)\,dx + Q(x,y)\,dy$ is an exact differential of f, then $df = P(x,y)\,dx + Q(x,y)\,dy$. But by the chain rule, $df = \frac{\partial f}{\partial x}dx + \frac{\partial f}{\partial y}dy$ and this equality holds for any displacements dx, dy. So

$$\frac{\partial f}{\partial x} = P, \quad \frac{\partial f}{\partial y} = Q.$$

From this, we have

$$\frac{\partial^2 f}{\partial y \partial x} = \frac{\partial P}{\partial y}, \quad \frac{\partial^2 f}{\partial x \partial y} = \frac{\partial Q}{\partial x}.$$

We know that the two mixed second derivatives are equal. So

$$\frac{\partial P}{\partial y} = \frac{\partial Q}{\partial x}.$$

The converse is not necessarily true. Even if this equation holds, the differential need not be exact. However, it is true if the domain is *simply-connected*.

Definition 1.2 (Simply-connected domain) *A domain $\mathscr{D}$ is simply-connected if it is connected and any closed curve in $\mathscr{D}$ can be shrunk to a point in $\mathscr{D}$ without leaving $\mathscr{D}$.*

Example 1.8 A disc in 2D is simply-connected. A disc with a "hole" in the middle is not simply-connected because a loop around the hole cannot be shrunk into a point. Similarly, a sphere in 3D is simply-connected but a torus is not. □

Theorem 1.2 *If $\frac{\partial P}{\partial y} = \frac{\partial Q}{\partial x}$ throughout a simply-connected domain $\mathscr{D}$, then*

$$P(x,y)\,dx + Q(x,y)\,dy = 0 \tag{1.13}$$

is an exact differential of a single-valued function in $\mathscr{D}$.

If the equation is exact, then the solution is simply $f = $ constant, and we can find f by integrating $\frac{\partial f}{\partial x} = P$ and $\frac{\partial f}{\partial y} = Q$.

Example 1.9 Consider

$$\bigl(y\cos(xy) + 1\bigr)dx + \bigl(x\cos(xy) + e^y\bigr)dy = 0.$$

We have

$$P = y\cos(xy) + 1, \quad Q = x\cos(xy) + e^y$$

Then $\frac{\partial P}{\partial y} = \frac{\partial Q}{\partial x} = \cos(xy) - xy\sin(xy)$. Hence, we are dealing with an exact equation. To find the function f we integrate either

$$\frac{\partial f}{\partial x} = y\cos(xy) + 1, \text{ or } \frac{\partial f}{\partial y} = x\cos(xy) + e^y.$$

Integrating the first equation gives

$$f(x,y) = \sin(xy) + x + g(y).$$

Note that since it was a partial derivative with respect to x holding y constant, the "constant" term can be any function of y. Differentiating the derived f with respect to y, we have

$$\frac{\partial f}{\partial y} = x\cos(xy) + g'(y) = x\cos(xy) + e^y.$$

Thus $g'(y) = e^y$ and $g(y) = e^y$. The constant of integration need not be included in the preceding line since the solution is $f(x,y) = c$. This is due to the fact that $df(x,y) = 0$ implies that $f(x,y) = $ constant. Thus, the final solution is

$$\sin(xy) + x + e^y = c.$$

If we are given an initial condition, say $y(\pi/2) = 1$, then $c = 1 + \pi/2 + e$. □

1.3.1   Integrating factor

In some cases, it is possible to convert a first-order differential equation of the form (1.13) that is not exact into an exact one. In such a scenario, it is possible to multiply (1.13) by a function $\mu(x,y)$, called the integrating factor. Thus we make the following definition.

Definition 1.3 *An integrating factor $\mu(x,y)$ of (1.13) is a function that makes*

$$\mu(x,y)P(x,y)\,dx + \mu(x,y)Q(x,y)\,dy = 0 \tag{1.14}$$

an exact differential equation. That is

$$\frac{\partial\mu(x,y)P}{\partial y} = \frac{\partial\mu(x,y)Q}{\partial x}.$$

We have the following theorem.

Theorem 1.3 *Equation (1.13) has*

(a) an integrating factor of x alone, $\mu(x) = e^{f(x)}$ if and only if

$$\frac{1}{Q}\left(P_y - Q_x\right) = f(x), \tag{1.15}$$

(b) an integrating factor of y alone, $\mu(y) = e^{g(y)}$ if and only if

$$\frac{1}{P}\left(Q_x - P_y\right) = g(y). \tag{1.16}$$

Proof *Suppose (1.15) holds with $\mu(x) = e^{f(x)}$. Then*

$$
\begin{aligned}
\frac{\partial(\mu Q)}{\partial x} &= \frac{\partial\mu}{\partial x}Q + \mu\frac{\partial Q}{\partial x} \\
&= \frac{\mu}{Q}(P_y - Q_x)Q + \mu\frac{\partial Q}{\partial x} \quad \left(\text{since } \mu'(x) = \mu(x)f(x)\right) \\
&= \mu\frac{\partial P}{\partial y} \\
&= \frac{\partial(\mu P)}{\partial y}.
\end{aligned}
$$

The rest of the proof will be in the Exercises.

Example 1.10 Consider the differential equation

$$6xy\,dx + (4y + 9x^2)dy = 0, \quad y > 0$$

which is not exact. Now

$$\frac{1}{Q}\left(P_y - Q_x\right) = \frac{1}{4y + 9x^2}\left(6x - 18x\right),$$

which is not a function of one variable alone. However,

$$\frac{1}{P}(Q_x - P_y) = \frac{1}{6xy}(18x - 6x) = \frac{2}{y}$$

and so by (b) of Theorem 1.3, the equation has the integrating factor

$$\mu(y) = e^{\int \frac{2}{y}dy} = y^2.$$

Hence the new differential equation

$$y^2\left[6xydx + (4y + 9x^2)dy\right] = 0$$

is exact and has the solution

$$3x^2y^3 + y^4 = c.$$

$\square$

1.3.2 Exercises

In Exercises 1.17–1.22 show the differential equation is exact and then find its solution.

Exercise 1.17

$$(3x^2y + y)dx + (x^3 + x + 1 + 2y)dy = 0, \quad y(0) = 2.$$

Exercise 1.18

$$(e^x \sin(y) + 3y)dx + (3x + e^x \cos(y))dy = 0.$$

Exercise 1.19

$$(6xy - y^3)dx + (4y + 3x^2 - 3xy^2)dy = 0, \quad y(1) = -1.$$

Exercise 1.20

$$(x^3 + \frac{y}{x})dx + (y^2 + \ln(x))dy = 0.$$

Exercise 1.21

$$(x + \arctan(y))dx + \frac{x+y}{1+y^2}dy = 0.$$

Exercise 1.22

$$(y^2 \cos(x) - 3x^2y - 4x)dx + (2y\sin(x) - x^3 + \ln(y) + 1)dy = 0, \quad y(0) = e.$$

Exercise 1.23 *Suppose $\mu(x)$ is a function of only x and is an integrating factor of $P(x,y)dx + Q(x,y)dy = 0$. Show that $\frac{1}{Q}(P_y - Q_x)$ is a function of x alone.*

Exercise 1.24 *Prove part (b) of Theorem 1.3.*

Exercise 1.25 *Solve the given differential equation by finding an appropriate integrating factor.*

(a) $(xy + y^2 + y)dx + (x + 2y)dy = 0$.

(b) $(2y^2 + 3x)dx + 2xydy = 0$.

(c) $(y\ln(y) + ye^x)dx + (x + y\cos(y))dy = 0$.

(d) $(4xy^2 + y)dx + (6y^3 - x)dy = 0$.

Exercise 1.26 *For appropriate integers r and q show that $\mu(x, y) = x^r y^q$ is an integrating factor for the differential equation and then solve it,*

$$(3y^2 + 10xy)dx + (5xy + 12x^2)dy = 0.$$

1.4 Linear Differential Equations

Consider the differential equation

$$y'(x) + a(x)y(x) = g(x, y(x)), \quad y(x_0) = y_0, \ x \geq x_0 \qquad (1.17)$$

where $g \in C(\mathbb{R} \times \mathbb{R}, \mathbb{R})$ and $a \in C(\mathbb{R}, \mathbb{R})$. Note that (1.17) can be nonlinear or linear, which depends on the function g. If $g(x, y) = g(x)$, a function of x alone then (1.17) is said to be linear (linear in y.) To obtain a formula for the solution, we multiply both sides of (1.17) by the integrating factor $e^{\int_{x_0}^{x} a(u)du}$. Observing that

$$\frac{d}{dt}\left(y(x)e^{\int_{x_0}^{x} a(u)du}\right) = y'(x)e^{\int_{x_0}^{x} a(u)du} + a(x)y(x)e^{\int_{x_0}^{x} a(u)du},$$

we arrive at

$$\frac{d}{dx}\left(y(x)e^{\int_{x_0}^{x} a(u)du}\right) = g(x, y(x))e^{\int_{x_0}^{x} a(u)du}.$$

An integration of the above expression from x_0 to x and using $y(x_0) = y_0$ yields

$$y(x)e^{\int_{x_0}^{x} a(u)du} = y_0 + \int_{x_0}^{x} g(s, y(s))e^{\int_{x_0}^{s} a(u)du}ds$$

from which we get

$$y(x) = y_0 e^{-\int_{x_0}^{x} a(u)du} + \int_{x_0}^{x} g(s, y(s))e^{-\int_{s}^{x} a(u)du}ds, \ x \geq x_0. \qquad (1.18)$$

It can be easily shown that if $y(x)$ satisfies (1.18), then it satisfies (1.17). Expression (1.18) is known as the variation of parameters formula. We note that (1.18) is a functional equation in y since the integrand is a function of y. If we replace the function g with a function $h(x)$ where $h \in C(\mathbb{R}, \mathbb{R})$, then (1.18) takes the special form

$$y(x) = y_0 e^{-\int_{x_0}^x a(u)du} + \int_{x_0}^x h(s)e^{-\int_s^x a(u)du}ds, \quad x \ge x_0. \tag{1.19}$$

Another special form of (1.18) is that, if the function $a(x)$ is constant for all $x \ge x_0$ and g is replaced with $h(x)$ as before, then we have from (1.19) that

$$y(x) = y_0 e^{-a(x-x_0)} + \int_{x_0}^x e^{-a(x-s)}h(s)ds, \quad x \ge x_0. \tag{1.20}$$

It is easy to compute, using (1.20) that the differential equation

$$y'(x) - 3y = e^{2x}, \quad y(0) = 3$$

has the solution

$$y(x) = 4e^{3x} - e^{2x}.$$

Remark 1 *If no initial conditions are assigned, then* (1.18) *takes the form*

$$y(x) = e^{-\int a(x)dx}\left(C + \int g(x, y(x))e^{\int a(x)dx}dx\right).$$

1.4.1 Exercises

Exercise 1.27 *Solve each of the given differential equation.*

(a) $x\dfrac{dy}{dx} + 2y = 5xy, \quad y(1) = 0.$

(b) $\dfrac{dy}{dx} + 2y = e^x, \quad y(1) = 0.$

(c) $(x+1)\dfrac{dy}{dx} + (x+2)y = 2xe^{-x}.$

(d) $\dfrac{dy}{dx} + \dfrac{e^x}{1+x^2} = y, \quad y(1) = 0.$

(e) $x\dfrac{dy}{dx} = y\ln(x), \quad y(1) = 2.$

Exercise 1.28 *Find a continuous solution satisfying*

$$y'(x) + y(x) = f(x), \quad y(0) = 0$$

where $f(x) = \begin{cases} 1, & 0 \le x \le 1 \\ 0, & x > 1. \end{cases}$

Is the solution differentiable at $x = 1$?

Exercise 1.29 *Find a continuous solution satisfying*

$$y'(x) + 2xy(x) = f(x), \quad y(0) = 2$$

where $f(x) = \begin{cases} x, & 0 \le x < 1 \\ 0, & x \ge 1. \end{cases}$

Is the solution differentiable at $x = 1$?

1.5 Homogeneous Differential Equations

In this section we look at first-order differential equations that can be written in the form

$$P(x,y)\, dx + Q(x,y)\, dy = 0. \tag{1.21}$$

We begin with the following definition.

Definition 1.4 *A function* $g(x,y)$ *of two variables is called homogeneous of degree* α *if for all* x, y *where* g *is defined and for all positive constants* λ *we have*

$$g(\lambda x, \lambda y) = \lambda^\alpha g(x,y). \tag{1.22}$$

For example the function $h(x,y) = xy + y^2$ is homogeneous of degree $\alpha = 2$, since

$$h(\lambda x, \lambda y) = \lambda^2 (xy + y^2) = \lambda^2 h(x,y).$$

On the other hand, $h(x,y) = xy + y^2 + 1$ is not homogeneous of any degree α since

$$h(\lambda x, \lambda y) = \lambda^2 (xy + y^2) + 1 \ne \lambda^\alpha h(x,y),$$

for any α.

Definition 1.5 *The differential equation* (1.21) *is called a homogeneous differential equation if both* P *and* Q *satisfy condition* (1.22) *for the same* α.

Method of finding solutions

Suppose (1.21) is a homogeneous differential equation. Then you may use either transformation

(a) $y = ux$ or

(b) $x = vy$

where u and v are continuous and are functions of the variable x. The choice of either (a) or (b) depends on the number of terms that multiply P and Q. For example, if P is multiplied by fewer terms, then go with (b). Similarly, if Q is multiplied by fewer terms, then go with (a). You may use either (a) or (b) if they are multiplied by the same number of terms.

Say we go with $y = ux$. Then compute $\dfrac{dy}{dx} = x\dfrac{du}{dx} + u$. Multiplying both sides by dx we get

$$dy = x\, du + u\, dx. \tag{1.23}$$

Next substitute $y = ux$ and (1.23) into (1.21) and the resulting differential equation is separable in x and u and can be easily solved. If we go with (b), then use

$$dx = y\, dv + v\, dy \tag{1.24}$$

and then substitute back into (1.21) to obtain a separable equation in terms of v and y.

Example 1.11 Consider

$$x\frac{dy}{dx} + x - 3y = 0. \tag{1.25}$$

Then the equation (1.25) takes the form

$$xdy + (x - 3y)dx = 0,$$

which is homogeneous of degree 1. Since dy is multiplied by one term only, we use $y = ux$. Using the above procedure, the differential equation reduces to

$$x(udx + xdu) + (x - 3xu)dx = 0.$$

Simplifying by x and then regrouping we arrive at the separable equation

$$\frac{dx}{x} = \frac{du}{2u - 1}.$$

Integrating both sides and then substituting $u = y/x$ we arrive at the solution of the original problem

$$\ln|x| = \frac{1}{2}\ln\left|\frac{2y}{x} - 1\right| + c,$$

for some constant c. $\qquad\qquad\square$

Notice that (1.25) can be written as

$$\frac{dy}{dx} = 1 + 3\frac{y}{x} = f\left(1, \frac{y}{x}\right),$$

and this hints at another way of defining homogeneous differential equations. Thus we make another alternate definition.

Definition 1.6 *A differential equation*

$$\frac{dy}{dx} = g(x, y) \tag{1.26}$$

is called a homogeneous differential equation if g is a homogeneous function of degree 0, that is

$$g(\lambda x, \lambda y) = \lambda^0 g(x, y) = g(x, y).$$

If we let $\lambda = \dfrac{1}{x}$, then

$$g(\lambda x, \lambda y) = \lambda^0 g(x,y) = g(x,y) = g(1, \tfrac{y}{x}).$$

Hence, if we let $F(\tfrac{y}{x}) = g(1, \tfrac{y}{x})$, then (1.26) can be put in the form

$$\frac{dy}{dx} = F(\tfrac{y}{x}). \tag{1.27}$$

This suggests making the substitution $u = \dfrac{y}{x}$. In this case the differential equation given by (1.27) is reduced to

$$\frac{du}{dx} = \frac{F(u) - u}{x},$$

which is separable.

Another type of first-order differential equation that requires transformations in both the dependent and independent variables is of the form

$$\frac{dy}{dx} = \frac{ax + by + c}{dx + ey + g}, \tag{1.28}$$

with $ae \neq bd$. To solve (1.28) we propose the substitutions

$$x = t - p, \quad y = w - k,$$

where the constants p and k must be carefully chosen so that the resulting equation is homogeneous. Using the chain rule, we see that

$$\frac{dy}{dx} = \frac{d}{dx}(w - k) = \frac{dw}{dx} = \frac{dw}{dt}\frac{dt}{dx} = \frac{dw}{dt}(1) = \frac{dw}{dt}.$$

Moreover,

$$ax + by + c = at + bw + (c - ap - bk).$$

Similarly,

$$dx + ey + g = dt + ew + (g - dp - ek).$$

In order for the resulting equation to be homogeneous we require

$$c - ap - bk = 0, \quad g - dp - ek = 0,$$

which has a unique solution p and k since $ae \neq bd$. This results in the new differential equation

$$\frac{dw}{dt} = \frac{at + bw}{dt + ew}, \tag{1.29}$$

which is homogeneous in w and t.

Example 1.12 Consider

$$\frac{dy}{dx} = \frac{x+2y-5}{-2x-y+4}.$$

It is clear that $ae \neq bd$. We use the substitutions

$$x = t - p, \quad y = w - k$$

and solve for the unique solution of

$$-5 - p - 2k = 0, \quad 4 + 2p + k = 0$$

and obtain $p = -1, k = -2$. Using (1.29) we arrive at the homogeneous differential equation

$$(2t + w)dw + (t + 2w)dt = 0.$$

This equation can be easily solved by letting $w = ut$, and obtaining the separable differential equation

$$\frac{dt}{t} = -\frac{2+u}{u^2 + 4u + 1}du.$$

An integration of both sides gives

$$\ln|t| = -\frac{1}{2}\ln|u^2 + 4u + 1| + C.$$

Or,

$$\ln|x-1| = -\frac{1}{2}\ln\left|\frac{(y-2)^2}{(x-1)^2} + 4\frac{y-2}{x-1} + 1\right| + C.$$

$\square$

1.5.1 Exercises

In Exercises 1.30–1.34 Show the given differential equations are homogeneous and solve them.

Exercise 1.30

$$xdy + (2y + 5x)dx = 0, \quad y(1) = 2.$$

Exercise 1.31

$$x\frac{dy}{dx} - 3y = \frac{y^2}{x}, \quad y(1) = 5.$$

Exercise 1.32

$$x^2\frac{dy}{dx}xy = x^2 + y^2, \quad y(1) = 2.$$

Exercise 1.33

$$2xy\frac{dy}{dx} = 4x^2 + 3y^2.$$

Exercise 1.34

$$x\frac{dy}{dx} = y + \sqrt{x^2 - y^2}, \quad y(1) = 0.$$

Exercise 1.35 *Solve*

$$\frac{dy}{dx} = \frac{4x - 3y + 13}{x - y + 3}, \quad y(0) = 1.$$

Exercise 1.36 *Solve*

$$\frac{dy}{dx} = -\frac{4x + 3y + 11}{2x + y + 5}.$$

Exercise 1.37 *Find a substitution that reduces the differential equation*

$$\frac{dy}{dx} = f(ax + by + c)$$

to a separable equation, where a, b and c are constants, with $b \neq 0$.

Exercise 1.38 *Use Exercise 1.37 to solve*

$$\frac{dy}{dx} = -1 + \frac{x + y + 1}{\ln(x + y + 1)}, \quad x + y + 1 > 0.$$

1.6   Bernoulli Equation

In the previous section we used substitutions to solve first-order differential equations. The same intuition can be applied to the Bernoulli differential equation

$$y' + P(x)y = Q(x)y^n. \tag{1.30}$$

Note that if $n = 0$, then (1.30) is a linear differential equation and it can be solved by the method of Section 1.4. Similarly, if $n = 1$, then (1.30) is a separable differential equation that can be solved by the method of Section 1.2. Thus, we consider (1.30) only for

$$n \neq 0, 1.$$

We make the substitution

$$W = y^{1-n},$$

so that

$$y' = \frac{1}{n-1} W^{\frac{n}{1-n}} W'.$$

Substituting into (1.30) we arrive at the linear differential equation in W and x

$$W' + (1 - n)P(x)W = (1 - n)Q(x) \tag{1.31}$$

that can be now solved by the method of Section 1.4.

Example 1.13 Consider

$$x\frac{dy}{dx} - y = e^x y^3, \quad x > 0.$$

Then the differential equation is Bernoulli with $n = 3$, $P(x) = -\frac{1}{x}$, and $Q(x) = \frac{e^x}{x}$. Thus, by (1.31) we have the new linear differential equation

$$W' + \frac{2}{x}W = -2\frac{e^x}{x},$$

which has the solution

$$W = cx^{-2} - 2\frac{e^x}{x} + 2\frac{e^x}{x^2}.$$

Since $y = W^{-1/2}$, the solution of the Bernoulli equation is given by

$$y = \left(cx^{-2} - 2\frac{e^x}{x} + 2\frac{e^x}{x^2}\right)^{-1/2}.$$

□

1.6.1   Exercises

Exercise 1.39 *Find a general solution of the Bernoulli equation.*

(a) $3(1+x^2)\dfrac{dy}{dx} = 2xy(y^3 - 1)$.

(b) $y^3\dfrac{dy}{dx} + \dfrac{y^4}{x} = \dfrac{\cos(x)}{x^4}$.

(c) $(x^2+1)\dfrac{dy}{dx} + 3x^3y = 6xe^{-3/x^2}$.

(d) $x\dfrac{dy}{dx} + 2y + \sin(x)\sqrt{y} = 0$.

(e) $\dfrac{dy}{dx} + 2y = x^3y^2\sin(x)$.

1.7   Higher-Order Differential Equations

Now is the time to consider higher-order linear differential equations. In particular, we consider the general nth order linear differential equation

$$a_n(x)\frac{d^n y}{dx^n} + a_{n-1}(x)\frac{d^{n-1}y}{dx^{n-1}} + \ldots + a_1(x)\frac{dy}{dx} + a_0(x)y = F(x). \tag{1.32}$$

Unless otherwise noted, we always assume that the coefficients $a_i(x)$, $i = 1, 2, \ldots, n$ and the function $F(x)$ are continuous on some open interval I. The interval I may be unbounded. If the function $F(x)$ vanishes for all $x \in I$, then we call (1.32) a *homogeneous* linear equation; otherwise, it is *nonhomogeneous*. Thus the *homogeneous* linear equation associated with (1.32) is

$$a_n(x)\frac{d^n y}{dx^n} + a_{n-1}(x)\frac{d^{n-1} y}{dx^{n-1}} + \ldots + a_1(x)\frac{dy}{dx} + a_0(x)y = 0. \qquad (1.33)$$

Definition 1.7 Let f_1, f_2, ..., f_n be set of functions defined on an interval I. We say the set $\{f_1, f_2, \ldots, f_n\}$ is linearly dependent on I if there exists constants $c_1, c_2, \ldots, c_n$ not all zero, such that

$$c_1 f_1 + c_2 f_2 + \ldots + c_n f_n = 0$$

for every $x \in I$. If the set of functions is not linearly dependent on the interval I, it is said to be linearly independent.

Definition 1.8 (Fundamental set of solutions). A set of n solutions of the linear differential system (1.33) all defined on the same open interval I, is called a fundamental set of solutions on I if the solutions are linearly independent functions on I.

We have the following corollary.

Corollary 2 Let the coefficients $a_i(x)$, $i = 1, 2, \ldots, n$ be continuous on an interval I. If $\{\phi_1(x), \phi_2(x), \ldots, \phi_n(x)\}$ form a fundamental set of solutions on I, then the general solution of (1.33) is given by

$$y(x) = c_1 \phi_1(x) + c_2 \phi_2(x) + \ldots + c_n \phi_n(x),$$

for constants c_i, $i = 1, 2, \ldots, n$.

Example 1.14 The second-order differential equation

$$y''(x) + y'(x) - 2y(x) = 0, \qquad (1.34)$$

has the two solutions $\varphi_1(x) = e^x$ and $\varphi_2(x) = e^{-2x}$ and they are linearly independent on $I = (-\infty, \infty)$ and hence the fundamental solution is

$$y(x) = c_1 \varphi_1(x) + c_2 \varphi_2(x) = c_1 e^x + c_2 e^{-2x},$$

for constants c_1 and c_2. $\qquad \square$

To completely describe the solution of either (1.32) or (1.33), we impose the initial conditions

$$y(x_0) = d_0, \quad y'(x_0) = d_1, \quad \ldots, \quad y^{(n-1)}(x_0) = d_{n-1}, \qquad (1.35)$$

for an initial point x_0, and constants d_i, $i = 0, 1, 2, \ldots, n-1$,

Theorem 1.4 [Existence and Uniqueness] Consider the (IVP) defined by (1.32) and (1.35), where F and $a_i(x)$, $i = 1, 2, \ldots, n$ are continuous on some open interval I. Assume $a_n(x) \neq 0$ for all $x \in I$. For $x_0 \in I$ and given the constants d_i, $i = 0, 1, 2, \ldots, n-1$, the differential equation (1.32) has a unique solution on the entire interval I satisfying the initial condition (1.35).

Another way of determining whether a set of functions is linearly independent or not is to look at the *Wronskian*.

Definition 1.9 (Wronskian) *Given two functions f and g, the Wronskian of f and g is the determinant*

$$W = \begin{vmatrix} f & g \\ f' & g' \end{vmatrix} = fg' - f'g.$$

We write $W(f,g)$ to emphasize the functions. Consider the two functions given in Example 1.14. Then

$$W(\varphi_1, \varphi_2) = \begin{vmatrix} e^x & e^{-2x} \\ e^x & -2e^{-2x} \end{vmatrix} = -2e^{-x} - e^{-x} = -3e^{-x} \neq 0, \text{ for all } x \in (-\infty, \infty).$$

This is an example of a linearly independent pair of functions. Note that the *Wronskian* is everywhere nonzero. On the other hand, if the functions f and g are linearly dependent, with $g = kf$ for a nonzero constant k, then

$$W(f,g) = \begin{vmatrix} f & kf \\ f' & kf' \end{vmatrix} = kff' - kff' = 0.$$

Thus the Wronskian of two linearly dependent functions is zero. This will be made formal in Theorem 1.5. The above Wronskian discussion can be easily extended to the set of functions $f_1, f_2, \ldots, f_n$, where

$$W(f_1, f_2, \ldots, f_n) = \begin{vmatrix} f_1 & f_2 & \cdots & f_n \\ f_1' & f_2' & \cdots & f_n' \\ \vdots & \vdots & \ddots & \vdots \\ f_1^{(n-1)} & f_2^{(n-1)} & \cdots & f_n^{(n-1)} \end{vmatrix}$$

For better illustration, we consider the second-order differential equation

$$y'' + b_1(x)y' + b_0(x)y = 0, \tag{1.36}$$

where the functions b_0 and b_1 are continuous on some fixed interval I.

Theorem 1.5 *Suppose y_1 and y_2 are solutions of (1.36) on the interval I. Then y_1 and y_2 are linearly independent if and only if*

$$W(y_1(x), y_2(x)) \neq 0, \quad \text{for all } x \in I.$$

Proof *Let y_1 and y_2 be linearly independent. Then,*

$$c_1 y_1(x) + c_2 y_2(x) = 0, \quad \text{for all } x \in I,$$

is true only when $c_1 = c_2 = 0$. Then a simple differentiation yields

$$c_1 y_1(x) + c_2 y_2(x) = 0 \text{ and } c_1 y_1'(x) + c_2 y_2'(x) = 0 \tag{1.37}$$

for all $x \in I$. Now (1.37) can only be true for some c_1, and c_2 not both zero if and only if $W(y_1(x), y_2(x)) = 0$, for all $x \in I$. If (1.37) holds for some point $x_0 \in I$, then the function $y = c_1 y_1 + c_2 y_2$ is a solution (1.36) and satisfies the initial conditions, $y(x_0) = y'(x_0) = 0$. On the other hand the zero function; $y = 0$ is also a solution and satisfies the initial conditions. This violates the uniqueness of the solution unless y and the zero solution, $y = 0$ are the same. Now $y = 0$ implies (1.37) is true for all $x \in I$. This shows that $W(y_1(x), y_2(x)) = 0$, for all $x \in I$ if and only if $W(y_1(x), y_2(x)) = 0$, for at least one $x_0 \in I$. This completes the proof.

Next, we define the general solution of the nonhomogeneous differential equation (1.32). The proof of the next theorem will be left as an exercise.

Theorem 1.6 *Let the coefficients $a_i(x)$, $i = 1, 2, \ldots, n$ and the function $F(x)$ be continuous on an interval I. Suppose $\{\phi_1(x), \phi_2(x), \ldots, \phi_n(x)\}$ form a fundamental set of solutions on I of the homogeneous differential equation (1.33), Denote such solution with*

$$y_h(x) = c_1 \phi_1(x) + c_2 \phi_2(x) + \ldots + c_n \phi_n(x),$$

for constants c_i, $i = 1, 2, \ldots, n$. Let $y_p(x)$ be a particular solution of the nonhomogeneous differential equation (1.32). Then the general solution of (1.32) on I is given by

$$y(x) = y_h(x) + y_p(x).$$

Example 1.15 Consider the second-order differential equation

$$y'' + 3y' + 2y = 6, \quad y(0) = 1, \quad y'(0) = -3. \tag{1.38}$$

Clearly, each of the functions $\varphi_1(x) = e^{-x}$, and $\varphi_2(x) = e^{-2x}$ is a solution of the homogeneous equation $y'' + 3y' + 2y = 0$. Also, they are linearly independent since

$$W(\varphi_1, \varphi_2) = \begin{vmatrix} e^{-x} & e^{-2x} \\ -e^{-x} & -2e^{-2x} \end{vmatrix} = -3e^{-3x} \neq 0, \text{ for all } x \in (-\infty, \infty).$$

Thus, the homogeneous solution of $y'' + 3y' + 2y = 0$ is given by

$$y_h(x) = c_1 e^{-x} + c_2 e^{-2x},$$

for constants c_1 and c_2. Moreover, $y_p(x) = 3$ is a particular solution of (1.38), since

$$y_p'' + 3y_p' + 2y_p = 6.$$

Thus, the general solution of (1.38) is

$$y(x) = y_h(x) + y_p(x) = c_1 e^{-x} + c_2 e^{-2x} + 3.$$

Applying the initial conditions we arrive at $c_1 = -7$, and $c_2 = 5$.

Later on in the chapter we will look at different techniques for finding the particular solution $y_p(x)$.   □

1.7.1 Exercises

Exercise 1.40 *Use Definition 1.7 to show that for any nonzero constant r the set*

$$\{e^{rx}, xe^{rx}, x^2e^{rx}, \ldots, x^ne^{rx}\}$$

is linearly independent.

Exercise 1.41 *Decide whether or not the solutions given determine a fundamental set of solutions for the equation.*

(a) $y''' - 3y'' - y' + 3y = 0$, $y_1 = e^{3x} + e^x$, $y_2 = e^x - e^{-x}$, $y_3 = e^{3x} + e^{-x}$.

(b) $y''' - 2y'' - y' + 2y = 0$, $y_1 = e^x$, $y_2 = e^{-x}$, $y_3 = e^{2x}$.

(c) $x^2y''' + xy'' - y' = 0$, $y_1 = 1 + x^2$, $y_2 = 2 + x^2$, $y_3 = \ln(x)$.

(d) $x^3y''' + 2x^2y'' + 3xy' - 3y = 0$, $y_1 = x$,
$y_2 = \cos(\sqrt{3}\ln(x))$, $y_3 = \sin(\sqrt{3}\ln(x))$, $x > 0$.

1.8 Equations with Constant Coefficients

We consider the nth order differential equation with constant coefficients

$$a_ny^{(n)}(x) + a_{n-1}y^{(n-1)}(x) + \ldots + a_2y''(x) + a_1y'(x) + a_0y(x) = 0 \qquad (1.39)$$

and try to find its solution. In the previous section we noticed that solutions to some of the differential equations that were considered, were exponential functions. Thus, we search for solutions of (1.39) of the form

$$y = e^{rx},$$

where r is a parameter to be determined. We begin with the observation that

$$\frac{d^k}{dx^k}(e^{rx}) = r^k e^{rx} \quad \text{for} \quad k = 0, 1, 2, \ldots, \qquad (1.40)$$

A substitution of (1.40) into (1.39) leads to

$$e^{rx}\left(a_nr^n + a_{n-1}r^{n-1} + \ldots + a_2r^2 + a_1r + a_0\right) = 0.$$

Since, $e^{rx} \neq 0$ for any finite r or x, we must have that

$$a_nr^n + a_{n-1}r^{n-1} + \ldots + a_2r^2 + a_1r + a_0 = 0. \qquad (1.41)$$

Equation (1.41) is referred to as the *characteristic equation* or the *auxiliary equation*.

Distinct Roots

Now suppose the roots of (1.41) can be found. Then we can always write the fundamental solution or general solution of (1.39). The easiest case is when all the roots r_i, $i = 1, 2, \ldots, n$ are real and distinct. That is, no two roots are the same, or

$$r_i \neq r_j, \quad i, j = 1, 2, \ldots, n.$$

We have the following theorem.

Theorem 1.7 (Distinct Real Roots) *Suppose all the roots of (1.41) r_i, $i = 1, 2, \ldots, n$ are real and distinct. Then the general solution of (1.39) is given by*

$$y = \sum_{k=1}^{n} c_k e^{r_k x}, \tag{1.42}$$

for constants c_k, $k = 1, 2, \ldots, n$.

Proof *Since the roots are real and distinct, the set $\{e^{r_k x}, k = 1, 2, \ldots, n\}$ is linearly independent. Moreover, each function in the set is a solution of (1.39) and hence they form a fundamental set of solutions. Then, by Theorem 1.6, the solution is given by (1.42).*

Example 1.16 Consider the third order differential equation

$$y''' + 2y'' - y' - 2y = 0.$$

Its characteristic equation is found to be $r^3 + 2r^2 - r - 2 = 0$, which factors into $(r^2 - 1)(r + 2) = 0$. Thus the three roots are $-2, -1$, and 1 and they are real and distinct and hence by Theorem 1.7 the general solution is

$$y = c_1 e^{-2x} + c_2 e^{-x} + c_3 e^x.$$

$\square$

Repeated Roots

Now we turn our attention to the case when the characteristic equation (1.41) has some of its roots repeated. In such cases, we are not able to produce n linearly independent solutions using Theorem 1.7. For example, if the characteristic equation of a given differential equation has the roots -1, 1, 2, and 2, then we can only produce the three linearly independent functions e^{-x}, e^x, and e^{2x}. The problem is then to find a way to obtain the linearly independent solutions. To that end we introduce the symbol $\mathcal{L}$ to represent a linear operator in the sense that for functions y_1 and y_2 in an appropriate space

$$\mathcal{L}(c_1 y_1 + c_2 y_2) = c_1 \mathcal{L} y_1 + c_2 \mathcal{L} y_2, \quad \text{for constants } c_1, c_2.$$

Thus, in terms of the operator $\mathcal{L}$, equation (1.39) can take the form $\mathcal{L} y = 0$, where

$$\mathcal{L} = a_n \frac{d^n}{dx^n} + a_{n-1} \frac{d^{n-1}}{dx^{n-1}} + \ldots + a_1 \frac{d}{dx} + a_0. \tag{1.43}$$

In addition, we introduce the term $D = \dfrac{d}{dx}$ and hence the notations,

$$Dy = y', \quad D^2y = y'', \quad \ldots, \quad D^ny = y^{(n)}.$$

For example, if y is a function of x and d is a constant,

$$(D-d)y = Dy - dy = y' - dy.$$

Thus, in terms of the operator $\mathscr{L}$, and D we may rewrite (1.43)

$$\mathscr{L} = a_nD^n + a_{n-1}D^{n-1} + \ldots + a_1D + a_0. \tag{1.44}$$

Now suppose that (1.41) has a simple root r_0 and another root r_1 with multiplicity k, where k is an integer such that $k > 1$. Then by Exercise 1.42, equation (1.44) reduces to

$$\mathscr{L} = (D-r_1)^k(D-r_0) = (D-r_0)(D-r_1)^k. \tag{1.45}$$

Then setting (1.45) equal to zero, which corresponds to the differential $\mathscr{L}y = 0$, we arrive at the two solutions $y_0 = e^{r_0x}$, and $y_1 = e^{r_1x}$. Remember, we need to find $k+1$ linearly independent solutions for the construction of the general solution. Thus, there are $k-1$ missing linearly independent solutions. Applying y to the operator in (1.45), yields

$$\mathscr{L}y = (D-r_0)(D-r_1)^ky = (D-r_0)\left[(D-r_1)^ky\right].$$

By setting $\mathscr{L}y = 0$ we arrive at

$$(D-r_1)^ky = 0. \tag{1.46}$$

Every solution of the kth order differential equation in (1.46) will also be a solution of the original differential equation $\mathscr{L}y = 0$. Since e^{r_1x} is already a known solution, we search for other solutions of the form

$$y(x) = u(x)e^{r_1x}, \tag{1.47}$$

where the function u is to be determined. Using the product rule we obtain

$$(D-r_1)\left(u(x)e^{r_1x}\right) = (Du(x))e^{r_1x} + r_1u(x)e^{r_1x} - r_1u(x)e^{r_1x} = (Du(x))e^{r_1x}.$$

Or,

$$(D-r_1)\left(u(x)e^{r_1x}\right) = (Du(x))e^{r_1x}.$$

Applying $u(x)e^{r_1x}$ to (1.46) and by an induction argument on k (see Exercise 1.43) it can be shown that

$$(D-r_1)^k\left(u(x)e^{r_1x}\right) = (D^ku(x))e^{r_1x}.$$

Thus $y = u(x)e^{r_1 x}$ is a solution of (1.46) if and only if $(D^k u(x))e^{r_1 x} = 0$. But this holds if and only if $(D^k u(x)) = 0$. Since $(D^k u(x)) = u^k(x) = 0$, the solution is

$$u(x) = c_0 + c_1 x + c_2 x^2 + \ldots + c_k x^{k-1},$$

which is a polynomial of degree at most $k - 1$. We arrived at the following theorem.

Theorem 1.8 (Repeated Roots) *Suppose a root r of the characteristic equation (1.41) has multiplicity $k > 1$. Then the root r contributes to the general solution of (1.39) the term*

$$\left(c_0 + c_1 x + c_2 x^2 + \ldots + c_k x^{k-1} \right) e^{rx},$$

for constants c_i, $i = 0, 1, 2, \ldots, k$.

To see the set

$$\{ e^{rx}, \ xe^{rx}, \ x^2 e^{rx}, \ \ldots, \ x^n e^{rx} \}$$

is linearly independent, we refer to Exercise 1.40.

Example 1.17   Consider the fourth order differential equation

$$y^{(4)} - 7y^{(3)} + 18y'' - 20y' + 8y = 0.$$

Its characteristic equation is found to be $r^4 - 7r^3 + 18r^2 - 20r + 8 = 0$, which factors into $(r - 2)^3 (r - 1) = 0$. Thus, we have a simple root 1 and another root 2 of multiplicity 3. Now by Theorem 1.8 the root 2 contributes the term $\left(c_0 + c_1 x + c_2 x^2 \right) e^{2x}$. Consequently, the general solution is

$$y = \left(c_0 + c_1 x + c_2 x^2 \right) e^{2x} + c_3 e^x.$$

$\square$

Complex Roots

We discuss the situation when one of the roots is complex. That is, if (1.41) has a simple complex root then, it appears in complex conjugate pairs $\alpha \pm i\beta$, where α, and β are real and $i = \sqrt{-1}$. Recall

$$e^t = \sum_{n=0}^{\infty} \frac{t^n}{n!} = 1 + t + \frac{t^2}{2!} + \frac{t^3}{3!} + \frac{t^n}{n!} + \ldots.$$

If we let $t = ix$, then the above series becomes

$$
\begin{aligned}
e^{ix} &= \sum_{n=0}^{\infty} \frac{(ix)^n}{n!} = 1 + ix - \frac{x^2}{2!} - \frac{ix^3}{3!} + \frac{x^4}{4!} + \frac{ix^5}{5!} - \ldots \\
&= \left(1 - \frac{x^2}{2!} + \frac{x^4}{4!} - \ldots \right) + i\left(x - \frac{x^3}{3!} + \frac{x^5}{5!} + \ldots \right) \\
&= \cos(x) + i\sin(x).
\end{aligned}
$$

From Euler's formula we know that

$$e^{(\alpha \pm i\beta)x} = e^{\alpha x}\Big(\cos(\beta x) \pm i\sin(\beta x)\Big),$$

and moreover, by Exercise 1.44 that for any complex number r,

$$De^{rx} = re^{rx}.$$

For emphasis, e^{rx} will be a solution of the differential equation given by (1.39) if and only if r is a root of its characteristic equation given by (1.41). Thus, if the conjugate compelx pair of roots $r_1 = \alpha + i\beta$, and $r_2 = \alpha - i\beta$, are simple (nonrepeated), then the corresponding part of the general solution is

$$
\begin{aligned}
K_1 e^{(\alpha+i\beta)x} + K_2 e^{(\alpha-i\beta)x} &= K_1 e^{\alpha x}\Big(\cos(\beta x) + i\sin(\beta x)\Big) \\
&+ K_2 e^{\alpha x}\Big(\cos(\beta x) - i\sin(\beta x)\Big) \\
&= e^{\alpha x}\Big(c_1 \cos(\beta x) + c_2 \sin(\beta x)\Big),
\end{aligned}
$$

where $c_1 = K_1 + K_2$, and $c_2 = (K_1 - K_2)i$. It is easy to verify that $e^{\alpha x}\cos(\beta x)$, and $e^{\alpha x}\sin(\beta x)$, are linearly independent. As a consequence we have the following theorem.

Theorem 1.9 (Complex Simple Roots) *Suppose the characteristic equation (1.41) has a nonrepeated pair of complex conjugate roots $\alpha \pm i\beta$. Then the corresponding part of the general solution of (1.39) is*

$$e^{\alpha x}\Big(c_1 \cos(\beta x) + c_2 \sin(\beta x)\Big)$$

for constants c_1, and c_2.

The next example summarizes all three cases.

Example 1.18 Suppose the characteristic equation of a given differential equation is found to be

$$(r^2 - 2r + 5)(r^2 - 9)(r + 2)^2 = 0.$$

We are interested in finding its general solution. The term $r^2 - 2r + 5 = 0$, has the pairs of complex conjugate roots $1 - 2i$, and $1 + 2i$, and therefore its contribution to the general solution is given by $e^x\Big(c_1 \cos(2x) + c_2 \sin(2x)\Big)$. Similarly, $r^2 - 9 = 0$ makes the contribution $c_3 e^{3x} + c_4 e^{-3x}$. Finally, $c_5 e^{-2x} + c_6 x e^{-2x}$ is the contribution corresponding to $(r + 2)^2 = 0$. Hence, the general solution is

$$
\begin{aligned}
y &= e^x\Big(c_1 \cos(2x) + c_2 \sin(2x)\Big) + c_3 e^{3x} + c_4 e^{-3x} \\
&+ c_5 e^{-2x} + c_6 x e^{-2x}.
\end{aligned}
$$

$\square$

1.8.1   Exercises

Exercise 1.42 *Show that for constants a, and b and a function y(x) that is differentiable,*

$$(D-a)(D-b)y = (D-b)(D-a)y.$$

Exercise 1.43 *Use an induction argument to show that*

$$(D-r_1)^k \left(u(x)e^{r_1 x} \right) = (D^k u(x))e^{r_1 x}.$$

Exercise 1.44 *Show that for any complex number r,*

$$De^{rx} = re^{rx}.$$

In Exercises 1.45- 1.46 the characteristic equation of a certain differential equation is given. Find the corresponding general solution.

Exercise 1.45

$$(r^2 + 4)(r^2 - 9)^2(r+2)^2 = 0.$$

Exercise 1.46

$$(r^2 + 4)^2(r^2 + 3r + 2)^3(r+5)^2 = 0.$$

In Exercises 1.47- 1.51 solve the given differential equation.

Exercise 1.47

$$y'' + y' - 2y = 0, \ \ y(0) = 1, \ \ y'(0) = 4.$$

Exercise 1.48

$$y''' - y'' - 4y' + 4y = 0,$$

Exercise 1.49

$$y^{(4)} - y''' + y'' - 3y' + 5y = 0.$$

Exercise 1.50

$$y^{(4)} + 2y''' + 3y'' + 2y' + y = 0.$$

Exercise 1.51

$$y''' + 10y'' + 25y' = 0, \ \ y(0) = 3, \ \ y'(0) = 4, \ y''(0) = 5.$$

1.9   Nonhomogeneous Equations

In this section we consider the nonhomogeneous *n*th order differential equation with constant coefficients

$$a_n y^{(n)}(x) + a_{n-1} y^{(n-1)}(x) + \ldots + a_2 y''(x) + a_1 y'(x) + a_0 y(x) = f(x), \qquad (1.48)$$

and the associated homogeneous equation

$$a_n y^{(n)}(x) + a_{n-1} y^{(n-1)}(x) + \ldots + a_2 y''(x) + a_1 y'(x) + a_0 y(x) = 0, \qquad (1.49)$$

where $f(x)$ is continuous on some interval I. In terms of the operator $\mathscr{L}$ Equations (1.48) and (1.49) take the form

$$\mathscr{L}y = f, \quad \mathscr{L}y = 0,$$

respectively. Let y_p be a given particular solution of (1.48). Then $\mathscr{L}y_p = f$. In addition, assume z is any other solution of (1.48). Then $\mathscr{L}z = f$, too. Due to the linearity of $\mathscr{L}$ we have that

$$\mathscr{L}(y_p - z) = \mathscr{L}y_p - \mathscr{L}z = f - f = 0.$$

Thus, $y_h = z - y_p$ is a solution of the associated homogeneous equation given by (1.49). It follows from Theorem 1.6 that,

$$y_h(x) = c_1 \phi_1(x) + c_2 \phi_2(x) + \ldots + c_n \phi_n(x),$$

for constants c_i, $i = 1, 2, \ldots, n$, where the functions $\varphi_i(x)$, $i = 1, 2, \ldots, n$ are linearly independent solutions of (1.49). Finding the particular solution y_p depends on two things:

(a) The type of function $f(x)$ in (1.48), and

(b) the nature of the homogeneous solution y_h of (1.49).

The method of this section only applies to functions $f(x)$ that are polynomial in x, combinations of *sine* or *cosine*, exponentials in x or combinations of the aftermentioned forms of $f(x)$. We illustrate the idea by displaying a few examples.

Example 1.19 The differential equation

$$y'' - 3y' + 2y = 4$$

has the homogenous solution $y_h(x) = c_1 e^x + c_2 e^{2x}$. Since $f(x) = 4$, is a constant we consider a particular solution of the form $y_p = A$, where A is to be determined. Substituting y_p into the differential equation and solving for A gives, $A = 2$. Thus the general solution is

$$y(x) = y_h(x) + y_p = c_1 e^x + c_2 e^{2x} + 2.$$

$\square$

Example 1.20 For

$$y'' - 3y' + 2y = 2e^{3x}$$

we have $y_h(x) = c_1 e^x + c_2 e^{2x}$. Since $f(x) = 2e^{3x}$, we consider a particular solution of the form $y_p = Ae^{3x}$, where A is to be determined. Substituting y_p into the differential equation we arrive at the relation $2Ae^{3x} = 2e^{3x}$. This gives $A = 1$, and hence the general solution is

$$y(x) = y_h(x) + y_p = c_1 e^x + c_2 e^{2x} + e^{3x}.$$

$\square$

Example 1.21 The equation

$$y'' - 3y' + 2y = 2e^x$$

has $y_h(x) = c_1 e^x + c_2 e^{2x}$. Since $f(x) = e^x$, we consider a particular solution of the form $y_p = Ae^x$, where A is to be determined. Substituting y_p into the differential equation we arrive at the relation $(A - 3A + 2A)e^x = 2e^x$. Or, $0e^x = 2e^x$, which can only imply that

$$0 = 2.$$

So what went wrong? Well, we said in the beginning that $f(x)$ depends on the form of y_h too. Let's start over. Now $f(x) = e^x$, and so we try a particular solution of the form $y_p = Ae^x$. But the term e^x is already present in y_h and so we try to multiply Ae^x by x. Thus, we end up with the particular solution of the form $y_p = Axe^x$. Substituting y_p into the differential equation we arrive at $A = -2$, and hence the general solution is

$$y(x) = y_h(x) + y_p = c_1 e^x + c_2 e^{2x} - 2xe^x.$$

$\square$

The table below provides guidance on how to construct y_p in the case that none of the terms present in y_h are parts of the forcing function $f(x)$.

1.9.1   Exercises

In Exercises 1.52–1.53 the characteristic equation and the forcing function of a certain differential equation are given. Write down the particular solution without solving for the coefficients.

Exercise 1.52

$$(r^2 + 4)(r^2 - 9)^2 (r + 2)^2 = 0; \quad f(x) = \sin(2x) + e^{-5x} + 10.$$

Exercise 1.53

$$r(r + 4)(r^2 + 3r + 2)^3 (r + 5)^2 = 0; \quad f(x) = \cos(2x) + xe^{-5x} + 1.$$

solve the differential equations.

Exercise 1.54 *Solve each of the given differential equations.*

(a) $y'' + y' + y = e^x \cos(2x) - 2e^x \sin(2x)$.

(b) $y'' - 5y' + 6y = xe^x \cos(2x)$.

TABLE 1.1
Shows how to find y_p.

$f(x)$	y_p
Constant C	A
e^{ax}	Ae^{ax}
Cx^n, $n = 0, 1, 2, \ldots$	$A_0 + A_1 x + A_2 x^2 + \ldots + A_n x^n$
$\cos(bx)$, or $\sin(bx)$	$A_1 \cos(bx) + A_2 \sin(bx)$
$x^n \cos(bx)$, or $x^n \sin(bx)$	$(A_0 + A_1 x + A_2 x^2 + \ldots + A_n x^n) \cos(bx)$ $+ (B_0 + B_1 x + B_2 x^2 + \ldots + B_n x^n) \sin(bx)$
$x^n e^{ax}$	$(A_0 + A_1 x + A_2 x^2 + \ldots + A_n x^n) e^{ax}$

(c) $y^{(4)} + 5y'' + 4y = \sin(x) + \cos(2x)$.

(d) $y^{(4)} - 2y'' + y = x^2 \cos(x)$.

(e) $y''' - y'' - 12y' = x - 2xe^{-3x}$.

1.10   Wronskian Method

Suppose we are to solve the differential equation

$$y'' + 4y = \sec(2x). \tag{1.50}$$

Then the method of Section 1.8 is not of much help here in constructing the particular solution y_p. This is due to the fact that $f(x) = \sec(2x)$ does not fit any of the forms given in Table 1.1. To find y_p we use the *Method of variation of constants* that we call here *the Wronskian method*. It is a general method and applies to any linear equation whether the coefficients are constants or not. Thus, we begin by considering the general linear second-order differential equation

$$y'' + P(x)y' + Q(x)y = f(x), \tag{1.51}$$

where the functions $P(x)$, $Q(x)$, and $f(x)$ are all continuous on some interval I. Assume $y_1(x)$, and $y_2(x)$ are known solutions on the interval I of the corresponding homogeneous equation

$$y'' + P(x)y' + Q(x)y = 0. \tag{1.52}$$

Then the homogeneous solution of (1.52) is

$$y_h(x) = c_1 y_1 + c_2 y_2,$$

for constants c_1, and c_2. Assume a particular solution $y_p(x)$ of (1.51) of the form

$$y_p(x) = u_1(x)y_1(x) + u_2(x)y_2(x) \tag{1.53}$$

where the functions u_1, and u_2 are to be found and continuous on the interval I. For the rest of this section we suppress the independent variable x in (1.53). Differentiating (1.53) with respect to x we obtain

$$y'_p = u'_1 y_1 + u_1 y'_1 + u'_2 y_2 + u_2 y'_2.$$

We assume u_1 and u_2 satisfy the natural condition

$$u'_1 y_1 + u'_2 y_2 = 0. \tag{1.54}$$

Substituting y_p and y'_p into (1.51) and making use of the fact that $y_1(x)$, and $y_2(x)$ are known solutions of the corresponding homogeneous equation (1.52), we arrive at the relation

$$u'_1 y'_1 + u'_2 y'_2 = f(x). \tag{1.55}$$

Solving (1.54) and (1.55) by using the process of elimination we get

$$u'_1 = \frac{f(x) y_2}{y_2 y'_1 - y_1 y'_2} \quad \text{and } u'_2 = \frac{f(x) y_1}{y_1 y'_2 - y_2 y'_1}. \tag{1.56}$$

Using Wronskian notations, and hence the name of this section, we arrive at the easy formulae to remember

$$u'_1 = \frac{W_1}{W} \quad \text{and} \quad u'_2 = \frac{W_2}{W}, \tag{1.57}$$

where

$$W = \begin{vmatrix} y_1 & y_2 \\ y'_1 & y'_2 \end{vmatrix}, \quad W_1 = \begin{vmatrix} 0 & y_2 \\ f(x) & y'_2 \end{vmatrix}, \quad \text{and} \quad W_2 = \begin{vmatrix} y_1 & 0 \\ y'_1 & f(x) \end{vmatrix}.$$

Before we go for an example, we briefly discuss how the method can be extended to nonhomogenous nth order differential equations of the form

$$y^{(n)}(x) + P_{n-1}(x) y^{(n-1)}(x) + \ldots + P_2(x) y''(x) + P_1(x) y'(x) + P_0(x) y(x) = f(x). \tag{1.58}$$

If its corresponding homogeneous equation has the homogeneous solution $y_h(x) = c_1 y_1 + c_2 y_2 + \ldots + c_n y_n$, then the particular solution y_p is of the form

$$y_h(x) = u_1 y_1 + u_2 y_2 + \ldots + u_n y_n,$$

where

$$u'_i = \frac{W_i}{W}, \quad i = 1, 2, \ldots, n.$$

Here,

$$W(y_1, y_2, \ldots, y_n) = \begin{vmatrix} y_1 & y_2 & \cdots & y_n \\ y'_1 & y'_2 & \cdots & y'_n \\ \vdots & \vdots & \ddots & \vdots \\ y_1^{(n-1)} & y_2^{(n-1)} & \cdots & y_n^{(n-1)} \end{vmatrix}$$

and W_i is the determinant obtained by replacing the ith column of the Wronskian by

$$\begin{pmatrix} 0 \\ 0 \\ \vdots \\ f(x) \end{pmatrix}.$$

We provide the following example.

Example 1.22 We consider (1.50). First, the homogeneous solution is

$$y_h = c_1 \sin(2x) + c_2 \cos(2x),$$

and so we may set $y_1 = \sin(2x)$, and $y_2 = \cos(2x)$. Moreover,

$$W = \begin{vmatrix} \sin(2x) & \cos(2x) \\ 2\cos(2x) & -2\sin(2x) \end{vmatrix} = -2, \ W_1 = \begin{vmatrix} 0 & \cos(2x) \\ \sec(2x) & -2\sin(2x) \end{vmatrix} = -1$$

and

$$W_2 = \begin{vmatrix} \sin(2x) & 0 \\ 2\cos(2x) & \sec(2x) \end{vmatrix} = \tan(2x).$$

Thus,

$$u_1' = \frac{1}{2}, \quad u_2' = -\frac{1}{2}\tan(2x).$$

An integration gives

$$u_1 = \frac{x}{2}, \quad u_2 = \frac{1}{4}\ln\left(\cos(2x)\right).$$

Hence

$$y_p = \frac{x}{2}\sin(2x) + \frac{1}{4}\ln\left(\cos(2x)\right)\cos(2x).$$

Finally, the general solution is $y = y_h + y_p$. □

1.10.1 Exercises

In Exercises 1.55- 1.60 solve the given differential equation.

Exercise 1.55
$$y'' + 9y = \csc(3x).$$

Exercise 1.56
$$y'' + 2y' + y = \frac{\ln(x)}{e^x}, \quad x > 0.$$

Exercise 1.57
$$y'' + y' + 2y = e^{-x}\sin(2x).$$

Exercise 1.58
$$y'' - y = \frac{2e^x}{e^x + e^{-x}}.$$

Exercise 1.59

$$y'' - 6y' + 9y = \frac{1}{x}, \quad x > 0.$$

Exercise 1.60

$$e^{4x}\left(y'' + 8y' + 16y\right) = \frac{1}{x^2}, \quad x > 0.$$

1.11 Cauchy-Euler Equation

We end this chapter by looking at Cauchy-Euler equations. An nth order homogeneous Cauchy-Euler equation is of the form

$$a_n x^n y^{(n)}(x) + a_{n-1}x^{n-1}y^{(n-1)}(x) + \ldots + a_2 x^2 y''(x) + a_1 xy'(x) + a_0 y(x) = 0, \quad x > 0$$

where a_j, $j = 1, 2, \ldots, n$, are constants. Euler equations are important since they pop up in many applications and partial differential equations. In addition, they make their presence in Chapter 4. We concentrate on the second-order Cauchy-Euler homogeneous equation

$$ax^2 y'' + dxy' + ky = 0,$$

and write it in the form

$$y'' + \frac{b}{x}y' + \frac{c}{x^2}y = 0, \quad x > 0. \tag{1.59}$$

We note that the coefficients of (1.59) are continuous everywhere except at $x = 0$. However, we shall consider the equation over the interval $(0, \infty)$. We solve the Cauchy-Euler equation (1.59) by making the substitution $x = e^t$, or equivalently $t = \ln(x)$. Once we rewrite (1.59) in terms of the new variables y, and t, then it is possible to use the method of Section 1.8 to find its general solution. Let

$$x = e^t, \text{ or equivalently } t = \ln(x).$$

Then,

$$\frac{dy}{dx} = \frac{dy}{dt}\frac{dt}{dx} = \frac{dy}{dt}\frac{1}{x} = e^{-t}\frac{dy}{dt}.$$

Moreover,

$$\begin{aligned}
y'' = \frac{d^2y}{dx^2} &= \frac{d}{dx}\left(\frac{dy}{dx}\right) = e^{-t}\frac{d}{dt}\left(e^{-t}\frac{dy}{dt}\right) \\
&= e^{-t}\left[-e^{-t}\frac{dy}{dt} + e^{-t}\frac{d^2y}{dt^2}\right] \\
&= -e^{-2t}\frac{dy}{dt} + e^{-2t}\frac{d^2y}{dt^2}.
\end{aligned}$$

Substituting into (1.59) and noting that $x^2 = e^{2t}$, we arrive at the second-order differential equation

$$\frac{d^2y}{dt^2} + (b-1)\frac{dy}{dt} + cy = 0. \tag{1.60}$$

We remark that on the interval $(-\infty, 0)$ we make the substitution $|x| = e^t$, or equivalently, $x = -e^t$, which will again reduce (1.59) to (1.60). Thus, once (1.60) is solved, we use the inverse substitution $t = \ln|x|$ to obtain solutions of the original equation (1.59). Recall the three cases that we discussed in Section 1.8.

(a) **(Distinct roots)** Let r_1 and r_2 be the two distinct roots of the auxiliary equation of (1.60). Then, the general solution is given by

$$y(t) = c_1 e^{r_1 t} + c_2 e^{r_2 t} = c_1(e^t)^{r_1} + c_2(e^t)^{r_2}.$$

Letting $t = \ln|x|$, we obtain the general equation of our original equation to be

$$y(x) = c_1|x|^{r_1} + c_2|x|^{r_2}.$$

(b) **(Repeated roots)** Let r be a repeated root of the auxiliary equation of (1.60). Then, the general solution is given by

$$y(t) = c_1 e^{rt} + c_2 t e^{rt} = c_1(e^t)^r + c_2 t(e^t)^r.$$

Letting $t = \ln|x|$, we obtain the general solution of our original equation to be

$$y(x) = c_1|x|^r + c_2|x|^r \ln|x|$$

for constants c_1, and c_2.

(c) **(Complex roots)** Let r be a complex root of the auxiliary equation of (1.60) that has a pair of complex conjugate roots $\alpha \pm i\beta$, then the solution is

$$y(t) = e^{\alpha t}\left(c_1 \cos(\beta t) + c_2 \sin(\beta t)\right)$$

for constants c_1, and c_2. Letting $t = \ln|x|$, we obtain the general solution of our original equation to be

$$y(x) = |x|^{\alpha}\left(c_1 \cos(\beta \ln|x|) + c_2 \sin(\beta \ln|x|)\right).$$

Example 1.23 Consider

$$x^2 y'' - xy' + y = \ln(x), \quad x > 0.$$

We need to put the equation in the standard form, by dividing with x^2 and arrive at

$$y'' - \frac{1}{x}y' + \frac{1}{x^2}y = \frac{\ln(x)}{x^2}, \quad x > 0.$$

Next we find y_h of

$$y'' - \frac{1}{x}y' + \frac{1}{x^2}y = 0, \quad x > 0.$$

Here $b = -1$ and $c = 1$. The auxiliary equation is

$$r^2 - 2r + 1 = 0,$$

with the repeated root $r = 1$. Thus, the homogeneous solution is

$$y_h = c_1 x + c_2 x \ln(x).$$

To find y_p we use the method of Section 1.50. Let $y_1 = x$, and $y_2 = x \ln(x)$. Set $f(x) = \frac{\ln(x)}{x^2}$. Then

$$W = x, \quad W_1 = -\frac{\ln^2(x)}{x}, \quad \text{and} \quad W_2 = \frac{\ln(x)}{x}.$$

As a consequence, we have

$$u_1' = \frac{W_1}{W} = -\frac{\ln^2(x)}{x^2} \quad \text{and} \quad u_2' = \frac{W_2}{W} = \frac{\ln(x)}{x^2}.$$

We carry out the integrations since they require techniques that students may benefit from. Let $u = \ln(x)$. Then

$$u_1 = -\int \frac{\ln^2(x)}{x^2} dx = -\int \frac{\ln^2(x)}{x} \frac{1}{x} dx = -\int u^2 e^{-u} du.$$

After an integration by parts twice in a row we arrive at

$$u_1 = u^2 e^{-u} + 2u e^{-u} + 2e^{-u} = \frac{\ln^2(x)}{x} + \frac{2\ln(x)}{x} + \frac{2}{x}.$$

Similarly, if we let $u = \ln(x)$, then

$$u_2 = \int \frac{\ln(x)}{x^2} dx = \int \frac{\ln(x)}{x} \frac{1}{x} dx = \int u e^{-u} du.$$

An integration yields,

$$u_2 = -u e^{-u} - e^{-u} = -\frac{\ln(x)}{x} - \frac{1}{x}.$$

Then, after some calculations we arrive at

$$y_p = u_1 y_1 + u_2 y_2 = 2 + \ln(x).$$

Finally,

$$y = y_h + y_p = c_1 x + c_2 x \ln(x) + 2 + \ln(x),$$

is the general solution.   □

1.11.1 Exercises

Exercise 1.61 *Solve each of the given differential equation.*

(a) $x^2y'' - 4xy' + 6y = 0$, $y(-2) = 8$, $y'(-2) = 0$.

(b) $x^2y'' - xy' + y = 4x\ln(x)$, $x > 0$.

(c) $x^2y'' + xy' + y = \sec(\ln(x))$, $x > 0$.

(d) $x^2y'' + 3xy' + y = \sqrt{x}$, $x > 0$.

(e) $x^2y'' - 2xy' + 2y = 2x^3$, $x > 0$.

(f) $x^2y'' + 2xy' + y = \ln(x)$, $y(1) = y'(1) = 0$.

Exercise 1.62 *Let α be a constant. Use the substitution $x - \alpha = e^t$ to reduce*

$$(x - \alpha)^2 y'' + b(x - \alpha)y' + cy = 0$$

to the equation

$$\frac{d^2y}{dt^2} + (b - 1)\frac{dy}{dt} + cy = 0.$$

In Exercises 1.63-1.64 use the results of Exercise 1.62 to solve each of the given differential equation.

Exercise 1.63

$$(x + 1)^2 y'' - (x + 1)y' + y = 0, \quad y(0) = 1, \ y'(0) = 0.$$

Exercise 1.64

$$(x - 2)^2 y'' + y = 0, \quad y(1) = 3, \ y'(1) = 1.$$

2

Partial Differential Equations

This chapter is intended to serve as an introduction to the topic of partial differential equations. We will discuss basic and fundamental topics that are suitable for this course. We mainly discuss first-order partial differential equations and Burger's equation using the method of characteristics. We move on to the study of second-order partial differential equations and their classifications. We consider the wave and heat equations on bounded and unbounded domains.

2.1 Introduction

A *partial differential equation*, short PDE, is an equation that contains the unknown function u and its partial derivatives. Recall from Chapter 1, Section 1.1 that given a function of two variables, $f(x,y)$, the *partial derivative* of f with respect to x is the rate of change of f as x varies, keeping y constant and it is given by

$$\frac{\partial f}{\partial x} = \lim_{h \to 0} \frac{f(x+h,y) - f(x,y)}{h}.$$

Similarly, the *partial derivative* of f with respect to y is the rate of change of f as y varies, keeping x constant and it is given by

$$\frac{\partial f}{\partial y} = \lim_{h \to 0} \frac{f(x,y+h) - f(x,y)}{h}.$$

More often we write f_x, f_y to denote $\frac{\partial f}{\partial x}$ and $\frac{\partial f}{\partial y}$, respectively. Similar notations will be used to denote higher partial derivatives and mixed partial derivatives. Let D be a subset of $\mathbb{R}^2$ and $u = u(x,y)$ such that $u : D \to \mathbb{R}$. Then we may denote the general first oder PDE in $u(x,y)$ by

$$F(x,\, y,\, u(x,y),\, u_x(x,y),\, u_y(x,y)) = 0,$$

or

$$F(x,\, y,\, u,\, u_x,\, u_y) = 0, \tag{2.1}$$

for some function F. In this Chapter, we limit our discussion to PDEs in two independent variables. Below we list some important PDEs.

$$u_t + cu_x = 0, \quad c \in \mathbb{R} \quad \text{(Transport equation)}.$$

DOI: 10.1201/9781003449881-2

$$u_{xx} + u_{yy} = 0 \quad \text{(Laplace's equation)}.$$

$$u_t + uu_x = 0 \quad \text{(Burger's equation)}.$$

$$u_{tt} - c^2 u_{xx} = 0 \quad \text{(Wave equation)}.$$

$$u_t - ku_{xx} = 0 \quad \text{(Heat equation)}.$$

Loosely speaking, by a solution of (2.1), we mean a function $u(x,y) = \varphi(x,y)$ such that

$$F\left(x, y, \varphi, \varphi_x, \varphi_y\right) = 0$$

for all $x, y \in D \subseteq \mathbb{R}^2$.

Example 2.1 For an arbitrary function f, $\varphi(x,t) = f(x - ct)$ is a general solution of the transport equation $u_t + cu_x = 0$ since

$$\varphi_t + c\varphi_x = -c\varphi'(x - ct) + c\varphi'(x - ct) = 0.$$

$\square$

Order and linearity are two of the main properties of PDEs.

Definition 2.1 *The order of a partial differential equation is the highest order derivative in the given PDE.*

Definition 2.2 *A given partial differential equation is said to be linear if the unknown function and all of its derivatives enter linearly.*

For example, all the equations listed above are linear except Burger's equation. Now to better understand linearity we utilize the operator concept $\mathscr{L}$ on an appropriate space, where $\mathscr{L}$ is a *differential operator*. Recall from Chapter 1, that an operator is really just a function that takes a function as an argument instead of numbers as we are used to dealing with in functions. For example, $\mathscr{L}u$ assigns u a new function $\mathscr{L}u$. Another example if we take

$$\mathscr{L} = \frac{\partial^2}{\partial t^2} - \frac{\partial^2}{\partial x^2},$$

then

$$\mathscr{L}u = u_{tt} - u_{xx}.$$

The next definition gives a precise and convenient way to test for linearity.

Definition 2.3 *An operator $\mathscr{L}$ is said to be linear if is satisfies*

(a)

$$\mathscr{L}(u_1 + u_2) = \mathscr{L}u_1 + \mathscr{L}u_2,$$

(b)

$$\mathscr{L}(cu_1) = c\mathscr{L}u_1,$$

for any functions u_1, u_2 and constant c. Moreover, the equation $\mathcal{L}u = 0$ is said to be linear if the operator $\mathcal{L}$ is linear.

Example 2.2 Consider $u_{tt} - u_{xx} = 0$. To show it is linear we let

$$\mathcal{L}u = u_{tt} - u_{xx}.$$

Then for any functions u, v, we have

$$\mathcal{L}(u+v) = (u+v)_{tt} - (u+v)_{xx} = [u_{tt} - u_{xx}] + [v_{tt} - v_{xx}] = \mathcal{L}u + \mathcal{L}v,$$

$$\mathcal{L}(cu) = (cu)_{tt} - (cu)_{xx} = c(u_{tt}) - c(u_{xx}) = c[u_{tt} - u_{xx}] = c\mathcal{L}u.$$

This shows the PDE in question is linear.   □

Example 2.3 The following PDE $uu_t - u_x = 0$ is not a linear. To see this, we let

$$\mathcal{L}u = uu_t - u_x.$$

Then for any functions u, v and a constant c, we have

$$\mathcal{L}(u+v) = (u+v)(u+v)_t - (u+v)_x = [uu_t - u_x] + [vv_t - v_x] + uv_t + vu_t \neq \mathcal{L}u + \mathcal{L}v.$$

This shows the PDE in question is not linear.   □

Let $\mathcal{L}u = 0$ and assume $\mathcal{L}$ is a linear operator. If u is a solution, then so is cu, for some constant c since $\mathcal{L}(cu) = c\mathcal{L}u = 0$. Similarly, if v is another solution then so is $c_1 u + c_2 v$, since for any constants c_1 and c_2 we have $\mathcal{L}(c_1 u + c_2 v) = c_1 \mathcal{L}u + c_2 \mathcal{L}v$. Thereupon, we have the following theorem.

Theorem 2.1 *[Principle of Superposition] Assume $\mathcal{L}$ is a linear operator. If $u_1(x,y)$, $u_2(x,y)$, $\ldots$, $u_n(x,y)$ are solutions of $\mathcal{L}u = 0$, then so is*

$$c_1 u_1(x,y) + c_2 u_2(x,y) + \ldots + c_n u_n(x,y) = \sum_{i=1}^{n} c_i u_i(x,y).$$

Example 2.4 In this example, we show that each of

$$u_n(x,t) = e^{-n^2 \pi^2 t} \sin(n\pi x), \quad n = 1, 2, \ldots, N$$

is a solution of the heat equation $u_{xx} - u_t = 0$, $\quad t > 0$, for the temperature $u = u(x,t)$ in a rod, considered as a function of the distance x measured along the rod and of the time t. In addition, we show for constants c_i, $i = 1, 2, \ldots, N$ that

$$u(x,t) = \sum_{n=1}^{N} c_n u_n(x,t)$$

is also a solution.

To do so, we let $\mathcal{L}$ be the differential operator given by $\mathcal{L} = \frac{\partial^2}{\partial x^2} - \frac{\partial}{\partial t}$. It is clear that $\mathcal{L}$ is linear. Now for any fixed n, $n = 1,2,\ldots,N$ we have

$$
\begin{aligned}
\mathcal{L}u_n &= \frac{\partial^2}{\partial x^2}\left(e^{-n^2\pi^2 t}\sin(n\pi x)\right) - \frac{\partial}{\partial t}\left(e^{-n^2\pi^2 t}\sin(n\pi x)\right) \\
&= \frac{\partial}{\partial x}\left(n\pi e^{-n^2\pi^2 t}\cos(n\pi x)\right) + n^2\pi^2 e^{-n^2\pi^2 t}\sin(n\pi x) \\
&= -n^2\pi^2 e^{-n^2\pi^2 t}\sin(n\pi x) + n^2\pi^2 e^{-n^2\pi^2 t}\sin(n\pi x) \\
&= 0.
\end{aligned}
$$

This shows $\mathcal{L}u_n(x,t) = 0$, $n = 1,2,\ldots,N$. As for the rest of the work, we apply u to $\mathcal{L}$.

$$
\begin{aligned}
\mathcal{L}u &= \mathcal{L}\left(\sum_{n=1}^{N} c_n u_n(x,t)\right) \\
&= \mathcal{L}(c_1 u_1(x,t)) + \mathcal{L}(c_2 u_2(x,t)) + \mathcal{L}(c_N u_N(x,t)) \\
&= c_1\mathcal{L}u_1(x,t) + c_2\mathcal{L}u_2(x,t) + c_N\mathcal{L}u_N(x,t) \\
&= 0.
\end{aligned}
$$

$\square$

Let $\mathcal{L}$ be a differential operator. Then the partial differential equation

$$\mathcal{L}u = f \tag{2.2}$$

for continuous function f in x and y is called *nonhomogeneous partial differential equation*. Its corresponding *homogeneous partial differential equation* is

$$\mathcal{L}u = 0. \tag{2.3}$$

If we assume $\mathcal{L}$ is linear, then the construction of the general solution of the non-homogeneous PDE given by (2.2) is similar to its counterpart in ordinary differential equations. Let u_p be a *particular solution* of (2.2) and u_h be the *homogeneous solution* of (2.3). Then, due to the linearity of the differential operator $\mathcal{L}$ we see that

$$\mathcal{L}(u_h + u_p) = \mathcal{L}u_h + \mathcal{L}u_p = 0 + f = f.$$

Thus, it suffices to find u_p of (2.2) and add it to the homogeneous solution u_h of (2.3) to get the general solution

$$u = u_h + u_p.$$

When the PDE is linear and involves only simple derivatives of only one variable, it is more likely that it can be solved along the lines of an ordinary differential equation, as the next example shows.

Example 2.5 In this example we display various forms of the solution $u = u(x,y)$ for the following PDEs

(a) $u_y = 0$,

(b) $u_{yy} + u = 0$,

(c) $u_{yyx} = 0$.

We begin with (a). The given PDE, $u_y = 0$ has no partial derivatives with respect to x, which indicates that the solution u is a function of x only. Thus, the solution is $u(x,y) = g(x)$, for some function g. Suppose we impose the initial condition $u(x,a) = e^{2x}$, then

$$e^{2x} = u(x,a) = g(x),$$

which uniquely determines the function g.

(b) Now the PDE, $u_{yy} + u = 0$ can be thought of as the second-order ODE $z'' + z = 0$, which has the solution $z = c_1 \cos(y) + c_2 \sin(y)$. Or, $u(x,y) = g(x)\cos(y) + h(x)\sin(y)$, since the constants c_1 and c_2 may depend on the other variable x, where the functions g and h are differentiable.

(c) Let f be twice differentiable function in y. Then, we integrate with respect to x and get

$$u_{yy} = f_{yy}(y),$$

and from which we have $u_y = f_y(y) + g(x)$. Integrating one more time we arrive at the general solution

$$u(x,y) = f(y) + yg(x) + h(x).$$

□

Here is another example.

Example 2.6 Consider the PDE $u_{yy} + u = 3 + y$.

Since it is nonhomogeneous we need to find u_h and u_p. From Example 2.5, we have

$$u_h(x,y) = g(x)\cos(y) + h(x)\sin(y).$$

Since $f = 3 + y$, we look for a particular solution in the form $u_p = A + By$, where A and B are to be determined by substituting, $(u_p)_{yy}$, and u_p into the original PDE. This implies that $A + By = 3 + y$, and therefore, $A = 3$, and $B = 1$. As a result, we obtain the general solution

$$u(x,y) = g(x)\cos(y) + h(x)\sin(y) + 3 + y.$$

□

2.1.1 Exercises

Exercise 2.1 *Determine the order and use the operator $\mathcal{L}$ to decide linearity, non-linearity of the given equations.*

(a) $u_{xx} + 3u_y + 4u = 0$.

(b) $u_{xy} + 3u_x = x$.

(c) $u_x + 3u_{yyyy} + u^{1/2} = 0$.

(b) $u_x = 1$.

Exercise 2.2 *Show that $u(x,t) = e^{t-x}$ is a solution of $u_{tt} - u_{xx} = 0$.*

Exercise 2.3 *Show that $u(x,y) = f(bx - ay)$ for a differentiable function f is a solution of $au_x + bu_y = 0$.*

Exercise 2.4 *Show that $u(x,t) = \dfrac{1}{6}t^3 + (x - t^2/2)t + (x - t^2/2)^2$ solves $u_t + tu_x = x$, $-\infty < x < \infty$, $t > 0$, and satisfies the initial data $u(x,0) = x^2$.*

Exercise 2.5 *Solve $2u_x - 3u_y = 0$, subject to the condition $u(0,y) = e^y$.*

Exercise 2.6 *Find the general solution of each of the PDEs.*

(a) $u_{xx} + 4u = 0$.

(b) $u_{xy} = 0$.

(c) $u_{xy} + u_y = 0$.

(d) $u_{xx} + 3u_x + 2u = e^{5x}$.

(e) $(x - y)u_{xy} - u_x + u_y = 0$. *(Hint: let $v = (x - y)u_x + u$.)*

(f) $u_{xxy} = 0$.

(g) $u_{xy} + u_y = e^{2x+3y}$.

2.2 Linear Equations

Now is the time to learn how to solve first-order linear partial differential equations with constant or variable coefficients. We begin with linear equations with constant coefficients.

2.2.1 Linear equations with constant coefficients

We consider the first-order nonhomogeneous general partial differential equation

$$Au_x(x,y) + Bu_y(x,y) + Cu(x,y) = G(x,y), \tag{2.4}$$

where A, B, and C are constants such that $A^2 + B^2 \neq 0$, and G is a given continuously differentiable function in x and y. We begin by examining the homogeneous solution of the corresponding homogenous equation

$$Au_x + Bu_y + Cu = 0, \tag{2.5}$$

where we have suppressed the independent variables x and y. Note that the requirement $A^2 + B^2 \neq 0$, implies that A and B are not both zero at the same time; otherwise, we would not have a differential equation to solve. Our aim is to use a linear transformation and transform (2.5) into an ordinary differential equation in a single independent variable, say, x, and the dependent variable u. We begin by letting $\xi = \xi(x,y)$ and $\eta(x,y) = \eta$, where

$$\xi = c_{11}x + c_{12}y \tag{2.6}$$

and

$$\eta = c_{21}x + c_{22}y \tag{2.7}$$

where the constants c_{11}, c_{12}, c_{21}, and c_{12} are to be appropriately chosen in order to reduce the PDE into an ODE. Using the chain rule, we have

$$u_x = u_\xi \xi_x + u_\eta \eta_x = c_{11}u_\xi + c_{21}u_\eta, \tag{2.8}$$

and

$$u_y = u_\xi \xi_y + u_\eta \eta_y = c_{12}u_\xi + c_{22}u_\eta. \tag{2.9}$$

Now substituting (2.8) and (2.9) into (2.5) and rearranging the terms we arrive at

$$\left(Ac_{11} + Bc_{12}\right)u_\xi + \left(Ac_{21} + Bc_{22}\right)u_\eta + Cu = 0. \tag{2.10}$$

Assume $A \neq 0$ and choose $c_{11} = 1$, $c_{12} = 0$, $c_{21} = B$, and $c_{22} = -A$. Then

$$\xi = x \text{ and } \eta = Bx - Ay.$$

Hence (2.10) becomes

$$Au_\xi + Cu = 0,$$

which is a separable ODE. Separating the variables while fixing η we arrive at

$$\frac{u_\xi}{u} = -\frac{C}{A},$$

from which we get $\ln|u| = -\frac{C}{A}\xi + g(\eta)$, for some function g. Taking exponential on both sides we obtain the homogeneous solution

$$u(\xi,\eta) = e^{-\frac{C}{A}\xi}f(\eta),$$

where f is an arbitrary continuously differentiable function. Thus, the general solution of the homogenous equation (2.5) is

$$u_h(x,y) = e^{-\frac{C}{A}x}f(Bx - Ay). \tag{2.11}$$

If u_p is a particular solution of (2.4), then the general solution of (2.4) is

$$u(x,y) = u_h(x,y) + u_p(x,y),$$

where $u_h(x,y)$ is given by (2.11). Now if $B \neq 0$ then u_h of (2.4) can be found to be

$$u_h(x,y) = e^{-\frac{C}{B}y} f(Bx - Ay). \tag{2.12}$$

Remark 2 *In the case that both A and B are not zero, you may use either* (2.11) *or* (2.12).

Example 2.7 Solve

$$u_x + u_y + u = x + y, \tag{2.13}$$

subject to the initial condition

$$u(0,y) = y^2.$$

First we find u_h and since neither A nor B is zero we have the luxury to use either (2.11) or (2.12). Using (2.12) with $A = B = C = 1$, we obtain

$$u_h(x,y) = e^{-y} f(x - y).$$

Using the concept of Chapter 1 of obtaining the particular solution, we try

$$u_p = a + bx + cy.$$

Then substituting into (2.13) we get

$$a + b + c + bx + cy = x + y.$$

It follows from the above expression that $a = -2$, $b = c = 1$. Then the general solution is

$$u(x,y) = e^{-y} f(x - y) + x + y - 2.$$

Our next task is to use the initial data to uniquely determine the function f. The initial data implies that when $x = 0$ and $y = y$ we have $u = y^2$. Thus, it follows from the general solution that

$$y^2 = e^{-y} f(-y) + y - 2.$$

This gives $f(-y) = e^y(y^2 - y + 2)$. Let $k = -y$. Then $f(k) = e^{-k}(k^2 + k + 2)$, from which we get

$$f(x - y) = e^{y-x}[(x - y)^2 + (x - y) + 2].$$

It follows from

$$u(x,y) = e^{-y} f(x - y) + x + y - 2$$

that

$$u(x,y) = e^{-x}[(x - y)^2 + (x - y) + 2] + x + y - 2.$$

□

Remark 3 *Note that if $C = 0$ in (2.5), then the homogeneous solution is reduced to*

$$u_h(x,y) = f(Bx - Ay),$$

for an arbitrary continuously differentiable function f.

Example 2.8 Solve the transport equation

$$u_t + cu_x = 0, \quad -\infty < x < \infty, \ t > 0, \tag{2.14}$$

subject to

$$u(x,0) = f(x). \tag{2.15}$$

Using the above remark, we immediately obtain the solution to be

$$u(x,t) = f(x - ct).$$

Now suppose we are given

$$f(x) = \begin{cases} 1, & x \le 0 \\ 1 - x, & 0 < x \le 1 \\ 0, & x > 1. \end{cases}$$

The easiest way is to replace x in f with $x - ct$ and rearrange the domains. Thus,

$$u(x,t) = f(x - ct) = \begin{cases} 1, & x - ct \le 0 \\ 1 - x + ct, & 0 < x - ct \le 1 \\ 0, & x - ct > 1. \end{cases}$$

It follows that

$$u(x,t) = \begin{cases} 1, & x \le ct \\ 1 - x + ct, & ct < x \le 1 + ct \\ 0, & x > 1 + ct. \end{cases}$$

$\square$

Next, we give a geometrical interpretation of the characteristic lines and solutions. Consider the simpler form of (2.5) with $C = 0$. That is,

$$Au_x + Bu_y = 0, \tag{2.16}$$

where $A^2 + B^2 \ne 0$. That is A and B can not both be zero. Let $\cdot$ denote the inner product in $\mathbb{R}^2$. Then (2.16) is equivalent to

$$< A, B > \cdot < u_x, u_y > = 0.$$

If we let $\vec{v} = < A, B >$, then the left side of the equation is the directional derivative of u in the direction of the vector $\vec{v}$. That is, the solution u of (2.16) must be constant in the direction of the vector $\vec{v} = Ai + Bj$. The lines parallel to the vector $\vec{v}$ have the equation for an arbitrary constant constant K

$$Bx - Ay = K, \tag{2.17}$$

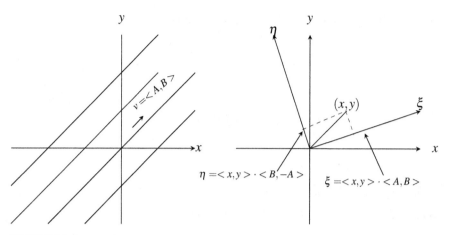

FIGURE 2.1
Characteristic lines; change of coordinates.

since the vector $< B, -A >$ is orthogonal to $\vec{v}$, and as such is a normal vector to all lines that are parallel to $\vec{v}$. Thus, (2.17) provides a family of lines and each one of them is uniquely determined by the specific value of K. The family of lines given by (2.17) are called *characteristic lines* for the equation (2.16). We conclude that the solution $u(x,y)$ is constant in the direction of $\vec{v}$, which is true also along the family of lines given by (2.17). Such any line is determined by $K = Bx - Ay$ and hence u will depend only on $Bx - Ay$. It follows that

$$u(x,y) = f(Bx - Ay),$$

as we have obtained in Remark 3, for an arbitrary function f. We refer to Fig. 2.1.

2.2.2 Exercises

Exercise 2.7 *Give all the details in arriving at Equation (2.12).*

Exercise 2.8 *Redo Example 2.7 using (2.11).*

Exercise 2.9 *For constants A and B use the transformation*

$$\xi = Ax + By, \quad \eta = Bx - Ay$$

and obtain the general solution

$$u(\xi, \eta) = e^{-\frac{C}{A^2+B^2}\xi} \left(f(\eta) + \int g(\xi, \eta) e^{\frac{C}{A^2+B^2}\xi} d\xi \right),$$

of the nonhomogeneous equation

$$Au_x(x,y) + Bu_y(x,y) + Cu(x,y) = g(x,y).$$

Exercise 2.10 *Use the results of Exercise 2.9 to solve*

$$-2u_x + 4u_y + 5u = e^{x+3y},$$

subject to

$$u(0,y) = 2 + y.$$

Exercise 2.11 *Solve each of the given PDEs.*

(a) $u_x + u_y + u = \sin(x)$.

(b) $u_x + u = x$, $\quad u(0,y) = y^2$.

(c) $2u_x + u_y = x + e^y$, $\quad u(0,y) = y^2$.

Exercise 2.12 *Solve*

(a) $u_x - 3u_y = \sin(x) + \cos(y)$, $\quad u(x,0) = x$.

(b) $2u_x + u_y = x + e^y$, $\quad u(0,y) = y^2$.

Exercise 2.13 *Solve*

(a) $u_x - 5u_y = 0$, $\quad u(x,0) = \dfrac{1}{1+x^2}$.

(b) $u_x + u_t = -3$, $\quad u(x,0) = e^{3x}$.

Exercise 2.14 *Solve for $u(x,t)$ and sketch $u(x,t)$ at $t = 1,2$.*

$$u_t - u_x = 0, \ -\infty < x < \infty, \ t > 0,$$

subject to

$$u(x,0) = f(x), \tag{2.18}$$

where

$$f(x) = \begin{cases} 1, & x \le 0 \\ 1-x, & 0 < x \le 1 \\ 0, & x > 1. \end{cases}$$

2.2.3   Equations with variable coefficients

We turn our attention to the general first-order linear partial differential equation with variable coefficients

$$A(x,y)u_x(x,y) + B(x,y)u_y(x,y) + C(x,y)u(x,y) = G(x,y), \tag{2.19}$$

where A and B are continuously differentiable functions in the variables in the open set $D \subset \mathbb{R}^2$ and that A and B do not simultaneously identically vanish in D. In addition, the functions C and G are continuous in D. We begin by examining the homogeneous equation

$$Au_x + Bu_y + Cu = 0, \tag{2.20}$$

where we have suppressed the independent variables x and y. Let

$$\xi = \xi(x,y), \quad \eta = \xi(x,y)$$

be a transformation on D with

$$J = \begin{vmatrix} \xi_x & \eta_x \\ \xi_y & \eta_y \end{vmatrix} = \xi_x \eta_y - \xi_y \eta_x \neq 0. \tag{2.21}$$

Note that condition (2.21) is necessary so that the transformation is invertible. Then by (2.8) and (2.9)

$$u_x = u_\xi \xi_x + u_\eta \eta_x,$$

and

$$u_y = u_\xi \xi_y + u_\eta \eta_y.$$

Substituting into (2.20) and rearranging the terms we obtain

$$(A\xi_x + B\xi_y)u_\xi + (A\eta_x + B\eta_y)u_\eta + Cu = 0, \tag{2.22}$$

where A, B, and C are functions of ξ and η. As before, our ultimate goal is to reduce (2.22) into an ODE and so we ask that

$$A\eta_x + B\eta_y = 0. \tag{2.23}$$

With this particular choice, the PDE given by (2.22) is reduced into an ODE in the variables ξ, and u. Let's assume $A \neq 0$. Let $y(x)$ be a curve with slope $\dfrac{B}{A}$, that forms the surface that is a solution to the PDE. Then,

$$\frac{d}{dx}\eta(x,y(x)) = \eta_x + \eta_y \frac{dy}{dx} = \eta_x + \frac{B}{A}\eta_y.$$

This implies that the characteristic lines given by $\eta(x,y) = k$, for constant k is a solution to the characteristic equation $\dfrac{dy}{dx} = \dfrac{B}{A}$, with $\eta_y \neq 0$, otherwise $\eta_x = 0$ as well, and this will not be a solution. Thus, $\eta(x,y) = k$ determines η. We may choose the other variable $\xi(x,y) = x$. We observe that for this change of variables

$$J = \begin{vmatrix} \xi_x & \eta_x \\ \xi_y & \eta_y \end{vmatrix} = \eta_y \neq 0,$$

and hence the transformation is invertible. Finally, our PDE reduces to

$$A(\xi,\eta)u_\xi(\xi,\eta) + C(\xi,\eta)u(\xi,\eta) = 0.$$

The above equation is a separable ODE and has the general solution

$$u_h(x,y) = f(\eta)e^{-\int \frac{C(\xi,\eta)}{A(\xi,\eta)}d\xi} = f(\eta)\Psi(\xi,\eta), \tag{2.24}$$

where f is an arbitrary function and Ψ is known. If u_p is a particular solution of (2.19), then the general solution of (2.19) is

$$u(x,y) = u_h(x,y) + u_p(x,y),$$

where u_h is given by (2.24).

Remark 4 *In some cases, it is difficult to compute the particular solution u_p of (2.19) and then add it to the homogeneous solution u_h to obtain the general solution. However, it might be beneficial to transform the whole equation of (2.19) into a PDE in terms of ξ and η and then use methods of ODEs to solve it. In doing so, the transformation turns (2.19) into the new PDE*

$$a(\xi,\eta)u_\xi + b(\xi,\eta)u_\eta + c(\xi,\eta)u = g(\xi,\eta). \tag{2.25}$$

This can be achieved by making use of

$$u_x = u_\xi \xi_x + u_\eta \eta_x,$$

and

$$u_y = u_\xi \xi_y + u_\eta \eta_y.$$

Example 2.9 Consider

$$x^2 u_x - xy u_y + yu = 0, \quad x \neq 0, \tag{2.26}$$

subject to

$$u(1,y) = e^{2y}.$$

Based on the above discussion, we have $A = x^2$, $B = -xy$, and $C = y$. The characteristic equation

$$\frac{dy}{dx} = \frac{-xy}{x^2},$$

has the characteristic lines as its solution given by $xy = k$. Thus, we define $\eta(x,y) = xy$. Let $\xi(x,y) = x$. Then the Jacobian

$$J = \begin{vmatrix} \xi_x & \eta_x \\ \xi_y & \eta_y \end{vmatrix} = x \neq 0.$$

Hence the transformation is invertible. Moreover,

$$u_x = u_\xi + yu_\eta, \quad u_y = xu_\eta.$$

Substituting into (2.26), we obtain

$$x^2 u_\xi + yu = 0.$$

In terms of ξ and η the equation is reduced to

$$u_\xi = \frac{-\eta}{\xi^3} u.$$

An integration with respect to ξ yields the solution

$$\begin{aligned} u(\xi,\eta) &= f(\eta)e^{-\int \frac{\eta}{\xi^3} d\xi} \\ &= f(\eta)e^{-\frac{\eta}{2\xi^2}}. \end{aligned}$$

In terms of x and y we have

$$u(x,y) = f(xy)e^{-\frac{y}{2x}}.$$

Using the initial data we obtain

$$e^{2y} = f(y)e^{-\frac{y}{2}},$$

from which we have $f(y) = e^{\frac{5y}{2}}$. Thus, $f(xy) = e^{\frac{5xy}{2}}$, and the solution is

$$u(x,y) = e^{\frac{5xy}{2}} e^{-\frac{y}{2x}}.$$

$\square$

2.2.4 Exercises

Exercise 2.15 *Solve*

(a) $yu_x + xu_y = 0$, $u(0,y) = e^{-y^2}$.

(b) $xu_x - u_y = 0$, $u(x,0) = x$.

Exercise 2.16 *Solve*

(a) $3u_x - 2u_y + u = x$, $u(x,0) = \dfrac{1}{1+x^2}$.

(b) $u_x + u_t = -3$, $u(x,0) = e^{3x}$.

Exercise 2.17 *Use Remark 4 to solve*

(a) $xu_x + u_y = 1$, $u(x,0) = \cos(3x)$.

(b) $xu_x - yu_y + y^2 u = y^2$, $x, y \neq 0$.

2.3 Quasi-Linear Equations

We now investigate the general first-order *quasi-linear* partial differential equation

$$A(x,y,u)u_x(x,y) + B(x,y,u)u_y(x,y) = C(x,y,u), \qquad (2.27)$$

where A, B and C are continuously differentiable functions of the variables x, y and u in some open set $\mathbb{D} \subset \mathbb{R}^3$. We represent the function $u(x,y)$ by a surface $z = u(x,y)$, in the xyz-space, see Fig. 2.2. Surfaces corresponding to solutions of a PDE are called *integral surfaces* of the PDE. Suppressing the argument, and using inner product, the left side of (2.27) can be written as

$$< A, B, C > \cdot < u_x, u_y, -1 > = 0.$$

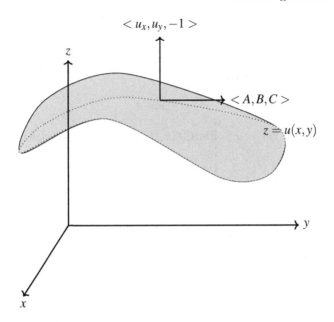

FIGURE 2.2
Surface $z = u(x,y)$.

Hence, the normal vector $< u_x,\ u_y,\ -1 >$ to the surface at a given point is orthogonal to the vector $< A,\ B,\ C >$ at that point. It follows that the vector $< A,\ B,\ C >$ must be tangent to the surface $u(x,y) - z = 0$ and therefore *integral surfaces* must be formed from the *integral curves* of the vector field $< A,\ B,\ C >$. Thus, the integral curves are given as solutions to the system of ODEs

$$\frac{dx}{dt} = A(x,y,z), \quad \frac{dy}{dt} = B(x,y,z), \quad \frac{dz}{dt} = C(x,y,z). \tag{2.28}$$

Note that the choice of parameter t in (2.28) is artificial and it can be suppressed to write (2.28) in the form

$$\frac{dx}{A(x,y,z)} = \frac{dy}{B(x,y,z)} = \frac{dz}{C(x,y,z)}. \tag{2.29}$$

Either of the systems (2.28) or (2.29) is called the characteristic system associated with (2.27). Characteristics curves are solutions of either system (2.28) or (2.29).

If a surface $S : z = u(x,y)$ is a union of characteristic curves, then S is an integral surface. We have the following theorem.

Theorem 2.2 *Assume the point $P = (x_0, y_0, z_0)$ is a point on the integral surface $S : z = u(x,y)$. Let γ be the characteristic curve through P. Then γ lies entirely on S.*

Proof *Assume* $\gamma = (x(t),\ y(t),\ z(t))$ *is a solution of* (2.29) *and satisfies* $\gamma = (x(t_0),\ y(t_0),\ z(t_0)) = (x_0,\ y_0,\ z_0)$ *at some initial time* t_0. *Let*

$$W(t) = z(t) - u(x(t), y(t)).$$

Then $W(t_0) = z(t_0) - u(x_0, y_0) = 0$, *since the point P lies on the surface S. It follows from the chain rule and* (2.28) *that*

$$
\begin{aligned}
\frac{dW}{dt} &= \frac{dz}{dt} - u_x(x,y)\frac{dx}{dt} - u_y(x,y)\frac{dy}{dt} \\
&= C(x,y,z) - u_x(x,y)A(x,y,z) - u_y(x,y)B(x,y,z),
\end{aligned}
$$

which can be written in the form

$$
\begin{aligned}
\frac{dW}{dt} &= C(x,y,W+u(x,y)) - u_x(x,y)A(x,y,W+u(x,y)) \\
&- u_y(x,y)B(x,y,W+u(x,y)). \tag{2.30}
\end{aligned}
$$

Since $u(x,y)$ *satisfies* (2.27) *we see that* $W = 0$ *is a particular solution of* (2.30). *By uniqueness of solutions, (uniqueness theorem from ODEs) we know that* $W = 0$ *only hold at* $t = t_0$. *Thus, the function* $W(t) = z(t) - u(x(t),\ y(t))$ *vanishes identically. This implies that the curve gamma lies entirely on S. This completes the proof.*

Next, we discuss the *Cauchy Problem for quasi-linear equation* (2.27), which says, find the *integral surface* $z = u(x,y)$ of (2.27) containing an initial curve Γ.

Method of solution:

Let Γ be *non-characteristic* contained in the surface $u(x,y) - z = 0$; (that is the tangent to Γ is nowhere tangent to the characteristic vector $<A,\ B,\ C>$ along Γ), be given initial curve. Parametrize Γ by

$$x = f(s), \quad y = g(s), \quad z = h(s), \quad \text{at initial time} \quad t = 0.$$

To construct an integral surface containing Γ, one can proceed as follow;

1) Solve either (2.28) or (2.29) and then use the data given by the initial curve Γ to obtain the constants of integrations. This should produce two independent integral curves,

$$u_1\Big(f(s), h(s), h(s)\Big) = c_1, \quad u_2\Big(f(s), h(s), h(s)\Big) = c_2.$$

2) Eliminate s from the two equations in 1) and obtain the functional relation $F(c_1, c_2) = 0$, between c_1 and c_2.

3) Then the solution to the Cauchy problem is

$$F\big[u_1(x,y,z),\ u_2(x,y,z)\big] = 0.$$

4) If we are given an initial curve, then utilize it to compute the arbitrary function in 3).

Remark 5 *(1) You may obtain different looking solutions, but this depends on whether you use (2.28) or (2.29). However, once you apply the initial data given by the initial curve Γ, then solutions should match.*

(2) The method can be easily extended to PDEs of multiple variables by adapting the relations (2.28) and (2.29). We shall explain this in one of the examples below.

Example 2.10 Find $u(x,y)$ satisfying

$$xu_x + yu_y = u + 1,$$

and

$$u(x,1) = 3x.$$

Let $z = u(x,y)$. First we parametrize the initial curve Γ given by the data $u(x,1) = 3x$. Let $x = s$ at $t = 0$. Then

$$\Gamma : x = s, \quad y = 1, \quad z = 3s, \quad \text{for} \quad t = 0.$$

Using system (2.28) we have

$$\frac{dx}{dt} = x, \quad \frac{dy}{dt} = y, \quad \frac{dz}{dt} = z + 1,$$

with corresponding solutions

$$x = c_1 e^t, \quad y = c_2 e^t, \quad z = c_3 e^t - 1.$$

Applying the initial data given by the parametrized curve Γ, we obtain $c_1 = s$, $c_2 = 1$, and $c_3 = 3s + 1$. Thus, we have the solutions

$$x = se^t, \quad y = e^t, \quad z = (3s+1)e^t - 1.$$

Eliminating s in the two equations $x = se^t$, $y = e^t$ gives $s = \frac{x}{e^t} = \frac{x}{y}$. Finally, substituting $s = \frac{x}{y}$ in $z = (3s+1)e^t - 1$ yields,

$$z = (3\frac{x}{y} + 1)y - 1 = 3x + y - 1.$$

So the solution is

$$u(x,y) = 3x + y - 1.$$

Fig. 2.3 shows the characteristic lines given by $y = \dfrac{x}{s}$ intersecting the initial curve $y = 1$ at exactly one point (they are not in the same direction) and hence, as the curves cover all the plane, the solution is defined everywhere. □

In the next example, we revisit the transport equation with incompatible data.

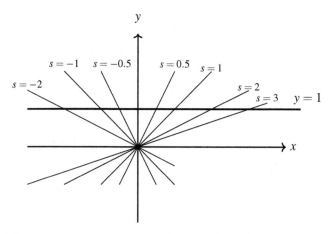

FIGURE 2.3
Characteristic lines $y = \frac{x}{s}$ intersecting initial curve $y = 1$.

Example 2.11 Consider the transport equation

$$u_x + cu_y = 0, \quad -\infty < x < \infty, \ y > 0, \qquad (2.31)$$

subject to

$$u(x, cx) = f(x). \qquad (2.32)$$

To parametrize the initial curve Γ we let $x = s$ at $t = 0$. Then

$$\Gamma : x = s, \quad y = cs, \quad z = f(s), \quad \text{for } t = 0.$$

Using system (2.28), we have

$$\frac{dx}{dt} = 1, \quad \frac{dy}{dt} = c, \quad \frac{dz}{dt} = 0,$$

and corresponding solutions along Γ are

$$x = t + c_1, \quad y = ct + c_2, \quad z = c_3.$$

Applying the initial data given by the parametrized curve Γ, we obtain $c_1 = s$, $c_2 = cs$, and $c_3 = f(s)$. Thus, we have the solutions

$$x = s + t, \quad y = ct + cs, \quad z = f(s).$$

We see that it is not feasible to eliminate s and write the solution u in x and y. Notice that the characteristic line given by $y = c(s + t) = cx$, for a fixed value of c is the same as the equation of the initial curve $y = cx$. In other words, the direction of the characteristic lines are in the same direction as the initial curve. $\qquad \square$

Example 2.12 Find $u(x,y)$ satisfying

$$u_x + u_y + u - 1 = 0,$$

subject to

$$u(x, x + x^2) = \sin(x), \quad x > 0.$$

Let $z = u(x,y)$. By parametrizing the initial curve Γ we have

$$\Gamma : x = s, \quad y = s + s^2, \quad z = \sin(s), \quad s > 0 \quad \text{for } t = 0.$$

Using system (2.28), we have

$$\frac{dx}{dt} = 1, \quad \frac{dy}{dt} = 1, \quad \frac{dz}{dt} = 1 - z,$$

and corresponding solutions evaluated at the initial curve

$$x = s + t, \quad y = s + s^2 + t, \quad z = 1 - (1 - \sin(s))e^{-t}.$$

The two equations $x = s + t$, $y = s + s^2 + t$, give $y = x + s^2$. Then solving for s gives $s = \sqrt{y - x}$. Substituting s and $t = x - s$ into the equation $z = 1 - (1 - \sin(s))e^{-t}$ and using $z = u(x,y)$ we obtain the solution

$$u(x,y) = 1 - \left((1 - \sin(\sqrt{y-x}))\right)e^{(-x + \sqrt{y-x})},$$

which is defined for $y > x$. In Fig. 2.4, we plotted the traces of the characteristic curves given by $y = x + c$, where $c = s^2$ against the data from the initial curve $y = x + x^2$ and the two intersect in the region $y > x$, where the solution exists. □

Example 2.13 Consider the transport equation

$$u_x + x u_y = 0, \quad -\infty < x, y < \infty, \tag{2.33}$$

subject to

$$u(x,0) = f(x). \tag{2.34}$$

To parametrize the initial curve Γ we let $x = s$ at $t = 0$. Then

$$\Gamma : x = s, \quad y = 0, \quad z = f(s), \quad \text{for } t = 0.$$

Using system (2.28), we get

$$\frac{dx}{1} = \frac{dy}{x},$$

gives

$$y = \frac{x^2}{2} + \frac{s^2}{2}.$$

We also get $z = f(s)$ from the relation $\frac{dz}{dt} = 0$. Solving for s we arrive at $s = \pm\sqrt{x^2 - 2y}$. Thus, in order for a solution to exist we must require that f be an even

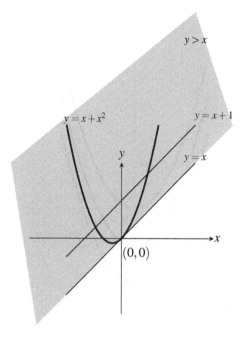

FIGURE 2.4
The solution exists in the shaded region $y > x$.

function. Since the solution is constant on the characteristic curves $y = \frac{x^2}{2} + \frac{s^2}{2}$, and the curves intersect the line $y = 0$ (the initial curve), the solution exists in the region where the traces of the characteristic lines given by $y = \frac{x^2}{2}$, intersects the y-axis. Moreover, we must make sure that $s^2 = 2y - x^2 > 0$. In conclusion, solutions exist in the region

$$y \le \frac{x^2}{2}$$

as depicted by Fig. 2.5.

□

The next example will show that based on the nature of the given equation, you will need to utilize either (2.28) or (2.29). But first, we make the following remark.

Remark 6 *A useful technique for integrating a system of first-order equations is that of multipliers. Recall from algebra that is $\frac{a}{b} = \frac{c}{d}$, then*

$$\frac{a}{b} = \frac{c}{d} = \frac{\lambda a + \mu c}{\lambda b + \mu d}$$

for arbitrary values of multipliers λ, μ. This can be generalized and one would have

$$\frac{a}{b} = \frac{c}{d} = \frac{e}{f} = \frac{\lambda a + \mu c + \nu e}{\lambda b + \mu d + \nu f} \tag{2.35}$$

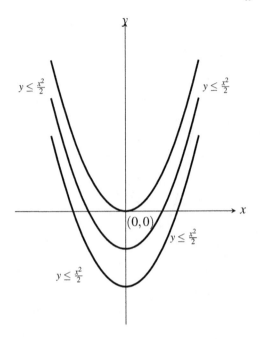

FIGURE 2.5
Feasible region for the solution.

for arbitrary multipliers λ, μ, and ν. Hopefully, with the right choices of the parameters λ, μ, and ν, expression (2.35) leads to simpler systems of ODEs that can be easily solved. In particular, if λ, μ, and ν, are chosen such that

$$\lambda A + \mu B + \nu C = 0,$$

then $\lambda dx + \mu dy + \nu dz = 0$. Now if there is a function u such that

$$du = \lambda dx + \mu dy + \nu dz = 0,$$

then $u(x,y,z) = c_1$ is an integral curve.

Example 2.14 Solve

$$x u_x + (x + u) u_y = y - x, \tag{2.36}$$

containing the curve

$$u(x, 2) = 1 + 2x.$$

A parametrization of the initial curve Γ gives

$$\Gamma : x = s, \quad y = 2, \quad z = 1 + 2s, \quad \text{for} \quad t = 0.$$

Here you can not use system (2.28) since the obtained ODEs will not be separable. Hence we resort to using (2.28), in combination with (2.35). From (2.28), we see that

$$\frac{dx}{x} = \frac{dy}{x+z} = \frac{dz}{y-x}.$$

Then, it follows from (2.35)

$$\frac{\lambda dx + \mu dy + v dz}{\lambda x + \mu(x+z) + v(y-x)} = \frac{dx}{x} = \frac{dy}{x+z} = \frac{dz}{y-x}. \tag{2.37}$$

In (2.37), set $\lambda = 0, \mu = 1$ and $v = 1$, and obtain

$$\frac{dy+dz}{y+z} = \frac{dx}{x}.$$

A direct integration yields, $\ln(y+z) = \ln(x) + k$, or $\ln\left(\frac{y+z}{x}\right) = k$. Taking exponential on both sides we arrive at the first integral curve,

$$c_1 = \frac{y+z}{x}, \tag{2.38}$$

where the constant k is replaced with the $c_1 = e^k$. Applying the initial data given by the initial curve we obtain $c_1 = \frac{3+2s}{s}$, or

$$s = \frac{3}{c_1 - 2}. \tag{2.39}$$

Similarly, if we set $\lambda = -1, \mu = 1$, and $v = 0$ in (2.37) we obtain $\dfrac{dy - dx}{z} = \dfrac{dz}{y-x}$, or

$$d(y-x)(y-x) = z dz.$$

An integration gives the second integral curve

$$(y-x)^2 - z^2 = c_2.$$

Applying the initial data we get the relation

$$(2-s)^2 - (1+2s)^2 = c_2. \tag{2.40}$$

Substituting the value of s given by (2.39) into (2.40) produces the expression

$$(2 - \frac{3}{c_1 - 2})^2 - (1 + 2\frac{3}{c_1 - 2})^2 = c_2. \tag{2.41}$$

Finally, to obtain a functional relation of the solution substitute c_1 and c_2 where c_1 is given by (2.38) and $c_2 = (y-x)^2 - z^2$ into (2.41). Don't forget to replace z by u for the final answer.

$\square$

2.3.1 Exercises

Exercise 2.18 *Find $u(x,y)$ satisfying*

$$u_x + u_y = 1 - u,$$

subject to

$$u(x, x+x^2) = e^x, \ x > 0.$$

Exercise 2.19 *Find $u(x,y)$ satisfying*

$$xuu_x - u_y = 0,$$

subject to

$$u(x,0) = x.$$

Exercise 2.20 *Find $u(x,y)$ satisfying*

$$xu_x + yuu_y = -xy, \ x > 0$$

subject to

$$u(x, \frac{1}{x}) = 5.$$

Exercise 2.21 *Find $u(x,y)$ satisfying*

$$uu_x + u_y = 3,$$

subject to

$$u(2x^2, 2x) = 0.$$

Exercise 2.22 *Find $u(x,y)$ satisfying*

$$xuu_x + y^2u_y = u^2, \quad x > 0, \ y > 0,$$

and

$$u = s, \ when \ x = \frac{1}{s}, \ y = 2s.$$

Exercise 2.23 *Find the general solution $u(x,y)$ of*

$$(x+u)u_x + (y+u)u_y = 0,$$

Exercise 2.24 *Find $u(x,y)$ satisfying*

$$xu_x + yu_y = xe^{-u},$$

subject to

$$u(x,x^2) = 0.$$

Exercise 2.25 *Find $u(x,y)$ satisfying*

$$xu_x + yu_y = 1 + y^2,$$

subject to

$$u(x,1) = x + 1.$$

Answer: $u(x,y) = ln(y) + \frac{y^2}{2} + \frac{1}{2} + \frac{x}{y}$.

Exercise 2.26 *Find $u(x,y)$ satisfying*

$$2xyu_x + (x^2 + y^2)u_y = 0, \ x \neq y, \ x > 0, \ y > 0,$$

subject to

$$u(x, 1-x) = \sin\left(\frac{x}{x-y}\right).$$

Exercise 2.27 *Find $u(x,y)$ satisfying*

$$x^2 u_x + uu_y = 1,$$

subject to

$$u(x, 1-x) = 0, \ x > 0.$$

Answer: $\dfrac{x}{1+xu} = \dfrac{u^2}{2} + 1 - y.$

Exercise 2.28 *Find $u(x,y)$ satisfying*

$$3u_x - 2u_y = x - u,$$

subject to

(a) $u(x,x) = 2x.$

(b) $u\left(x, \dfrac{1-2x}{3}\right) = 0.$

Exercise 2.29 *Use the idea of Remark 6 to solve*

(a) $(x+y)u_x + yu_y = x - y; \ u(x,1) = 1 - 3x.$

(b) $(x+y)u_x + (x+yu)u_y = u^2 - 1; \ u(x,1) = x^2.$

(c) $(y-u)u_x + (u-x)u_y = x - y; \ u(x,2x) = 0.$

(d) $(y-x)u_x + (y+x)u_y = \dfrac{x^2+y^2}{u}; \ u(x,0) = x.$

Exercise 2.30 *Find $u(x,y,z)$ satisfying*

(a) $xu_x + yu_y + u_z = u; \ u(x,y,0) = h(x,y), \text{ for a suitable function } h.$

(b) $u_x + 3u_y - 2u_z = u; \ u(0,y,z) = h(y,z), \text{ for a suitable function } h.$

(c) $yu_x + xu_y + u_z = 0; \ u(x,y,0) = h(x,y), \text{ for a suitable function } h.$

2.4 Burger's Equation

This section is devoted to the study of Burger's equation

$$u_y + a(u)u_x = 0, \quad -\infty < x < \infty, \ y > 0 \tag{2.42}$$

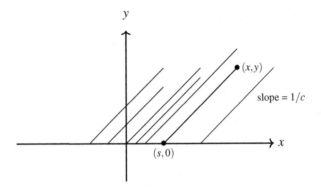

FIGURE 2.6
Characteristic lines $y = \frac{1}{c}x - \frac{s}{c}$ do not run into each others.

subject to the initial data

$$u(x,0) = h(x), \tag{2.43}$$

for a suitable function h. We seek a solution $u = u(x,y)$ that satisfies (2.42) and (2.43). Burger's equation (2.42) can be thought of as nonlinear one-wave equation, where $a(u)$ is the wave speed. We begin by considering the wave speed $a(u)$ as a constant c. In particular, we are analyzing

$$u_y + cu_x = 0, \quad -\infty < x < \infty, \tag{2.44}$$

along the parametrized initial curve

$$\Gamma : x = s, \quad y = 0, \quad z = h(s), \quad \text{at} \quad t = 0.$$

From Example 2.11, we have the characteristic lines (Fig. 2.6)

$$s = x - cy, \quad \text{or} \quad y = \frac{1}{c}x - \frac{s}{c}$$

and the solution u is constant along those lines. Thus, the solution is given by

$$u(x,y) = h(x - cy).$$

If we take the initial function $h(x) = e^{-x^2}$, then the solution becomes

$$u(x,y) = e^{-(x-cy)^2},$$

and are graphed in Fig. 2.8 for wave speed $c = 1$ and different values of y.

Now we turn our attention to equation (2.42) subject to the initial data given by (2.43). One of the characteristic equation is

$$\frac{dx}{dy} = a(u).$$

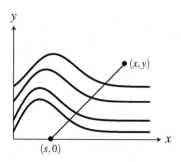

FIGURE 2.7
Wave propagation of the solutions $u(x+cy,y) = h(s-cy)$ by considering y as spacial time.

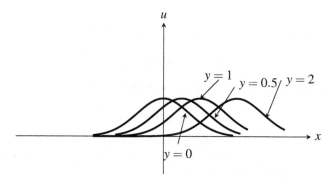

FIGURE 2.8
Wave propagation of the solutions $u(x,y) = e^{-(x-cy)^2}$ for wave speed $c = 1$.

If $x(t)$ is the corresponding characteristic curve, then

$$\frac{du(x,y)}{dy} = u_y + u_x\frac{dx}{dy} = u_y + a(u)u_x = 0.$$

This shows that the solution u is constant along the characteristics. Moreover, the characteristics are straight lines as it is obvious from the calculation below

$$\frac{d^2x}{dy^2} = \frac{d}{dy}(\frac{dx}{dy}) = a'\frac{du(x,y)}{dy} = 0.$$

Let

$$\Gamma: x = s, \quad y = 0, \quad z = h(s), \quad \text{at} \quad t = 0.$$

Using system (2.28), we have

$$\frac{dx}{dt} = a(z), \quad \frac{dy}{dt} = 1, \quad \frac{dz}{dt} = 0,$$

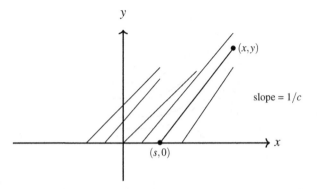

FIGURE 2.9
Characteristic lines running into each others in the nonlinear case, unlike Fig. 2.6.

and corresponding solutions along Γ are

$$y = t, \quad x = s + a(h(s))t, \quad z = h(s).$$

Replacing t with y we obtain the family of characteristic lines

$$x = s + a(h(s))y, \tag{2.45}$$

where s is the x-intercept of the characteristic curves. The characteristics are straight lines whose slopes are not constants but varies with s as Fig. 2.9 shows. Eliminating s and using $z = u$ we obtain the solution

$$u(x, y) = h(x - a(u)y). \tag{2.46}$$

We want a continuous solution with continuous u_x and u_y. Using (2.46) we have

$$u_x = h'(x - a(u)y)(1 - a'(u)u_x y),$$

or

$$u_x = \frac{h'(s)}{1 + a'(u)h'(s)y} \tag{2.47}$$

Similarly, taking partial derivative with respect to y we get

$$u_y = h'(x - a(u)y)(-a'(u)u_y y - a(u)).$$

Solving for u_y gives

$$u_y = \frac{h'(s)a(u)}{1 + a'(u)h'(s)y}. \tag{2.48}$$

Thus, along the characteristic $x = s + a(h(s))y$, we have u_x and u_y given by (2.47) and (2.48), respectively. Moreover, u_x and u_y become infinite at the positive time

$$y = -\frac{1}{a'(u)h'(s)}, \quad \text{provided that} \quad a'(u)h'(s) < 0. \tag{2.49}$$

If $a'(u) > 0$, then in order for solutions to exist, expression (2.49) implies that $h'(s) > 0$. In other words, $h(s)$ is an increasing function. Otherwise, the solutions will experience a "blow -up." For example, if $a(u) = u$, then condition (2.49) takes the form

$$y = \min\{-\frac{1}{h'(s)}\},$$

and solutions will experience blow-up at and beyond the time $y = -\frac{1}{h'(s_0)}$, where $h(s_0)$ is the minimum of $h(s)$ at $s = s_0$, and $h'(s) < 0$; that is h is non-increasing.

Example 2.15 Solve

$$u_y + uu_x = 0, \quad -\infty < x < \infty, y > 0 \tag{2.50}$$

subject to the initial data

$$u(x,0) = h(x) = \begin{cases} 1, & x \leq 0 \\ 1-x, & 0 < x \leq 1 \\ 0, & x > 1. \end{cases} \tag{2.51}$$

Parametrizing the initial curve Γ, we get

$$\Gamma : x = s, \quad y = 0, \quad z = \begin{cases} 1, & s \leq 0 \\ 1-s, & 0 < s \leq 1 \\ 0, & s > 1. \end{cases}$$

Using $a(u) = u$, it follows from the above discussion that

$$y = t, \quad x = zt + s, \quad z = h(s) = \begin{cases} 1, & s \leq 0 \\ 1-s, & 0 < s \leq 1 \\ 0, & s > 1. \end{cases}$$

Eliminating t from x and y, the characteristics satisfy

$$x = zy + s \quad \text{or} \quad s = x - zy. \tag{2.52}$$

We try to piece the solution together since our initial data is given piecewise.

1. For $s \leq 0$, $u = z = 1$. Moreover, we have from (2.52) that $s = x - y$. Since, $s \leq 0$, it follows that $x - y \leq 0$. We conclude that

$$u = 1, \quad \text{for} \quad x \leq y.$$

2. For $s > 1$, $u = z = 0$. Moreover, we have from (2.52) that $s = x$. We conclude that

$$u = 0, \quad \text{for} \quad x > 1.$$

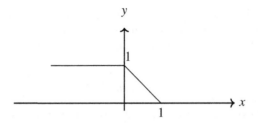

FIGURE 2.10
Initial wave profile.

3. For $0 < s \le 1$, $u = z = 1 - s$. Moreover, we have from (2.52) that $s = x - zy = x - (1-s)y$. Solving for s yields $s = \dfrac{x-y}{1-y}$. Now substituting this value of s into $u = 1 - s$ gives

$$u = 1 - \frac{x-y}{1-y} = \frac{1-x}{1-y}.$$

As for the domain, we have $0 < s \le 1$, which implies that $0 < \dfrac{x-y}{1-y} \le 1$. Rearranging the terms we arrive at $y < x \le 1$. We conclude that

$$u = \frac{1-x}{1-y}, \quad \text{for} \quad y < x \le 1.$$

Finally, the solution is

$$u(x,y) = \begin{cases} 1, & x \le y \\ \frac{1-x}{1-y}, & y < x \le 1 \\ 0, & x > 1. \end{cases} \tag{2.53}$$

$\square$

The obtained solution in (2.53) is valid for $0 \le y < 1$, and discontinuous at $y = 1$. The characteristics run into each others in the wedged region where $y > 1$. (See Fig. 2.11). Next, our goal is to extend the solution beyond $y \ge 1$. To do so, we introduce a curve starting at the discontinuity point $(1,1)$ and try to construct such curve (shock path) as shown in the next section.

Example 2.16 Find the blow-up time for

$$u_y + u u_x = 0, \quad -\infty < x < \infty, \ y > 0$$

subject to the initial data

$$u(x,0) = \frac{1}{1+x^2}.$$

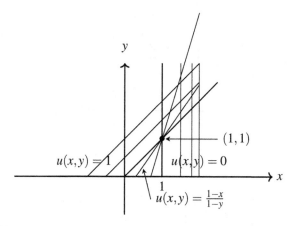

FIGURE 2.11
Characteristic lines running into each others in the nonlinear case, unlike Fig. 2.6.

According to (2.49), the blow-up time is $y = \min\{-\frac{1}{h'(\xi)}\}$, with $h(\xi) = \frac{1}{1+\xi^2}$.
Wherfore, we have $h'(\xi) = -\frac{2\xi}{(1+\xi^2)^2}$. We need to find the minimum of the function g where

$$g(\xi) = -\frac{1}{h'(\xi)} = \frac{(1+\xi^2)^2}{2\xi}.$$

After some calculations, we find that

$$g'(\xi) = \frac{3\xi^4 + 2\xi^2 - 1}{2\xi^2}.$$

Setting $g'(\xi) = 0$, it follows that the only feasible solution (positive and real time) is $\xi = \frac{1}{\sqrt{3}}$, which minimizes the function g. Thus, the blow-up time is

$$y = g(\frac{1}{\sqrt{3}}) = \frac{8\sqrt{3}}{9}.$$

$\square$

2.4.1 Shock path

Consider the first-order PDE

$$u_t + \big(F(u)\big)_x = 0, \quad -\infty < x < \infty, \ t > 0 \tag{2.54}$$

where F is continuously differentiable function. An equation of this form is called *conservation law* for the following reasons. Integrate (2.54) from $x = a$ to $x = b$ and get

$$\frac{d}{dt} \int_a^b u(x,t)dx + \int_a^b \big(F(u)\big)_x dx = 0.$$

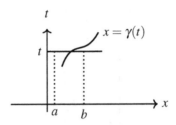

FIGURE 2.12
Shock path.

But,

$$\int_a^b \left(F(u)\right)_x dx = F\left(u(b,t)\right) - F\left(u(a,t)\right),$$

and so we have

$$\frac{d}{dt}\int_a^b u(x,t)dx = F\left(u(a,t)\right) - F\left(u(b,t)\right). \tag{2.55}$$

If u is the amount of a quantity per unit length, then the left of (2.55) is the time rate of change of the total amount of the quantity inside the interval $[a,b]$. If $F\left(u(x,t)\right)$ is the flux through x, that is, the amount of the quantity per unit time positively flowing across x, then (2.55) implies that the rate of the quantity in $[a,b]$ equals the flux in at $x = a$ minus the flux out through $x = b$.

As depicted in Fig. 2.12, let $x = \gamma(t)$ be a smooth curve across which u is discontinuous. Assume u is smooth on each side of the curve γ. Let u_0 and u_1 denote the right and left limits of u at $\gamma(t)$, respectively. That is,

$$\lim_{x \to \gamma^+(t)} u(x,t) = u_0 \quad \text{and} \quad \lim_{x \to \gamma^-(t)} u(x,t) = u_1.$$

From (2.55), we have

$$
\begin{aligned}
F\left(u(a,t)\right) - F\left(u(b,t)\right) &= \frac{d}{dt}\int_a^{\gamma^-(t)} u(x,t)dx + \frac{d}{dt}\int_{\gamma^+(t)}^b u(x,t)dx \\
&= \int_a^{\gamma^-(t)} u_t(x,t)dx + u(\gamma^-(t),t)\frac{d\gamma^-}{dt} \\
&\quad + \int_{\gamma^+(t)}^b u_t(x,t)dx - u(\gamma^+(t),t)\frac{d\gamma^+}{dt} \\
&= \int_a^{\gamma^-(t)} u_t(x,t)dx + u_1\frac{d\gamma^-}{dt} \\
&\quad + \int_{\gamma^+(t)}^b u_t(x,t)dx - u_0\frac{d\gamma^+}{dt}. \tag{2.56}
\end{aligned}
$$

As

$$a \rightarrow \gamma^-(t) \quad \text{and} \quad b \rightarrow \gamma^+(t),$$

we have

$$\int_a^{\gamma^-(t)} u_t(x,t)dx \rightarrow 0 \quad \text{and} \quad \int_{\gamma^+(t)}^b u_t(x,t)dx \rightarrow 0.$$

As a consequence, expression (2.56) reduces to

$$F(u_1) - F(u_0) = (u_1 - u_0)\frac{d\gamma}{dt}. \tag{2.57}$$

Adopting the notation

$$[F(u)] = F(u_1) - F(u_0) \quad \text{and} \quad [u] = u_1 - u_0,$$

equation (2.57) takes the form

$$\frac{d\gamma}{dt} = \frac{[F(u)]}{[u]}. \tag{2.58}$$

Using (2.58), the weak solution that evolves will be a piecewise smooth function with a discontinuity or shock wave, that propagates with shock speed.

Example 2.17 Consider the problem of Example 2.15. By setting $t = y$, then $u_t + uu_x = 0$ is equivalent to

$$u_t + \left(\frac{1}{2}u^2\right)_x = 0, \quad x \in \mathbb{R}, t > 0.$$

It follows that

$$F(u) = \frac{1}{2}u^2.$$

We know from Fig. 2.13 that the shock occurs at and beyond $(1,1)$. Also from Fig. 2.13, to the left of the shock we have $u_1 = 1$, and to the right of the shock, $u_0 = 0$. So, $[u] = u_1 - u_0 = 1$ and

$$[F(u)] = F(u_1) - F(u_0) = \frac{1}{2}u_1^2 - \frac{1}{2}u_0^2 = \frac{1}{2}.$$

Thus the path of shock, γ has the slope

$$\frac{d\gamma}{dt} = \frac{1}{2}.$$

Therefore, γ is of the form $2x = t + c$. Since this path passes through $(1,1)$, we see that $c = 1$. It follows that the shock line is given by

$$x(t) = \frac{t}{2} + \frac{1}{2}.$$

Therefore, for $t \geq 1$, the solution is given by

$$u(x,y) = \begin{cases} 1, & x(t) < \frac{t}{2} + \frac{1}{2} \\ 0, & x(t) > \frac{t}{2} + \frac{1}{2}. \end{cases} \tag{2.59}$$

□

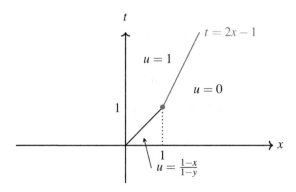

FIGURE 2.13
Characteristic lines intersecting the shock line $t = 2x - 1$.

2.4.2 Exercises

Exercise 2.31 *Find the breaking time for*

$$u_t + uu_x = 0, \quad u(x,0) = e^{-2x^2}, \quad x \in \mathbb{R}, \ t \geq 0.$$

Exercise 2.32 *Consider the PDE*

$$u_t + u^2 u_x = 0, \quad u(x,0) = \frac{1}{x^4 + 1}, \quad x \in \mathbb{R}, \ t \geq 0.$$

(a) Find and graph the characteristics.

(b) Determine the breaking time and find the shock line.

(c) Find the solution before the breaking time.

Exercise 2.33 *Find the breaking time for*

$$u_t + u^3 u_x = 0, \ u(x,0) = x^{1/3}, \quad x \in \mathbb{R}, \ t \geq 0$$

and then find the solution.

Exercise 2.34 *Solve and find the shock line of the traffic flow problem and explain the physical meaning of the solution*

$$u_t + (1 - 2u)u_x = 0,$$

subject to

$$u(x,0) = \begin{cases} 1/2, & x < 0 \\ 1, & x > 0. \end{cases}$$

Exercise 2.35 *Solve and find the shock line*

$$u_t + uu_x = 0, \quad u(x,0) = \begin{cases} 0, & x \leq 0 \\ x, & 0 \leq x \leq 2 \\ 2, & x > 2. \end{cases}$$

Exercise 2.36 *Solve and find the shock line*

$$u_t + u^2 u_x = 0, \quad u(x,0) = \begin{cases} 0, & x \le 0 \\ 2, & 0 \le x \le 1 \\ 0, & x \ge 1. \end{cases}$$

Exercise 2.37 *Solve*

$$u_t + u u_x = 0, \quad u(x,0) = \begin{cases} 0, & x \le -1 \\ 2, & -1 \le x \le 1 \\ 1, & x > 1. \end{cases}$$

Exercise 2.38 *Solve*

$$u_t + u u_x = 0, \quad u(x,0) = \begin{cases} 1, & x \le 0 \\ 1 - \frac{x}{a}, & 0 \le x < a \\ 1, & x \ge 1. \end{cases}$$

Exercise 2.39 *(a) Solve*

$$u_t + u u_x = 0, \quad u(x,0) = \begin{cases} 1, & x \le 0 \\ 2, & x \ge 0. \end{cases}$$

(b) Draw the characteristics.

2.5 Second-Order PDEs

In this section we consider general second-order partial differential equations. We begin by looking into their classifications and reduction of orders. Consider the general second-order linear PDE

$$A(x,y)u_{xx} + 2B(x,y)u_{xy} + C(x,y)u_{yy} + D(x,y)u_x + E(x,y)u_y + F(x,y)u = G(x,y), \tag{2.60}$$

where the function u and the coefficients are twice continuously differentiable in some domain $\Omega \subset \mathbb{R}^2$. We shall consider (2.60) along with the Cauchy conditions imposed on some curve Γ that is defined by $y = f(x)$. Imposing Cauchy conditions implies that u_x and u_y are known on Γ. That is,

$$u_x(x, y(x)) = f(x), \quad u_y(x, y(x)) = g(x),$$

where f and g are known functions. A differentiation of these relations with respect to x gives

$$u_{xx}(x, y(x)) + \frac{dy}{dx} u_{xy}(x, y(x)) = f_x(x) \tag{2.61}$$

$$u_{xy}(x, y(x)) + \frac{dy}{dx} u_{yy}(x, y(x)) = g_x(x). \tag{2.62}$$

In addition, along the curve Γ, equation (2.60) takes the form

$$A u_{xx}(x, y(x)) + 2B u_{xy}(x, y(x)) + C u_{yy}(x, y(x)) = H, \tag{2.63}$$

where H is a known function in x. Equations (2.61)–(2.63) determine u_{xx}, u_{yy}, and u_{xy} uniquely unless

$$\triangle = \begin{vmatrix} A & 2B & C \\ 0 & 1 & \frac{dy}{dx} \\ 1 & \frac{dy}{dx} & 0 \end{vmatrix} = -A\left(\frac{dy}{dx}\right)^2 + 2B\frac{dy}{dx} - C = 0.$$

Or,

$$A\left(\frac{dy}{dx}\right)^2 - 2B\frac{dy}{dx} + C = 0. \tag{2.64}$$

The above equation is quadratic in $\frac{dy}{dx}$ with solutions

$$\frac{dy}{dx} = \frac{B \pm \sqrt{B^2 - AC}}{A}. \tag{2.65}$$

When $B^2 - AC > 0$, there exists two families of curves such that no solution can be found when Cauchy conditions are imposed on them. The families of curves are known as the characteristics. On the other hand, there are no characteristics when $B^2 - AC < 0$, and one family of characteristics exists when $B^2 - AC = 0$. We call the initial curve Γ characteristic with respect to (2.60) and Cauchy conditions if $\triangle = 0$ along Γ, noncharacteristic if $\triangle \neq 0$ along Γ. When Γ is noncharacteristic, the Cauchy data uniquely determine the solution. However, in the case of a characteristic initial curve Γ, then (2.61)–(2.63) are inconsistent, unless more data is offered. Thus, when Cauchy data coincide with the initial curve Γ, the PDE (2.60) has no solution.

Definition 2.4 *The PDE in* (2.60) *has the following classifications: it is hyperbolic if* $B^2 - AC > 0$, *parabolic if* $B^2 - AC = 0$, *and elliptic if* $B^2 - AC < 0$.

We conclude from Definition 2.4, that the classification of the PDE (2.60) depends on the *highest order terms*. Our next task is to use transformations that will reduce the complicated PDE (2.60) to a simpler one that we can easily solve using the knowledge of the previous sections of this chapter. We introduce the transformations

$$\xi = \xi(x, y), \quad \eta = \eta(x, y), \tag{2.66}$$

where ξ and η are twice continuously differentiable and that the Jacobian

$$J = \begin{vmatrix} \xi_x & \xi_y \\ \eta_x & \eta_y \end{vmatrix} \neq 0 \tag{2.67}$$

in the region of interest. Then, x and y are uniquely determined from the system (2.66). With this in mind, using the chain rule, we obtain

$$u_x = u_\xi \xi_x + u_\eta \eta_x$$

$$u_y = u_\xi \xi_y + u_\eta \eta_y$$

$$\begin{aligned} u_{xx} &= (u_\xi)_x \xi_x + u_\xi \xi_{xx} + (u_\eta)_x \eta_x + u_\eta \eta_{xx} \\ &= u_{\xi\xi} \xi_x^2 + 2u_{\eta\xi} \xi_x \eta_x + u_{\eta\eta} \eta_x^2 + u_\xi \xi_{xx} + u_\eta \eta_{xx}. \end{aligned}$$

In a similar fasion, we obtain

$$u_{xy} = u_{\xi\xi} \xi_x \xi_y + u_{\xi\eta}(\xi_x \eta_y + \xi_y \eta_x) + u_{\eta\eta} \eta_x \eta_y + u_\xi \xi_{xy} + u_\eta \eta_{xy},$$

$$u_{yy} = u_{\xi\xi} \xi_y^2 + 2u_{\eta\xi} \xi_y \eta_y + u_{\eta\eta} \eta_y^2 + u_\xi \xi_{yy} + u_\eta \eta_{yy}.$$

Substituting into (2.60) we obtain the new and reduced PDE

$$\hat{A} u_{\xi\xi} + 2\hat{B} u_{\xi\eta} + \hat{C} u_{\eta\eta} + M(\xi, \eta, u, u_\xi, u_\eta) = 0, \tag{2.68}$$

where the new coefficients are known and we list the highest order terms. That is,

$$\hat{A} = A\xi_x^2 + 2B\xi_x \xi_y + C\xi_y^2,$$

$$\hat{B} = A\xi_x \eta_x + B(\xi_x \eta_y + \xi_y \eta_x) + C\xi_y \eta_y,$$

and

$$\hat{C} = A\eta_x^2 + 2B\eta_x \eta_y + C\eta_y^2.$$

Equation (2.68) is called the *canonical form* of (2.60). Thus, it can be easily shown that

$$\hat{B}^2 - \hat{A}\hat{C} = J^2(B^2 - AC),$$

which preserves the personification of the PDE (2.60) under the transformation (2.66). In the next discussion we explain how to find the transformations ξ and η. Suppose none of A, B, C is zero. Assume that under the transformations (2.66) $\hat{A}$ and $\hat{C}$ vanish. Let's consider $\hat{A} = 0$. Then it follows that

$$A\xi_x^2 + 2B\xi_x \xi_y + C\xi_y^2 = 0.$$

Divide by ξ_y^2 to obtain

$$A\left(\frac{\xi_x}{\xi_y}\right)^2 + 2B\left(\frac{\xi_x}{\xi_y}\right) + C = 0, \tag{2.69}$$

which is quadratic in $\dfrac{\xi_x}{\xi_y}$. Now along the curve $\xi = $ constant, we have

$$d\xi = \xi_x dx + \xi_y dy = 0,$$

or

$$\frac{dy}{dx} = -\frac{\xi_x}{\xi_y}. \tag{2.70}$$

Similarly, if we set $\hat{C} = 0$ then we obtain parallel equations to (2.69) and (2.70) in terms of η. A comparison of (2.65), (2.69), and (2.70) yields that ξ and η are the solutions to the ordinary differentials equations given by (2.65)

$$\frac{dy}{dx} = \frac{B \pm \sqrt{B^2 - AC}}{A}$$

along which $\xi = $ constant and $\eta = $ constant.

Remark 7 **1.** $B^2 - AC > 0.$ **(Hyperbolic equations)**
   In this case ξ and η are uniquely obtained from (2.65) and satisfy (2.67).

2. $B^2 - AC = 0.$ **(Parabolic equations)**
   In this case we can only determine either ξ or η from (2.65). Say ξ is a solution to $\frac{dy}{dx} = \frac{B}{A}$. Then $\xi = \frac{B}{A}x - y$. We may set $\eta = x$, and under this transformation of ξ and η (2.67) is satisfied. As a matter of fact, $J = 1 \neq 0$.

3. $B^2 - AC < 0.$ **(Elliptic equations)**

In this case we determine both ξ and η from (2.65) and arrive at the transformation

$$\xi = \frac{B}{A}x - y + i\frac{\sqrt{B^2 - AC}}{A}x \quad and \quad \eta = \frac{B}{A}x - y - i\frac{\sqrt{B^2 - AC}}{A}x.$$

One may show (see Exercise 2.43) that $J \neq 0$, and $\hat{A} \neq 0$, $\hat{C} \neq 0$, and $\hat{B} = 0$.

Example 2.18 Solve

$$u_{xx} - 2u_{xy} - 3u_{yy} = 0,$$

subject to the Cauchy conditions

$$u(x,0) = x^2, \quad u_y(x,0) = \frac{1}{3}.$$

From (2.64) we have

$$\left(\frac{dy}{dx}\right)^2 + 2\frac{dy}{dx} - 3 = 0,$$

with

$$\frac{dy}{dx} = \frac{-1 \pm \sqrt{4}}{1} = 1, -3.$$

The two differential equations have the solutions

$$y = x + c_1, \quad y = -3x + c_2.$$

Solving for c_1 and c_2, we get $c_1 = y - x$, and $c_2 = y + 3x$. As a consequence, we may set

$$\xi = y + 3x, \quad \eta = y - x.$$

We note that the Jacobian $J = 4 \neq 0$. It follows that

$$u_x = 3u_\xi - u_\eta$$

$$u_y = u_\xi + u_\eta$$

$$u_{xx} = 9u_{\xi\xi} - 6u_{\eta\xi} + u_{\eta\eta}$$

$$u_{xy} = 3u_{\xi\xi} + 2u_{\xi\eta} - u_{\eta\eta}$$

$$u_{yy} = u_{\xi\xi} + 2u_{\eta\xi} + u_{\eta\eta}.$$

Thus,

$$
\begin{aligned}
u_{xx} - 2u_{xy} - 3u_{yy} &= 9u_{\xi\xi} - 6u_{\eta\xi} + u_{\eta\eta} - 2(3u_{\xi\xi} + 2u_{\xi\eta} - u_{\eta\eta}) \\
&\quad - 3(u_{\xi\xi} + 2u_{\eta\xi} + u_{\eta\eta}) \\
&= -16u_{\xi\eta} = 0.
\end{aligned}
$$

Thus, under the transformation $\xi = y + 3x$, $\eta = y - x$, the original PDE is transformed to the canonical form

$$u_{\xi\eta} = 0,$$

which has the solution

$$u(\xi, \eta) = F(\xi) + G(\eta),$$

for some functions F and G. In terms of x and y the general solution is

$$u(x, y) = F(y + 3x) + G(y - x).$$

Applying the Cauchy condition $x^2 = u(x, 0)$, we arrive at

$$x^2 = F(3x) + G(-x). \tag{2.71}$$

To apply the second Cauchy condition, we notice that

$$u_y(x, y) = F'(y + 3x) + G'(y - x).$$

Thus,

$$1/3 = F'(3x) + G'(-x).$$

Integrate both sides and then multiply the resulting equation with 3, to get

$$x = F(3x) - 3G(-x). \tag{2.72}$$

Solving the system of equations given by (2.71) and (2.72) we obtain

$$F(3x) = \frac{3x^2 + x}{4}, \quad G(-x) = \frac{x^2 - x}{4}.$$

Let $z = 3x$, then $F(z) = \frac{z^2}{12} + \frac{z}{12}$, and we conclude that

$$F(y + 3x) = \frac{(y + 3x)^2}{12} + \frac{(y + 3x)}{12}.$$

Similarly, if we set $w = -x$, then $G(w) = \frac{w^2}{4} + \frac{w}{4}$ and as a consequence,

$$G(y-x) = \frac{(y-x)^2}{4} - \frac{(y-x)}{4}.$$

Finally, the solution is

$$
\begin{aligned}
u(x,y) &= F(y+3x) + G(y-x) \\
&= \frac{(y+3x)^2}{12} + \frac{(y+3x)}{12} + \frac{(y-x)^2}{4} - \frac{(y-x)}{4}.
\end{aligned}
$$

$\square$

2.5.1   Exercises

Exercise 2.40 *Find the characteristics and reduce to canonical form and then solve.*

$$u_{xx} - 2\sin(x)u_{xy} - \cos^2(x)u_{yy} - \cos(x)u_y = 0.$$

Exercise 2.41 *Find the characteristics and reduce to canonical form and then solve*

$$4u_{xx} + 5u_{xy} + u_{yy} + u_x + u_y = 0.$$

Exercise 2.42 *Show that in the hyperbolic case when $B^2 - AC > 0$, we have $\hat{A} = \hat{C} = 0$, and $\hat{B} \neq 0$.*

Exercise 2.43 *Show that in the elliptic case when $B^2 - AC < 0$, we have $\hat{A} \neq 0$, $\hat{C} \neq 0$, and $\hat{B} = 0$.*

Exercise 2.44 *Solve*

$$u_{xx} - 4u_{xy} + 4u_{yy} = 0,$$

subject to the Cauchy conditions

$$u(x,0) = e^{2x}, \; u_y(x,0) = 5.$$

Exercise 2.45 *Solve*

$$3u_{xx} + 10u_{xy} + 3u_{yy} = 0,$$

subject to the Cauchy conditions

$$u(x,0) = e^{-x}, \; u_y(x,0) = x.$$

Exercise 2.46 *Find the characteristics and reduce to canonical form and then find the general solution*

$$4u_{xx} + 5u_{xy} + u_{yy} + u_x + u_y = 3.$$

Exercise 2.47 *Find the characteristics and reduce to canonical form and then find the general solution*

$$x^2u_{xx} + 2xyu_{xy} + y^2u_{yy} + xyu_x + y^2u_y = 0.$$

Exercise 2.48 *Find the characteristics and reduce to canonical form and then solve*

$$u_{xx} + 2u_{xy} + 2u_{yy} = 0.$$

Exercise 2.49 *For constants a and b use the transformation*

$$v = ue^{-(a\xi + b\eta)}$$

to transform the PDE

$$u_{xx} - u_{yy} + 3u_x - 2u_y + u = 0,$$

into

$$v_{\xi\eta} = cv, \ c = \ constant.$$

Exercise 2.50 *Solve*

$$x^2 u_{xx} - y^2 u_{yy} = xy,$$

subject to the Cauchy conditions

$$u(x, 1) = x, \ u_y(x, 1) = 6.$$

Exercise 2.51 *Solve*

$$(x - y)u_{xy} - u_x + u_y = 0.$$

Hint: set $v = (x - y)u_x + u.$

Exercise 2.52 *Solve*

$$xyu_{xx} + x^2 u_{xy} - yu_x = 0,$$

subject to the Cauchy conditions

$$u(x, 0) = e^x, \ u_y(x, 0) = 5.$$

2.6 Wave Equation and D'Alembert's Solution

This section is about the wave equation. We will discuss the derivation of the wave equation, the D'Alembert solution, the domain of dependence of solutions, and solutions to the nonhomogeneous wave equation.

We begin this long section with the derivation of the wave equation. Let $l > 0$ be the length of a thin string that is stretched between two points on the x-axis. A string vibrates only if it is tightly stretched. Assume the string undergoes relatively small transverse vibrations (think of the string of a musical instrument, say a violin string). Let ρ be the linear density of the string, measured in units of mass per unit of length. We will assume that the string is made of homogeneous material. and its density is constant along the entire length of the string. The displacement of the string from its equilibrium state at time t and position x will be denoted by $u(t, x)$. We assume the string is positioned in such a way that its left endpoint coincides with the origin of the xu coordinate system. Consider the motion of a small portion of the string sitting atop the interval $[a,b]$. Then the corresponding mass is $\rho(b-a)$, and acceleration u_{tt}. Using Newton's second law of motion, we have

$$\rho(b-a) = \text{Total force.} \tag{2.73}$$

Since the mass of the string is negligible, we may discard the effect of gravity on the string. In addition, we may as well ignore air resistance, and other external forces. Thus, the only force that is acting on the string is the tension force $\mathbb{T}(x,t)$. Assuming that the string is perfectly flexible, the tension force will have the direction of the tangent vector along the string. At a fixed time t the position of the string is given by the parametric equations, $x = x$, $u = u(x,t)$, where x is a parameter. Then, the tangent vector is $< 1, u_x >$, with corresponding unit vector $< \dfrac{1}{\sqrt{1+u_x^2}}, \dfrac{u_x}{\sqrt{1+u_x^2}} >$. Under this set up the tension force takes the form

$$\mathbb{T}(x,t) = < \frac{T(x,t)}{\sqrt{1+u_x^2}}, \frac{T(x,t)u_x}{\sqrt{1+u_x^2}} > \tag{2.74}$$

where $T(x,t)$ is the magnitude of the tension force. Due to the assumption of a small vibration, it is safe to assume that u_x is small, and thus, via Taylor's expansion we have

$$\sqrt{1+u_x^2} = 1 + \frac{1}{2}u_x^2 + o(u_x^4) \approx 1.$$

Substituting this approximation into (2.74), we arrive at an equivalent form of the tension force

$$\mathbb{T}(x,t) = < T(x,t), T(x,t)u_x > .$$

Since there is no longitudinal displacement, we arrive at the following identities for the balances of forces (2.73) in the x, respectively u directions

$$0 = T(b,t) - T(a,t)$$

$$\rho(b-a)u_{tt} = T(b,t)u_x(b,t) - T(a,t)u_x(a,t).$$

Simply stated, the first equation shows that the tensions from the two edges of the little portion of the string balance each other out in the x direction (no longitudinal motion). From this, we can also infer that the position of the string has no impact on the tension force. Hence, the second equation might be rewritten as

$$\rho u_{tt} = T\frac{u_x(b,t) - u_x(a,t)}{b-a}.$$

Taking the limit in the above equation we arrive at the wave equation

$$\rho u_{tt} = \lim_{b \to a} T\frac{u_x(b,t) - u_x(a,t)}{b-a} = Tu_{xx},$$

or

$$u_{tt} - c^2 u_{xx} = 0,$$

with $c^2 = \frac{T}{\rho}$.

The above derivation can be generalized to incorporate effects of other forces, as displayed below. The wave equation

$$u_{tt} - c^2 u_{xx} + r u_t = 0, \quad r > 0,$$

reflects air resistance as a force proportional to the speed u_t. On the other hand, the wave equation

$$u_{tt} - c^2 u_{xx} + k u = 0, \quad k > 0,$$

incorporates transverse elastic force that is proportional to the displacement u. Finally, the wave equation

$$u_{tt} - c^2 u_{xx} = f(x, t), \quad r > 0,$$

incorporates externally applied forces. Such equation is refereed to as inhomogeneous wave equation, which we will discuss its solution at the end of this section.

A good and important application of a hyperbolic PDE with Cauchy conditions is the one-dimensional wave equation

$$u_{tt} - c^2 u_{xx} = 0 \tag{2.75}$$

$$u(x, 0) = f(x) \tag{2.76}$$

$$u_t(x, 0) = g(x) \tag{2.77}$$

where the function f is twice continuously differentiable and g is continuously differentiable. The function f is the initial displacement and g is the initial velocity. Using the method of the previous section, with $A = 1$, $B = 0$, and $C = -c^2$, we have the characteristics, which are solutions of the differential equations

$$\frac{dt}{dx} = \frac{\pm c}{c^2} = \pm \frac{1}{c}.$$

Thus, the corresponding characteristic lines (solutions) are

$$c_1 = x - ct, \quad c_2 = x + ct.$$

Let

$$\xi = x + ct, \quad \eta = x - ct.$$

Then,

$$u_{xx} = u_{\xi\xi} + 2u_{\xi\eta} + u_{\eta\eta},$$

and

$$u_{tt} = c^2 (u_{\xi\xi} - 2u_{\xi\eta} + u_{\eta\eta}).$$

Substituting into (2.75) we arrive at the canonical form $-4c^2 u_{\xi\eta} = 0$. Since $c \neq 0$, we must have

$$u_{\xi\eta} = 0,$$

which has the general solution

$$u(\xi, \eta) = F(\xi) + G(\eta),$$

where the functions F and G are arbitrary and required to be twice differentiable. In terms of t and x, the solution takes the form

$$u(x,t) = F(x+ct) + G(x-ct). \tag{2.78}$$

To determine the arbitrary functions F and G we apply the initial conditions or Cauchy conditions (2.76) and (2.77) and obtain

$$f(x) = F(x) + G(x), \tag{2.79}$$

and

$$g(x) = cF'(x) - cG'(x). \tag{2.80}$$

Integrating (2.80) from x_0 to x we arrive at

$$F(x) - G(x) = \frac{1}{c}\int_{x_0}^{x} g(s)ds + K, \tag{2.81}$$

where $x_0 \in \mathbb{R}$ and K are constants. Solving for F and G from (2.79) and (2.81) yields

$$F(x) = \frac{1}{2}[f(x) + \frac{1}{c}\int_{x_0}^{x} g(s)ds + K],$$

and

$$G(x) = \frac{1}{2}[f(x) - \frac{1}{c}\int_{x_0}^{x} g(s)ds - K].$$

Then by (2.78) the general solution takes the form

$$
\begin{aligned}
u(x,t) &= \frac{1}{2}[f(x+ct) + \frac{1}{c}\int_{x_0}^{x+ct} g(s)ds + K] \\
&+ \frac{1}{2}[f(x-ct) - \frac{1}{c}\int_{x_0}^{x-ct} g(s)ds - K] \\
&= \frac{1}{2}[f(x+ct) + f(x-ct)] \\
&+ \frac{1}{2c}\left[\int_{x_0}^{x+ct} g(s)ds - \int_{x_0}^{x-ct} g(s)ds\right].
\end{aligned}
$$

Combining the two integrals we arrive at the *D'Alembert solution*

$$u(x,t) = \frac{1}{2}[f(x+ct) + f(x-ct)] + \frac{1}{2c}\int_{x-ct}^{x+ct} g(s)ds. \tag{2.82}$$

It is simple to verify $u(x,t)$ given by (2.82) is a solution of the wave equation (2.75). Moreover, by a direct substitution into the solution (2.82), it is evident that the initial

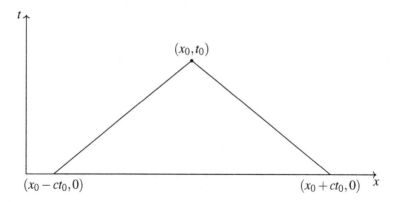

FIGURE 2.14
Nonhomogeneous wave equation.

conditions uniquely determine (2.82). According to (2.82), the value $u(x_0,t_0)$ depends on the initial data f and g in the interval $[x_0 - ct_0, x_0 + ct_0]$ which is cut out of the initial line by the the two characteristics lines with slopes $\pm\frac{1}{c}$ passing through the point (x_0,t_0). The interval $[x_0 - ct_0, x_0 + ct_0]$ on the line $t = 0$ is called the *domain of dependence*, as indicated in Fig. 2.14.

The next theorem is about stability; it says that for a small change in the initial data, only produces a small change in the solution.

Theorem 2.3 *Let $u^*(x,t)$ be another solution of (2.75)–(2.77) with initial data f^* and g^*. Define $|h| = \max_{-\infty<x<\infty} |h(x)|$, for $h : \mathbb{R} \to \mathbb{R}$ is continuous. Similarly, for $u = u(x,t)$, we define*

$$|u|_T = \max_{-\infty<x<\infty; |t|\leq T} |u(x,t)|.$$

Assume there is a small change in the initial data over a finite time T. That is, for small and positive ε we see that

$$|f - f^*| < \frac{\varepsilon}{2}, \quad |g - g^*| < \frac{\varepsilon}{2T}.$$

Then,

$$|u(x,t) - u^*(x,t)| < \varepsilon.$$

Proof *Using (2.82), we have*

$$
\begin{aligned}
|u(x,t) - u^*(x,t)| &= \left| \frac{1}{2}(f(x+ct) - f^*(x+ct) + f(x-ct) - f^*(x-ct)) \right.\\
&\quad + \left. \frac{1}{2c}\int_{x-ct}^{x+ct} (g(s) - g^*(s))ds \right|\\
&\leq \frac{1}{2}|f(x+ct) - f^*(x+ct)| + \frac{1}{2}|f(x-ct) - f^*(x-ct)|
\end{aligned}
$$

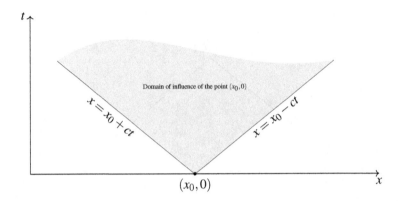

FIGURE 2.15
Domain of influence of the point $(x_0,0)$.

$$+ \quad \frac{1}{2c}\int_{x-ct}^{x+ct} |g(s)-g^*(s)|ds$$

$$\leq \quad \frac{1}{2}(\varepsilon/2+\varepsilon/2) + \frac{1}{2c}\int_{x-ct}^{x+ct} \frac{\varepsilon}{2T}ds$$

$$\leq \quad \varepsilon/2 + \frac{1}{2c}(2cT\frac{\varepsilon}{2T}) = \varepsilon.$$

Thus, for $|t| \leq T$, we have shown that

$$|u-u^*|_T < \varepsilon.$$

This completes the proof.

The D'Alambert solution given by (2.82) indicates that if an initial velocity, or initial displacement is in the neighborhood of (x_0,t_0), it can only influence the area $t > t_0$ bounded by the characteristic lines with slope $\pm\frac{1}{c}$ passing through the point (x_0,t_0), as shown in Fig. 2.15 with initial time $t_0 = 0$.

Example 2.19 Consider

$$4u_{tt} - 9u_{xx} = 0$$

subject to $u(x,0) = x^2$, $u_t(x,0) = \sin(x)$. Then $u(x,t)$ is given by the D'Alambert's solution (2.82) with $f = x^2$, $g = \sin(x)$, and $c^2 = \frac{9}{4}$. Thus,

$$\begin{aligned}
u(x,t) &= \frac{1}{2}[(x+\frac{3}{2}t)^2 + (x-\frac{3}{2}t)^2] + \frac{1}{3}\int_{x-\frac{3}{2}t}^{x+\frac{3}{2}t} \sin(s)ds \\
&= x^2 + \frac{9}{4}t^2 + \frac{2}{3}\sin(x)\sin(\frac{3}{2}t).
\end{aligned}$$

$\square$

Now that we displayed an example, let us take a closer look at the geometrical interpretation of (2.78). The term on the right-hand side of (2.78) is called the *progressive wave*. If we let $x^* = ct$, then the transformation $\xi = x + ct = x + x^*$ is a translation of the coordinate system to the left by x^*. Thus, $F(x + ct)$ is a wave that moves in the negative x direction with speed c without change in its shape. For example, $u(x,t) = \cos(x + ct)$ represents a cosine wave which moves in the negative x-direction with speed c without changing its shape. Similarly, $F(x - ct)$ is a wave which moves in the positive x-direction with speed c without change in its shape. Consequently, the solution

$$u(x,t) = F(x + ct) + G(x - ct)$$

is the sum of two waves traveling in opposite directions, and the shape of $u(x,t)$ will change with time.

Example 2.20 Consider the wave problem with zero initial velocity

$$u_{tt} - c^2 u_{xx} = 0,$$

subject to

$$u_t(x,0) = 0,$$

$$u(x,0) = \begin{cases} h, & |x| \le a \\ 0, & |x| > a. \end{cases}$$

This initial data corresponds to an initial disturbance of the string centered at $x = 0$ of height h. The solution is given by (2.82) with $g(x) = 0$. In other words,

$$u(x,t) = \frac{1}{2}[f(x + ct) + f(x - ct)].$$

We need to piece together the solution. Notice that

$$f(x + ct) = \begin{cases} h, & |x + ct| \le a \\ 0, & |x + ct| > a \end{cases}$$

and

$$f(x - ct) = \begin{cases} h, & |x - ct| \le a \\ 0, & |x - ct| > a. \end{cases}$$

As a consequence the solution is defined piecewise over four different regions. We will only consider all regions for $t \ge 0$. It is clear from the definitions of $f(x + ct)$ and $f(x - ct)$ that the four regions are:

$$I = \{|x + ct| \le a, \ |x - ct| \le a\},$$
$$II = \{|x + ct| \le a, \ |x - ct| > a\},$$
$$III = \{|x + ct| > a, \ |x - ct| \le a\},$$
$$IV = \{|x + ct| > a, \ |x - ct| > a\},$$

with

$$u_I(x,t) = h, \quad u_{II}(x,t) = \frac{h}{2}, \quad u_{III}(x,t) = \frac{h}{2}, \quad u_{IV}(x,t) = 0.$$

The notation $u_I(x,t)$ stands for the value of u in region I, and so on. See Fig. 2.16. $\square$

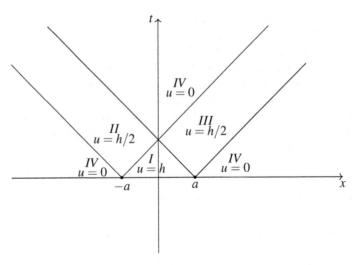

FIGURE 2.16
Different values of u.

Example 2.21 Consider the wave problem with zero initial displacement

$$u_{tt} - c^2 u_{xx} = 0,$$

subject to

$$u(x,0) = 0,$$

$$u_t(x,0) = \begin{cases} g_0, & |x| \le a \\ 0, & |x| > a. \end{cases}$$

This is similar to the previous example but we will have to adjust the interval of integrations. However, here we have six different regions that we list

$$I = \{x - ct < x + ct < -a < a\},$$

$$II = \{x - ct < -a < x + ct < a\},$$

$$III = \{x - ct < -a < a < x + ct\},$$

$$IV = \{-a < x - ct < x + ct < a\},$$

$$V = \{-a < x - ct < a < x + ct\},$$

$$VI = \{-a < a < x - ct < x + ct\},$$

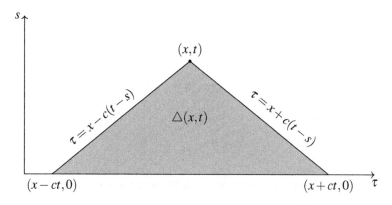

FIGURE 2.17
Nonhomogeneous wave equation.

with

$$u(x,t) = \begin{cases} 0 & \text{in } I \\[2mm] \frac{1}{2c}\int_{-a}^{x+ct} g_0 dx = \frac{g_0}{2c}(x+ct+a), & \text{in } II \\[2mm] \frac{1}{2c}\int_{-a}^{a} g_0 dx = g_0 t, & \text{in } III \\[2mm] \frac{1}{2c}\int_{x-ct}^{x+ct} g_0 dx = g_0 t, & \text{in } IV \\[2mm] \frac{1}{2c}\int_{x-ct}^{a} g_0 dx = \frac{g_0}{2c}(-x+ct+a), & \text{in } V \\[2mm] 0, & \text{in } VI \end{cases}$$

□

Next we consider the nonhomogeneous wave equation

$$u_{tt} - c^2 u_{xx} = h(x,t) \tag{2.83}$$

subject to

$$u(x,0) = f(x), \quad u_t(x,0) = g(x), \tag{2.84}$$

where the function h is assumed to be continuous with respect to both arguments. We show that the solution of (2.83) along with the initial data (2.84) is given by

$$\begin{aligned} u(x,t) &= \frac{1}{2}[f(x+ct)+f(x-ct)] + \frac{1}{2c}\int_{x-ct}^{x+ct} g(s)ds \\ &+ \frac{1}{2c}\int\int_{\triangle(x,t)} h(\tau,s)d\tau ds, \end{aligned} \tag{2.85}$$

where $\triangle(x,t)$ is shown in Fig. 2.17. We will do this by piecing together the solution of the homogeneous problem (2.75)–(2.77), which has the solution given by (2.82), and the solution u^p of

$$u_{tt}^p - c^2 u_{xx}^p = h(x,t) \tag{2.86}$$

subject to

$$u^P(x,0) = 0, \quad u_t^P(x,0) = 0. \tag{2.87}$$

We already have the transformation

$$\xi = x + ct, \quad \eta = x - ct.$$

Solving for x and t we get

$$x = \frac{\xi + \eta}{2}, \quad t = \frac{\xi - \eta}{2c}. \tag{2.88}$$

Under the same transformation, we saw the left side of (2.86) becomes

$$-4c^2 u_{\xi\eta} = 0.$$

Therefore, (2.86) takes the form

$$u_{\xi\eta}^P = -\frac{1}{4c^2} h(\xi, \eta). \tag{2.89}$$

Setting $t = 0$ in (2.88), we immediately have $\xi = \eta$. Thus, the first initial condition of (2.87) reduces to

$$u^P(\xi, \xi) = 0.$$

Using $u_x^P = u_\xi^P \xi_x + u_\eta^P \eta_x = u_\xi^P + u_\eta^P$ we have $u_x^P(x,0) = 0$, implies that

$$u_\xi^P(\xi, \xi) + u_\eta^P(\xi, \xi) = 0.$$

Similarly, $u_t^P = u_\xi^P \xi_t + u_\eta^P \eta_t = c u_\xi^P - c u_\eta^P$. Thus, the second boundary condition of (2.87) reduces to

$$c u_\xi^P(\xi, \xi) - c u_\eta^P(\xi, \xi) = 0.$$

From the last two equations above, it is immediate that $u_\xi^P(\xi, \xi) = u_\eta^P(\xi, \xi) = 0$. Fix a point (x_0, t_0). Then the corresponding point in the characteristic variables is (ξ_0, η_0). In order to find the value of the solution at this point we begin by integrating (2.89) in term of η from ξ to η_0 and obtain

$$\int_\xi^{\eta_0} u_{\xi\eta}^P d\eta = -\frac{1}{4c^2} \int_\xi^{\eta_0} h(\xi, \eta) d\eta.$$

However,

$$\int_\xi^{\eta_0} u_{\xi\eta}^P d\eta = u_\xi^P(\xi, \eta_0) - u_\xi^P(\xi, \xi) = u_\xi^P(\xi, \eta_0).$$

As a result, we have

$$u_\xi^P(\xi, \eta_0) = -\frac{1}{4c^2} \int_\xi^{\eta_0} h(\xi, \eta) d\eta = \frac{1}{4c^2} \int_{\eta_0}^\xi h(\xi, \eta) d\eta. \tag{2.90}$$

Similar to above, the integral,

$$\int_{\eta_0}^{\xi_0} u_\xi^p(\xi,\eta)d\xi = u^p(\xi_0,\eta_0) - u^p(\xi_0,\xi_0) = u^p(\xi_0,\eta_0). \tag{2.91}$$

Integrating (2.90) with respect to ξ from η_0 to ξ_0 and then using (2.91), we arrive at

$$
\begin{aligned}
u^p(\xi_0,\eta_0) &= \frac{1}{4c^2}\int_{\eta_0}^{\xi_0}\int_{\eta_0}^{\xi} h(\xi,\eta)d\eta d\xi \\
&= \frac{1}{4c^2}\int\int_\triangle h(\xi,\eta)d\xi d\eta, \tag{2.92}
\end{aligned}
$$

where the double integral is taken over the triangle of dependence of the point (x_0,t_0), as shown in Fig. 2.14. Left to transform the double integral in (2.92) to a double integral in terms of the variables (x,t). For $\xi = x+ct$, $\eta = x-ct$, we have

$$J = \begin{vmatrix} \xi_x & \xi_y \\ \eta_x & \eta_y \end{vmatrix} = \begin{vmatrix} 1 & c \\ 1 & -c \end{vmatrix} = -2c \neq 0.$$

Thus,

$$u^p(\xi,\eta) = \frac{1}{4c^2}\int\int_{\triangle(x,t)} h(\tau,s)|J|d\tau ds = \frac{1}{2c}\int\int_{\triangle(x,t)} h(\tau,s)d\tau ds, \tag{2.93}$$

where $\triangle(x,t)$ is shown in Fig. 2.17. Finally, adding (2.93) to (2.82), we obtain (2.85). For illustrational purpose, we provide the following two examples.

Example 2.22 Consider

$$4u_{tt} - 9u_{xx} = 4xt$$

subject to $u(x,0) = x^2$, $u_t(x,0) = \sin(x)$. Here $u(x,t)$ is given by the D'Alambert's solution (2.85) with $f = x^2$, $g = \sin(x)$, $h(x,t) = xt$ and $c^2 = \frac{9}{4}$. With the aid of Example 2.19 we have

$$
\begin{aligned}
u(x,t) &= x^2 + \frac{9}{4}t^2 + \frac{2}{3}\sin(x)\sin(\frac{3}{2}t) + \int_0^t\int_{x-\frac{3}{2}(t-s)}^{x+\frac{3}{2}(t-s)} \tau s\, d\tau ds \\
&= x^2 + \frac{9}{4}t^2 + \frac{2}{3}\sin(x)\sin(\frac{3}{2}t) + \frac{1}{3}\int_0^t \frac{s}{2}[x^2 - c^2(t-s)^2]ds \\
&= x^2 + \frac{9}{4}t^2 + \frac{2}{3}\sin(x)\sin(\frac{3}{2}t) + \frac{t^2x^2}{12} - \frac{3t^4}{96}.
\end{aligned}
$$

$\square$

Example 2.23 Consider

$$u_{tt} - 9u_{xx} = \frac{12}{t^2+1}$$

subject to $u(x,0) = x$, $u_t(x,0) = e^{-x}$.

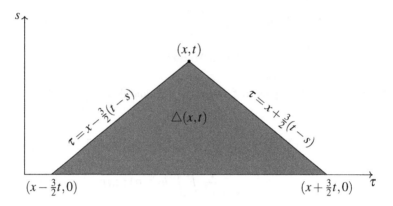

FIGURE 2.18
$\triangle(x,t)$.

(a) Find the solution.

(b) Determine the region where the solution is uniquely defined when $0 \le x \le 4$.

For (a) the solution is given with

$$
\begin{aligned}
u(x,t) &= \frac{1}{2}[(x+3t)+(x-3t)] + \frac{1}{6}\int_{x-3t}^{x+3t} e^{-s}\,ds \\
&\quad + 2\int_0^t \int_{x-3(t-s)}^{x+3(t-s)} \frac{1}{s^2+1}\,d\tau\,ds \\
&= x + \frac{1}{6}\left(e^{-(x-3t)} - e^{-(x+3t)}\right) + 12\int_0^t \frac{t-s}{s^2+1}\,ds \\
&= x + \frac{1}{6}\left(e^{-(x-3t)} - e^{-(x+3t)}\right) + 12t\tan^{-1}(t) - 6\ln(t^2+1).
\end{aligned}
$$

On the other hand, for (b) the initial data is prescribed at $t = 0$, and $0 \le x \le 4$, and so we need to work with the two points $(0,0)$ and $(4,0)$. The characteristic lines are

$$
x - 3t = c_1, \quad \text{and} \quad x + 3t = c_2.
$$

At $(0,0)$, we have $c_1 = c_2 = 0$ and at $(4,0)$ we have $c_1 = c_2 = 4$. Thus, the region of existence and uniqueness of the solution is the region bounded by the four lines,

$$
x - 3t = 0, \quad x + 3t = 0, \quad x - 3t = 4, \quad \text{and} \quad x + 3t = 4,
$$

as shown in Fig. 2.19. $\qquad\qquad\qquad\qquad\qquad\qquad\qquad\qquad\qquad\qquad\square$

2.6.1 Exercises

Exercise 2.53 *Solve*

$$
u_{tt} - 4u_{xx} = 0, \quad u(x,0) = \sin(x), \quad u_t(x,0) = \cos(x).
$$

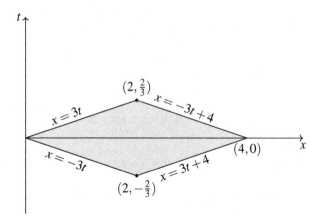

FIGURE 2.19
Region for existence and uniqueness.

Exercise 2.54 *At points in space where no sources are present the spherical wave equation satisfies*

$$u_{tt} = c^2 \left(u_{rr} + \frac{2}{r} u_r \right). \tag{2.94}$$

Equation (2.94) is obtained by writing the homogeneous wave equation in spherical coordinates r, θ, ϕ and neglecting the angular dependence. Assume the initial functions

$$u(r,0) = f(r), \ u_t(r,0) = g(r).$$

Make the change of variables $v = ru$ to transform (2.94) into the equation in v: $v_{tt} = c^2 v_{rr}$.

Solve for v and then find the general solution of (2.94) subject to the initial data.

Exercise 2.55 *Solve*

$$u_{tt} - 4u_{xx} = \cos(x)\sin(t), \quad u(x,0) = x, \ u_t(x,0) = \sin(x).$$

Exercise 2.56 *Consider*

$$2u_{tt} - 18u_{xx} = \frac{36}{t^2 + 1}, \quad u(x,0) = x^2, \ u_t(x,0) = e^{-2x}.$$

(a) Find the solution.

(b) Determine and sketch the region where the solution is uniquely defined when $0 \le x \le 6$.

Exercise 2.57 *[Semiinfinite string with a fixed end] Consider a semiinfinite vibrating string with a fixed end, that is*

$$u_{tt} - c^2 u_{xx} = 0, \quad x > 0 \ t > 0,$$

subject to

$$u(x,0) = f(x), \quad u_t(x,0) = g(x), \ x \geq 0,$$

$$u(0,t) = 0, \ t \geq 0.$$

Show that its solution is given by

$$u(x,t) = \frac{1}{2}[f(x+ct) + f(x-ct)] + \frac{1}{2c}\int_{x-ct}^{x+ct} g(s)ds, \ \text{for } x > ct,$$

$$u(x,t) = \frac{1}{2}[f(x+ct) - f(ct-x)] + \frac{1}{2c}\int_{ct-x}^{x+ct} g(s)ds, \ \text{for } x < ct.$$

Exercise 2.58 *Find the solution for*

$$u_{tt} - 4u_{xx} = 0, \quad x > 0, \quad t > 0,$$

subject to

$$u(x,0) = |\sin(x)|, \quad x > 0$$

$$u_t(x,0) = 0, \quad x \geq 0,$$

$$u(0,t) = 0, \quad t \geq 0.$$

Exercise 2.59 *Construct the solution for the wave problem with zero initial velocity*

$$u_{tt} - u_{xx} = 0, \quad u_t(x,0) = 0,$$

$$u(x,0) = \begin{cases} 1, & |x| \leq 1 \\ 0, & |x| > 1 \end{cases}$$

for

(a) $0 < t < 1$.

(b) $t > 1$.

Exercise 2.60 *Construct the solution for the wave problem with zero initial displacement.*

$$u_{tt} - u_{xx} = 0, \quad u(x,0) = 0,$$

$$u_t(x,0) = \begin{cases} 1-x^2, & |x| < 1 \\ 0, & |x| \geq 1. \end{cases}$$

Exercise 2.61 *Construct the solution for the wave problem with zero initial velocity.*

$$u_{tt} - u_{xx} = 0, \quad u_t(x,0) = 0,$$

$$u(x,0) = \begin{cases} 1-|x|, & |x| < 1 \\ 0, & |x| \geq 1. \end{cases}$$

Exercise 2.62 *Solve*

$$u_{tt} - u_{xx} = 1, \quad u(x,0) = \sin(x), \ u_t(x,0) = x.$$

Exercise 2.63 *Assume that* $u(x,t)$

$$u_{tt} - u_{xx} = 0, \ \ 0 < x < l, \ t \geq 0,$$

with Robin boundary conditions

$$u(0,t) - u_x(0,t) = 0, \ and \ u(l,t) + u_x(l,t) = 0.$$

Show that the energy function

$$E(t) = \frac{1}{2} \int_0^l \left(u_t^2(x,t) + u_x^2(x,t) \right) dx + \frac{1}{2} u^2(0,t) + \frac{1}{2} u^2(l,t)$$

is conserved. That is $\frac{d}{dt} E(t) = 0.$

2.6.2 Vibrating string with fixed ends

Obtaining the solution of a vibrating string with fixed ends is more complicated than solving a wave equation or an infinite string due to the repeated reflection of waves from boundaries. We want to find the solution of

$$
\begin{cases}
u_{tt} - c^2 u_{xx} = 0, & 0 < x < l, \ t > 0 \\
u(x,0) = f(x), & 0 \leq x \leq l \\
u_t(x,0) = g(x), & 0 \leq x \leq l \\
u(0,t) = 0 = u(l,t), & t \geq 0.
\end{cases} \tag{2.95}
$$

As before, the characteristic lines are given by

$$\xi = x + ct, \quad \eta = x - ct.$$

From (2.78), we have

$$u(x,t) = F(x+ct) + G(x-ct), \tag{2.96}$$

where the functions F and G are arbitrary and differentiable. Equation (2.96) is valid and over the domain

$$0 \leq x + ct \leq l, \quad \text{and} \quad 0 \leq x - ct \leq l.$$

Moreover, the solution is uniquely determined by the initial data in the the region

$$t \leq \frac{x}{c}, \quad t \leq \frac{l-x}{c}, \quad t \geq 0.$$

For the fixed end $u(0,t) = 0, \ t \geq 0$, we have

$$0 = u(0,t) = F(ct) + G(-ct). \tag{2.97}$$

Letting $\zeta = -ct$, we obtain from (2.97) that

$$G(\zeta) = -F(-\zeta), \quad \zeta \leq 0. \tag{2.98}$$

Equation (2.98) extends the range of G to negative values and can then be used to do the same for F. If we apply the initial data to (2.96), then it was obtained from Section 2.6 that

$$F(\xi) = \frac{1}{2}\left[f(\xi) + \frac{1}{c}\int_0^\xi g(s)ds + K\right], \quad 0 \le \xi = x + ct \le l \tag{2.99}$$

$$G(\eta) = \frac{1}{2}\left[f(\eta) - \frac{1}{c}\int_0^\eta g(s)ds - K\right], \quad 0 \le \eta = x - ct \le l \tag{2.100}$$

Using (2.98) in combination with (2.99) and (2.100) we arrive at by setting $G(\zeta) = -F(-\zeta)$ that

$$\frac{1}{2}\left[f(\zeta) - \frac{1}{c}\int_0^\zeta g(s)ds\right] = -\frac{1}{2}\left[f(-\zeta) + \frac{1}{c}\int_0^{-\zeta} g(s)ds\right]$$

$$= -\frac{1}{2}\left[f(-\zeta) - \frac{1}{c}\int_0^\zeta g(-s)ds\right].$$

By comparing both sides of the above expression, we immediately see that (2.98) is satisfied when

$$f(\zeta) = -f(-\zeta), \quad \text{and} \quad g(\zeta) = -g(-\zeta).$$

In other words, we must extend the functions f and g to be odd functions with respect to $x = 0$.

Now we turn our attention to the boundary condition $0 = u(l,t)$. As before, we have

$$0 = u(l,t) = F(l + ct) + G(l - ct).$$

Letting $\zeta = l + ct$ in the above equation we get

$$F(\zeta) = -G(2l - \zeta), \quad \zeta \ge l. \tag{2.101}$$

This equation extends the range of F to positive values $l \le \zeta \le 2l$. As before, setting $F(\zeta) = -G(2l - \zeta)$, we arrive at

$$\frac{1}{2}\left[f(\zeta) + \frac{1}{c}\int_0^\zeta g(s)ds\right] = -\frac{1}{2}\left[f(2l - \zeta) - \frac{1}{c}\int_0^{2l-\zeta} g(s)ds\right]$$

$$= -\frac{1}{2}\left[f(2l - \zeta) + \frac{1}{c}\int_{2l}^\zeta g(2l - \tau)d\tau\right].$$

By comparing both sides of the above expression, we immediately arrive at

$$f(2l - \zeta) = -f(\zeta), \quad \text{and} \quad \int_0^\zeta g(s)ds = -\int_{2l}^\zeta g(2l - \tau)d\tau.$$

Differentiating the second expression with respect to ζ, we obtain

$$g(2l - \zeta) = -g(\zeta).$$

These conditions on f and g means that we can extend these functions to $l \leq \zeta \leq 2l$ by performing an odd extension about $x = l$. In summary, the conditions on f and g are

$$\begin{cases} f(x) = -f(-x), & -l \leq x \leq 0 \\ f(2l-x) = -f(x), & l \leq x \leq 2l, \end{cases} \tag{2.102}$$

and

$$\begin{cases} g(x) = -g(-x), & -l \leq x \leq 0 \\ g(2l-x) = -g(x), & l \leq x \leq 2l. \end{cases} \tag{2.103}$$

To make some sense out of (2.102) and (2.103), we notice that

$$f(x+2l) = -f(2l - (x+2l)) = -f(-x) = f(x).$$

A similar situation occurs for the function g. Thus, f and g need to be periodic odd extensions of the original functions with period $2l$. Now we try to piece the solution together.

Let f_p and g_p denote the odd extensions of $2l$-periodic of f and g, respectively. Then,

$$f_p(x) = \begin{cases} f(x), & 0 < x < l, \\ -f(-x), & -l < x < 0, \end{cases} \quad g_p(x) = \begin{cases} g(x), & 0 < x < l, \\ -g(-x), & -l < x < 0 \end{cases}$$

Consider the wave problem on the whole real line with the extended initial data

$$\begin{cases} v_{tt} - c^2 u_{xx} = 0, & -\infty < x < \infty, \ t > 0 \\ v(x,0) = f_p(x), & 0 \leq x \leq l \\ v_t(x,0) = g_p(x), & 0 \leq x \leq l. \end{cases}$$

With this set up we automatically have $v(0,t) = v(l,t) = 0$, and the restriction

$$u(x,t) = v(x,t)\big|_{0 \leq x \leq l}$$

will solve (2.95).

Finally, using the D'Alembert solution, we see that

$$u(x,t) = \frac{1}{2}[f_p(x+ct) + f_p(x-ct)] + \frac{1}{2c} \int_{x-ct}^{x+ct} g_p(s)ds, \quad 0 < x < l. \tag{2.104}$$

Next, we discuss periodic odd extension, and for more on the subject we refer to Appendix A. Suppose we have a function f that is piecewise continuous on the interval $(0,l)$. We define the *Fourier sine series* of f by

$$f(x) = \sum_{n=1}^{\infty} b_n \sin\left(\frac{n\pi x}{l}\right), \quad 0 < x < l, \tag{2.105}$$

where the coefficients b_n, $n = 1, 2, \ldots$, are constants. To determine the coefficients b_n we make use of the orthogonality property

$$\int_0^l \sin(\frac{m\pi x}{l}) \sin(\frac{n\pi x}{l}) dx = \begin{cases} 0, & m \neq n, \\ \frac{l}{2}, & m = n. \end{cases}$$

With this in mind, by multiplying both sides of (2.105) by $\sin(\frac{m\pi x}{l})$ and integrating from $x = 0$ to $x = l$, we arrive at

$$b_n = \frac{2}{l} \int_0^l f(x) \sin(\frac{n\pi x}{l}) dx, \tag{2.106}$$

where we have assumed that term-by-term integration is valid. It can be shown that if

$$f(0) = 0, \quad \text{and} \quad f(\pi) = 0,$$

then series (2.105) converges uniformly to $f(x)$ for all $0 < x < l$. It is clear that the series converges to zero when $x = 0$ and when $x = l$. Moreover, the series in (2.105) converges to the *odd periodic extension,* with period $2l$, of f for all values of x. We have the following example.

Example 2.24 Let

$$f(x) = x, \quad 0 < x < \pi.$$

Then,

$$
\begin{aligned}
b_n &= \frac{2}{\pi} \int_0^\pi x \sin(nx) dx = \frac{2}{\pi} \left[-\frac{x\cos(nx)}{n} \Big|_0^\pi + \frac{1}{n} \int_0^\pi \cos(nx) dx \right] \\
&= 2\frac{(-1)^{n+1}}{n}, \quad n = 1, 2, \ldots
\end{aligned}
$$

Thus,

$$f_p = 2 \sum_{n=1}^\infty \frac{(-1)^{n+1}}{n} \sin(nx), \quad 0 < x < \pi. \tag{2.107}$$

It can be shown that this series converges to $f(x) = x$ when $0 < x < \pi$. □

Example 2.25 Consider

$$\begin{cases} u_{tt} = u_{xx}, & 0 < x < \pi, \; t > 0 \\ u(x,0) = x, & 0 \leq x \leq \pi \\ u_t(x,0) = 0, & 0 \leq x \leq \pi \\ u(0,t) = 0 = u(\pi,t), & t \geq 0. \end{cases}$$

Using (2.104) and (2.107), we arrive at

$$
\begin{aligned}
u(x,t) &= \frac{1}{2}[f_p(x+ct) + f_p(x-ct)] \\
&= \frac{1}{2} \sum_{n=1}^\infty 2\frac{(-1)^{n+1}}{n} \left[\sin(n(x+t)) + \sin(n(x-t)) \right]
\end{aligned}
$$

Clearly, the above solution satisfies $u(0,t) = 0$, $u(\pi,t) = 0$. □

Example 2.26 Consider

$$
\begin{cases}
u_{tt} = u_{xx}, & 0 < x < \pi, \ t > 0 \\
u(x,0) = 0, & 0 \le x \le \pi \\
u_t(x,0) = x, & 0 \le x \le \pi \\
u(0,t) = 0 = u(\pi,t), & t \ge 0.
\end{cases}
$$

Using (2.104) and (2.107), we arrive at

$$
\begin{aligned}
u(x,t) &= \frac{1}{2} \int_{x-t}^{x+t} g_p(s)\,ds, \\
&= \sum_{n=1}^{\infty} \frac{(-1)^{n+1}}{n} \int_{x-t}^{x+t} \sin(ns)\,ds \\
&= \sum_{n=1}^{\infty} \frac{(-1)^{n+1}}{n} \left[-\frac{\cos(ns)}{n} \Big|_{x-t}^{x+t} \right] \\
&= \sum_{n=1}^{\infty} \frac{(-1)^{n+1}}{n^2} \left[-\cos(n(x+t)) + \cos(n(x-t)) \right]
\end{aligned}
$$

Clearly, the above solution satisfies $u(0,t) = 0$, $u(\pi,t) = 0$. $\qquad\square$

2.6.3 Exercises

Exercise 2.64 *Consider*

$$
\begin{cases}
u_{tt} = u_{xx}, & 0 < x < \pi, \ t > 0 \\
u(x,0) = x^3, & 0 \le x \le \pi \\
u_t(x,0) = 0, & 0 \le x \le \pi \\
u(0,t) = 0 = u(\pi,t), & t \ge 0.
\end{cases}
$$

(a) Show that

$$
f_p(x) = 2 \sum_{n=1}^{\infty} (-1)^{n+1} \frac{(n\pi)^2 - 6}{n^3} \sin(nx), \quad 0 < x < \pi.
$$

(b) Find the solution $u(x,t)$.

Exercise 2.65 *Consider*

$$
\begin{cases}
u_{tt} = u_{xx}, & 0 < x < \pi, \ t > 0 \\
u(x,0) = 1, & 0 \le x \le \pi \\
u_t(x,0) = 0, & 0 \le x \le \pi \\
u(0,t) = 0 = u(\pi,t), & t \ge 0.
\end{cases}
$$

(a) Show that

$$f_p(x) = \frac{4}{\pi} \sum_{n=1}^{\infty} \frac{\sin(2n-1)x}{2n-1}, \quad 0 < x < \pi.$$

(b) Find the solution $u(x,t)$.

Exercise 2.66 *Consider*

$$\begin{cases} u_{tt} = u_{xx}, & 0 < x < \pi, \ t > 0 \\ u(x,0) = \pi - x, & 0 \le x \le \pi \\ u_t(x,0) = 0, & 0 \le x \le \pi \\ u(0,t) = 0 = u(\pi,t), & t \ge 0. \end{cases}$$

(a) Show that

$$f_p(x) = 2\sum_{n=1}^{\infty} \frac{\sin(nx)}{n}, \quad 0 < x < \pi.$$

(b) Find the solution $u(x,t)$.

Exercise 2.67 *Consider*

$$\begin{cases} u_{tt} = u_{xx}, & 0 < x < 1, \ t > 0 \\ u(x,0) = 1, & 0 \le x \le 1 \\ u_t(x,0) = 0, & 0 \le x \le 1 \\ u(0,t) = 0 = u(1,t), & t \ge 0. \end{cases}$$

(a) Show that

$$f_p(x) = \frac{2}{\pi} \sum_{n=1}^{\infty} \frac{(-1)^{n+1}}{n} \sin(n\pi x), \quad 0 < x < 1.$$

(b) Find the solution $u(x,t)$.

Exercise 2.68 *Consider*

$$\begin{cases} u_{tt} = u_{xx}, & 0 < x < 1, \ t > 0 \\ u(x,0) = x(1-x^2), & 0 \le x \le 1 \\ u_t(x,0) = 0, & 0 \le x \le 1 \\ u(0,t) = 0 = u(1,t), & t \ge 0. \end{cases}$$

(a) Show that

$$f_p(x) = \frac{12}{\pi^3} \sum_{n=1}^{\infty} \frac{(-1)^{n+1}}{n^3} \sin(n\pi x), \quad 0 < x < 1.$$

(b) Find the solution $u(x,t)$.

Exercise 2.69 *Solve*

$$\begin{cases} u_{tt} = u_{xx}, & 0 < x < 1, \ t > 0 \\ u(x,0) = 0, & 0 \le x \le 1 \\ u_t(x,0) = x(1-x^2), & 0 \le x \le 1 \\ u(0,t) = 0 = u(1,t), & t \ge 0. \end{cases}$$

Exercise 2.70 *Solve*

$$\begin{cases} u_{tt} = 4u_{xx}, & 0 < x < 1, \ t > 0 \\ u(x,0) = 0, & 0 \le x \le 1 \\ u_t(x,0) = x(1-x), & 0 \le x \le 1 \\ u(0,t) = 0 = u(1,t), & t \ge 0. \end{cases}$$

2.7   Heat Equation

We consider the heat conduction problem of a thin rod and look at the solution. Part of our presentation is inspired and influenced by Strauss [11]. Let $u(x,t)$ represent the temperature at position x in a thin insulated rod at time t. The vertical axis will measure the temperature, and the rod will be positioned along the x-axis in the xu-coordinate system. As in the wave equation, we consider a small portion of the rod over an interval $[a,b]$. The heat or thermal energy of the rod situated at the interval $[a,b]$ is given by

$$D(x,t) = \int_a^b c\rho u\, dx,$$

where c denotes the specific heat capacity of the material of the rod and ρ is the mass density of the rod. The instantaneous change of the heat with respect to time will be the time derivative of the above equation

$$\frac{d}{dt}D(x,t) = \frac{d}{dt}\int_a^b c\rho u\, dx = \int_a^b c\rho u_t\, dx. \tag{2.108}$$

The above expression is true since a and b are constants and u is continuous. Recall that the thin rod is insulated. The change in heat must be balanced by the heat flux across the cross-section of the cylindrical piece around the interval $[a,b]$, as the heat cannot be gained or lost in the absence of an external heat source. Fourier's law states that the heat flux across the boundary will be inversely proportional to the temperature derivative in the direction of the boundary's outward normal, in this instance the x-derivative. The second way to compute the time rate of change of D is to notice that, in the Absence of heat sources within the rod, the quantity of heat in u can change only through the flow of heat across the boundaries of u at $x = a$ and $x = b$. The heat flux through a section of the rod is called the heat flux through the section.

Let κ denote the *thermal conductivity* of the rod. Recall that the thermal conductivity of a material is a measure of its ability to conduct heat. Then the heat flux into u at $x = a$ and $x = b$ is

$$-\kappa u_x(a,t), \quad \text{and} \quad \kappa u_x(b,t),$$

respectively. Thus, the total time rate of change of D is the sum of the rates at the two ends. Using the fundamental theorem of calculus, we may write

$$\frac{d}{dt}D(x,t) = \kappa\left[u_x(b,t) - u_x(a,t)\right] = \kappa\int_a^b u_{xx}dx. \tag{2.109}$$

A quick comparison of equations (2.108) and (2.109) yields

$$\int_a^b c\rho u_t dx = \kappa\int_a^b u_{xx}dx,$$

or

$$\int_a^b \left(c\rho u_t - \kappa u_{xx}\right)dx = 0.$$

Since the above integral must hold for all $x \in [a,b]$ with $a < b$ we must have

$$c\rho u_t - \kappa u_{xx} = 0,$$

throughout the material. Dividing by $c\rho$ and setting $k = \frac{\kappa}{c\rho}$, we arrive at the one-dimensional *heat equation*

$$u_t - ku_{xx} = 0, \tag{2.110}$$

where k is called the *thermal diffusivity* of the material.

If we consider the heat equation in (2.110) on an interval $I \in \mathbb{R}$, then we have the heat problem, with initial and boundary conditions

$$\begin{cases} u_t = ku_{xx}, & x \in I,\, t > 0 \\ u(x,0) = f(x), & x \in I \\ u \text{ satisfies certain BCs} \end{cases} \tag{2.111}$$

Here, the initial condition $u(x,0) = f(x)$, means the lateral surface of the rod is insulated and parallel to the x-axis, and its initial temperatures are $f(x)$ for $x \in I$.

In practice, the most common boundary conditions are the following:

- $u(0,t) = 0 = u(l,t) : (I = (0,l)$, Dirichlet$)$. It is the case when both faces of the rod are kept at temperature zero.

- $u_x(0,t) = 0 = u_x(l,t) : (I = (0,l)$, Neumann$)$. It is the case when both faces of the rod are insulated.

- $u_x(0,t) - a_0u(0,t) = 0$ and $u_x(l,t) + a_lu(l,t) = 0 : (I = (0,l)$, Robin$)$.

- $u(-l,t) = u(l,t) = 0$ and $u_x(-l,t) = u_x(l,t) = 0 : (I = (-l,l)$, Periodic$)$.

Before we attempt to find the solution to the heat equation, we prove the uniqueness of the solution of the nonhomogeneous heat equation. We do so by defining an energy function V and show along the solutions of the heat equation, the energy function is nonnegative and its derivative is less or equal to zero. We begin by considering the nonhomogeneous heat equation with initial and boundary conditions

$$\begin{cases} u_t - ku_{xx} = f(x,t), & 0 \le x \le l, \ t > 0 \\ u(x,0) = \phi(x), & 0 \le x \le l, \\ u(0,t) = g(t), \ u(l,t) = h(t), \end{cases} \qquad (2.112)$$

for given functions f, ϕ, g, h.

Theorem 2.4 *The heat equation given by (2.112) has at most one solution.*

Proof *Assume (2.112) has two solutions u and v. Set $w = u - v$. Then*

$$w(x,0) = u(x,0) - v(x,0) = \phi(x) - \phi(x) = 0,$$

$$w(0,t) = u(0,t) - v(0,t) = g(t) - g(t) = 0,$$

$$w(l,t) = u(l,t) - v(l,t) = h(t) - h(t) = 0.$$

Moreover,

$$\begin{aligned} w_t - kw_{xx} &= (u-v)_t - k(u-v)_{xx} \\ &= u_t - ku_{xx} - (v_t - kv_{xx}) \\ &= f(x,t) - f(x,t) \\ &= 0. \end{aligned}$$

Thus, we arrive at the homogeneous heat equation in w,

$$\begin{cases} w_t - kw_{xx} = 0, & 0 \le x \le l, \ t > 0 \\ w(x,0) = 0, & 0 \le x \le l, \\ w(0,t) = 0, \ w(l,t) = 0, & t > 0. \end{cases} \qquad (2.113)$$

Define the energy function

$$V[w](t) = \frac{1}{2} \int_0^l [w(x,t)]^2 dx. \qquad (2.114)$$

Then it is clear that

$$V[w](0) = \frac{1}{2} \int_0^l [w(x,0)]^2 dx = \frac{1}{2} \int_0^l [0]^2 dx = 0,$$

and $V[w](t)$ is positive for $t > 0$. To obtain any meaningful information from the energy function, we must show it is decreasing in time along the solutions of (2.113). Thus, using the first equation in (2.113) we arrive at

$$\frac{d}{dt} V[w](t) = \int_0^l w(x,t)w_t(x,t)dx = k \int_0^l w(x,t)w_{xx}(x,t)dx.$$

An integration by parts yields

$$
\begin{aligned}
\frac{d}{dt}V[w](t) &= kw(x,t)w_x(x,t)\big|_0^l - k\int_0^l \left(w_x(x,t)\right)^2 dx \\
&= k\left[w(l,t)w_x(l,t) - w(0,t)w_x(0,t)\right] - k\int_0^l \left(w_x(x,t)\right)^2 dx \\
&= -k\int_0^l \left(w_x(x,t)\right)^2 dx \le 0.
\end{aligned}
$$

Since the energy function V is decreasing, we get

$$
0 \le V[w](t) \le V[w](0) = 0.
$$

Hence,

$$
V[w](t) = \frac{1}{2}\int_0^l [w(x,t)]^2 dx = 0, \quad \text{for all } t \ge 0,
$$

which implies $w \equiv 0$, for all $x \in [0,l]$, $t > 0$. Wherefore, $u - v = 0$ for all $x \in [0,l]$, $t > 0$. This shows

$$
u = v \quad \text{for all} \quad x \in [0,l], \, t > 0.
$$

This completes the proof.

Next we extend Theorem 2.4 to show uniqueness of solution for the heat equation on $\mathbb{R}$. Consider the heat equation

$$
\begin{cases}
u_t - ku_{xx} = f(x,t), & -\infty \le x \le \infty, \, t > 0 \\
u(x,0) = \phi(x), & -\infty \le x \le \infty, \\
\lim_{x\to\pm\infty} u = 0, \quad \lim_{x\to\pm\infty} u_x = 0, & t > 0,
\end{cases}
\tag{2.115}
$$

for a given function ϕ.

Theorem 2.5 *The heat equation given by (2.115) has at most one solution.*

Proof *Assume (2.115) has two solutions u and v. Set $w = u - v$. Then by similar arguments as in the proof of Theorem 2.4 w is a solution to the homogeneous heat equation*

$$
\begin{cases}
w_t - kw_{xx} = 0, & -\infty \le x \le \infty, \, t > 0 \\
w(x,0) = 0, & -\infty \le x \le \infty, \\
\lim_{x\to\pm\infty} w = 0, \quad \lim_{x\to\pm\infty} w_x = 0, & t > 0.
\end{cases}
\tag{2.116}
$$

Define the energy function for (2.116) by

$$
V[w](t) = \frac{1}{2}\int_{-\infty}^{\infty} [w(x,t)]^2 dx.
\tag{2.117}
$$

Then it is clear that

$$V[w](0) = \frac{1}{2} \int_{-\infty}^{\infty} [w(x,0)]^2 dx = 0,$$

and $V[w](t) \geq 0,$ *for* $t \geq 0.$ *Moreover, using (2.117) we get*

$$\frac{d}{dt} V[w](t) = \int_{-\infty}^{infty} w(x,t)w_t(x,t)dx = k \int_{-\infty}^{\infty} w(x,t)w_{xx}(x,t)dx.$$

An integration by parts yields

$$
\begin{aligned}
\frac{d}{dt} V[w](t) &= kw(x,t)w_x(x,t)|_{x=-\infty}^{x=\infty} - k \int_{-\infty}^{\infty} (w_x(x,t))^2 dx \\
&= -k \int_{-\infty}^{\infty} (w_x(x,t))^2 dx \leq 0.
\end{aligned}
$$

Since the energy function V is decreasing, we get

$$0 \leq V[w](t) \leq V[w](0) = 0.$$

Hence,

$$V[w](t) = \frac{1}{2} \int_{-\infty}^{\infty} [w(x,t)]^2 dx = 0, \quad \text{for all } t \geq 0,$$

which implies $w \equiv 0,$ *for all* $x \in (-\infty, \infty),$ $t > 0.$ *Consequently,* $u - v = 0$ *for all* $x \in (-\infty, \infty),$ $t > 0.$ *This shows*

$$u = v \quad \text{for all} \quad x \in (-\infty, \infty), t > 0.$$

This completes the proof.

The next theorem is about stability; it says that for a small change in the initial data, only produces a small change in the solution.

Theorem 2.6 *Let* $u^*(x,t)$ *be another solution of (2.115) with initial data* $g^*.$ *Define the* L^2 *norm of a function h as*

$$||h||_2 = \left(\int_{-\infty}^{\infty} h^2(x)dx \right)^{\frac{1}{2}}.$$

Assume there is a small change in the initial data. That is, for small and positive ε *we have that* $||g - g^*||_2 < \varepsilon.$ *Then,*

$$||u(x,t) - u^*(x,t)||_2 < \varepsilon.$$

Proof *The proof depends on the energy function. For simpler notation, we let*

$$w(x,t) = u(x,t) - u^*(x,t),$$

then w is a solution to the homogeneous heat problem

$$\begin{cases} w_t - kw_{xx} = 0, & -\infty \leq x \leq \infty, \, t > 0 \\ w(x,0) = g(x) - g^*(x), & -\infty \leq x \leq \infty, \\ \lim_{x \to \pm\infty} w = 0, \quad \lim_{x \to \pm\infty} w_x = 0 \end{cases} \qquad (2.118)$$

Define the energy function V for (2.118) by (2.117). Then by Theorem 2.5, we have V is decreasing along the solutions of (2.118) with $V[w](t) \geq 0$, for $t \geq 0$. Thus

$$V[w](t) \leq V[w](0) = \frac{1}{2}\int_{-\infty}^{\infty} [w(x,0)]^2 dx.$$

Accordingly, we have

$$\frac{1}{2}\|w\|_2^2 = V[w] \leq \frac{1}{2}\|w(x,0)\|_2^2,$$

or

$$\frac{1}{2}\|u - u^*\|_2^2 \leq \frac{1}{2}\|g - g^*\|_2^2,$$

which implies that

$$\|u - u^*\|_2 \leq \varepsilon, \, t \geq 0.$$

We have established stability for all $t \geq 0$ in terms of the square error. This completes the proof.

2.7.1 Solution of the heat equation

Finding the solution of the heat equation using the method of characteristics as we did for the wave equation will not be of much success. To see this, we consider the heat equation over $\mathbb{R}$,

$$u_t - ku_{xx} = 0. \qquad (2.119)$$

Then we have $A = k$, $B = C = 0$. Using (2.65) we obtain

$$\frac{dt}{dx} = \frac{B \pm \sqrt{B^2 - AC}}{A} = 0.$$

Wherefore, we only have one characteristic line given by

$$t = c,$$

for some constant c. Remember, we should be able to trace any point in the xt plane along the characteristic lines, which is not the case here since they are parallel to the x-axis.

Our aim then is to find another approach to establishing a bounded solution of the heat equation on an unbounded domain. We consider the heat problem with initial condition

$$\begin{cases} u_t - ku_{xx} = 0, & -\infty < x < \infty, \, t > 0 \\ u(x,0) = \phi(x), & -\infty < x < \infty. \end{cases} \qquad (2.120)$$

To arrive at the solution of (2.120), we begin by considering simple form of the initial condition. In particular, we first derive the solution of the heat problem with initial condition of the form

$$\begin{cases} V_t - kV_{xx} = 0, & -\infty < x < \infty, \, t > 0 \\ V(x,0) = H(x), \end{cases} \tag{2.121}$$

where $H(x)$ is the *Heaviside step function* defined by

$$H(x) = \begin{cases} 1, & x > 0 \\ 0, & x < 0. \end{cases} \tag{2.122}$$

The next lemma is needed in our future work.

Lemma 1 *For $x \in \mathbb{R}$,*

$$\int_0^\infty e^{-x^2} dx = \frac{\sqrt{\pi}}{2}.$$

Proof *Let*

$$I = \int_0^\infty e^{-x^2} dx.$$

Then, for $y \in \mathbb{R}$,

$$I^2 = I \cdot I = \int_0^\infty e^{-x^2} dx \int_0^\infty e^{-y^2} dy = \int_0^\infty \int_0^\infty e^{-(x^2+y^2)} dx dy.$$

We switch to polar coordinates by letting

$$x = r\cos(\theta), \, y = r\sin(\theta).$$

Then,

$$\begin{aligned} I^2 &= \int_0^{\pi/2} \int_0^\infty e^{-r^2} r \, dr \, d\theta \\ &= \int_0^{\pi/2} \left(\int_0^\infty e^{-r^2} r \, dr \right) d\theta \\ &= \int_0^{\pi/2} \left(-\frac{1}{2} e^{-r^2} \big|_0^\infty \right) d\theta \\ &= \int_0^{\pi/2} \frac{1}{2} d\theta = \frac{\pi}{4}. \end{aligned}$$

Taking the square root on both sides we arrive at

$$I = \int_0^\infty e^{-x^2} dx = \frac{\sqrt{\pi}}{2}.$$

This completes the proof.

In the next lemma we explore the invariance properties of the heat equation.

Lemma 2 *[Invariance properties of the heat equation] The heat equation (2.119) is invariant under these transformations.*

(a) *If $u(x,t)$ is a solution of (2.119), then so is $u(x-z,t)$ for any fixed z. (Spatial translation)*

(b) *If $u(x,t)$ is a solution of (2.119), then so are u_x, u_t, u_{xx}, and so on. (Differentiation)*

(c) *If $u_1, u_2, \ldots, u_n$ are solutions of (2.119), then so is $\sum_{i=1}^{n} c_i u_i$ for any constants $c_1, c_2, \ldots, c_n$. (Linear combinations)*

(d) *If $S(x,t)$ solves (2.119), then so is*

$$\int_{-\infty}^{\infty} S(x-y,t)g(y)dy$$

for function g as long as the integral converges.

(e) *If $u(x,t)$ is a solution of (2.119), then so is $u(\sqrt{a}x, at)$ for any constant $a > 0$. (Dilation, or scaling)*

Proof *The proof of parts (a)–(c) are straightforward and we refer to Exercise 2.71. To prove (d), we assume a finite interval $[-b,b]$ partitioned by points $\{y_i\}_{i=1}^n$ such that $-b = y_1 < y_2 < \cdots < y_n = b$ with equal length Δy. Then using (c) combined with representing the integral with the Riemann sum we may write*

$$\int_{-\infty}^{\infty} S(x-y,t)g(y)dy = \lim_{b\to\infty} \int_{-b}^{b} S(x-y,t)g(y)dy = \lim_{b\to\infty}\lim_{n\to\infty} \sum_{i=1}^{n} S(x-y,t)g(y_i)\Delta y.$$

As for the proof of (d), we make use of the chain rule. Let $v(x,t) = u(\sqrt{a}x, at)$. Then, $v_t = au_t(\sqrt{a}x, at)$, $v_x = \sqrt{a}u_x(\sqrt{a}x, at)$, and $v_{xx} = au_{xx}(\sqrt{a}x, at)$. Substituting into (2.120) we arrive at

$$au_t(\sqrt{a}x, at) - kau_{xx}(\sqrt{a}x, at) = 0,$$

or

$$u_t(\sqrt{a}x, at) - ku_{xx}(\sqrt{a}x, at) = 0.$$

This completes the proof.

Now we are in a good position to solve (2.120). As we have mentioned before, we will find a particular solution of (2.121) and then make use of it to obtain the solution of the heat problem given in (2.120). First we note that $H(x)$ is invariant under the dilation, since

$$H(\sqrt{a}x) = \begin{cases} 1, & \sqrt{a}x > 0 \\ 0, & \sqrt{a}x < 0 \end{cases} = \begin{cases} 1, & x > 0 \\ 0, & x < 0 \end{cases} = H(x),$$

as $a > 0$. Due to (e) and the fact that H is invariant under the dilation, we know that $V(\sqrt{a}x, at)$ solves (2.121). Due to the uniqueness of solutions, we must have

$V(\sqrt{a}x, at) = V(x,t)$, for all $x \in \mathbb{R}$ and $t > 0$. Thus, V is invariant under the dilation $(x,t) \rightarrow (\sqrt{a}x, at)$. Since our goal is to transfer (2.121) to an ODE, we want to eliminate one of the variables in V, which can be easily achieved by letting $a = \frac{1}{t}$. Then

$$V(x,t) = V(\frac{x}{\sqrt{t}}, \frac{1}{t}t) = V(\frac{x}{\sqrt{t}}, 1).$$

Let q be a function such that

$$V(x,t) = q(\frac{x}{\sqrt{t}}).$$

Thus, V is completely determined by the function of one variable q. Then,

$$q_t = -\frac{x}{2t^{\frac{3}{2}}} q'(\frac{x}{\sqrt{t}}), \quad q_{xx} = \frac{1}{t} q''(\frac{x}{\sqrt{t}}).$$

This reduces the PDE into,

$$V_t - kV_{xx} = -\frac{x}{2t^{\frac{3}{2}}} q'(\frac{x}{\sqrt{t}}) - \frac{k}{t} q''(\frac{x}{\sqrt{t}}) = 0,$$

which implies that

$$-\frac{x}{2t^{\frac{1}{2}}} q'(\frac{x}{\sqrt{t}}) - kq''(\frac{x}{\sqrt{t}}) = 0.$$

Setting, $z = \frac{x}{\sqrt{t}}$, the above second-order differential equation takes the form

$$q''(z) + \frac{z}{2k} q'(z) = 0.$$

Let $G(z) = q'(z)$, then the above equation reduces to the first-order differential equation

$$G'(z) + \frac{z}{2k} G(z) = 0,$$

which has the solution (see Chapter 1)

$$G(z) = q'(z) = Ce^{-\frac{z^2}{4k}}.$$

Integrating from 0 to z gives

$$q(z) = C \int_0^z e^{-\frac{s^2}{4k}} ds + D,$$

for unknown constants C and D. This yields

$$V(x,t) = q(\frac{x}{\sqrt{t}}) = C \int_0^{\frac{x}{\sqrt{t}}} e^{-\frac{s^2}{4k}} ds + D. \tag{2.123}$$

Note that by making a substitution in Lemma 1, one can show that

$$\int_0^\infty e^{-\frac{s^2}{4k}} ds = \sqrt{k\pi},$$

which we will need to compute the constants in (2.123). Since (2.123) is only valid for $t > 0$, so to check the initial condition in (2.121) we take the limit at $t \to 0^+$. Additionally, we observe that

$$\lim_{t\to 0+} \frac{x}{\sqrt{t}} = \begin{cases} \infty, & x > 0 \\ -\infty, & x < 0. \end{cases}$$

Thus, for $x > 0$, we have that

$$1 = \lim_{t\to 0+} V(x,t) = C \int_0^\infty e^{-\frac{s^2}{4k}} ds + D = C\sqrt{k\pi} + D.$$

On the other hand, for $x < 0$, we obtain

$$0 = \lim_{t\to 0+} V(x,t) = C \int_0^{-\infty} e^{-\frac{s^2}{4k}} ds + D = -C\sqrt{k\pi} + D.$$

By solving the system of equations

$$1 = C\sqrt{k\pi} + D, \quad 0 = -C\sqrt{k\pi} + D,$$

we obtain

$$C = \frac{1}{\sqrt{4k\pi}}, \quad D = \frac{1}{2}.$$

Plugging the constants into (2.123), we arrive at

$$V(x,t) = \frac{1}{\sqrt{4k\pi}} \int_0^{\frac{x}{\sqrt{t}}} e^{-\frac{s^2}{4k}} ds + \frac{1}{2}.$$

We try to put V in terms of the error function that we define below.

Definition 2.5 *The error function is the following improper integral considered as a real function* $erf : \mathbb{R} \to \mathbb{R}$, *such that*

$$erf(x) = \frac{2}{\sqrt{\pi}} \int_0^x e^{-z^2} dz,$$

where exponential is the real exponential function. In addition the complementary error function,

$$erfc(x) = 1 - erf(x) = \frac{2}{\sqrt{\pi}} \int_x^0 e^{-z^2} dz.$$

Let $z = \frac{s}{\sqrt{4k}}$. Then

$$\frac{1}{\sqrt{4k\pi}} \int_0^{\frac{x}{\sqrt{t}}} e^{-\frac{s^2}{4k}} ds = \frac{1}{\sqrt{\pi}} \int_0^{\frac{x}{\sqrt{4kt}}} e^{-z^2} dz = \frac{1}{2} erf\left(\frac{x}{\sqrt{4kt}}\right),$$

Hence the unique particular solution of (2.121) is given by

$$V(x,t) = \frac{1}{\sqrt{\pi}} \int_0^{\frac{x}{\sqrt{4kt}}} e^{-z^2} dz + \frac{1}{2}, \tag{2.124}$$

and in terms of the error function,

$$V(x,t) = \frac{1}{2} + \frac{1}{2} erf(\frac{x}{\sqrt{4kt}}). \tag{2.125}$$

Solving (2.120).
Now, we attempt to find the solution for the general heat problem given by (2.120). Define the function

$$S(x,t) = \frac{\partial V}{\partial x}(x,t). \tag{2.126}$$

By the invariance property (a), $S(x,t)$ solves (2.120). Therefore, by the invariance property (b),

$$u(x,t) = \int_{-\infty}^{\infty} S(x-y,t)\phi(y)dy, \quad \text{for } t > 0 \tag{2.127}$$

solves (2.120). We must show $u(x,t)$ given by (2.127) is the unique solution of (2.120). This can be accomplished by showing it satisfies the initial condition. Utilizing (2.126), we can write u as follows

$$u(x,t) = \int_{-\infty}^{\infty} \frac{\partial V}{\partial x}(x-y,t)\phi(y)dy = -\int_{-\infty}^{\infty} \frac{\partial}{\partial y}[V(x-y,t)]\phi(y)dy.$$

We note that $S(x-y,t)$ decays exponentially as $y-x$ grows larger. For now, we assume

$$\phi(\pm\infty) = 0,$$

so we may perform an integration by parts on the above integral. That is

$$u(x,t) = -V(x-y,t)\phi(y)\Big|_{-\infty}^{\infty} + \int_{-\infty}^{\infty} V(x-y,t)\phi'(y)dy = \int_{-\infty}^{\infty} V(x-y,t)\phi'(y)dy.$$

Setting $t = 0$, and noticing that $V(x-y,0)$ is $H(x)$ we obtain

$$u(x,0) = \int_{-\infty}^{\infty} V(x-y,0)\phi'(y)dy = \int_{-\infty}^{x} \phi'(y)dy = \phi(y)\Big|_{-\infty}^{x} = \phi(x).$$

Thus, $u(x,t)$ does satisfy the initial condition of (2.120). Left to compute $S(x,t)$ which can be easily done from (2.126) with V given by (2.124). That is, using (2.124) we obtain

$$S(x,t) = \frac{\partial V}{\partial x}(x,t) = \frac{1}{\sqrt{4\pi kt}} e^{-\frac{x^2}{4kt}}. \tag{2.128}$$

Finally, substituting S given by (2.128) into (2.127), we have the explicit form of the solution of (2.120)

$$u(x,t) = \frac{1}{\sqrt{4\pi kt}} \int_{-\infty}^{\infty} e^{-\frac{(x-y)^2}{4kt}} \phi(y)dy, \quad \text{for } t > 0. \tag{2.129}$$

The function $S(x,t)$ is known as the *heat kernel*. Other resources may refer to it as *fundamental solution, source function, Green's function,* or *propagator* of the heat equation

We present the following example.

Example 2.27 Consider the heat problem

$$\begin{cases} u_t - 4v_{xx} = 0, & -\infty < x < \infty, \, t > 0 \\ u(x,0) = \phi(x), & -\infty < x < \infty, \end{cases}$$

with initial data

$$\phi(x) = \begin{cases} 1 - |x|, & |x| < 1 \\ 0, & |x| \geq 1 \end{cases}$$

We are searching for a bounded solution $u(x,t)$. Substituting the initial data into (2.129) yields

$$
\begin{aligned}
u(x,t) &= \frac{1}{\sqrt{4\pi kt}} \int_{-\infty}^{\infty} e^{-\frac{(x-y)^2}{4kt}} \phi(y)dy \\
&= \frac{1}{\sqrt{16\pi t}} \left[\int_{-1}^{0} e^{-\frac{(x-y)^2}{16t}} (1+y)dy + \int_{0}^{1} e^{-\frac{(x-y)^2}{16t}} (1-y)dy \right].
\end{aligned}
$$

We make the change of variables

$$s = \frac{y-x}{\sqrt{16t}}$$

and transform the above integrals to obtain

$$
\begin{aligned}
u(x,t) &= \frac{1}{\sqrt{\pi}} \Bigg[\int_{-\frac{1+x}{\sqrt{16t}}}^{\frac{-x}{\sqrt{16t}}} e^{-s^2} (1+x+4s\sqrt{t})ds \\
&\quad + \int_{\frac{-x}{\sqrt{16t}}}^{-\frac{x-1}{\sqrt{16t}}} e^{-s^2} (1-x-4s\sqrt{t})ds \Bigg].
\end{aligned}
$$

The integrals with integrands in which e^{-s^2} is not multiplied by a term that has an s in it can be written in terms of the error function. The rest can be integrated out and at the end, we end up with the solution

$$
\begin{aligned}
v(x,t) &= \frac{1}{2}(1+x)\left(erf\left(\frac{1+x}{\sqrt{16t}}\right) - erf\left(\frac{x}{\sqrt{16t}}\right)\right) \\
&\quad + \frac{1}{2}(1-x)\left(erf\left(\frac{x}{\sqrt{16t}}\right) - erf\left(\frac{x-1}{\sqrt{16}}\right)\right) \\
&\quad + \frac{2\sqrt{t}}{\sqrt{\pi}}\left(e^{-\frac{(x+1)^2}{16t}} - e^{-\frac{x^2}{16t}}\right) + \frac{2\sqrt{t}}{\sqrt{\pi}}\left(e^{-\frac{(x-1)^2}{16t}} - e^{-\frac{x^2}{16t}}\right). \quad (2.130)
\end{aligned}
$$

$\square$

We end this section by looking into the solution of the nonhomogenous heat problem with initial condition,

$$\begin{cases} u_t - ku_{xx} = f(x,t), & -\infty < x < \infty, \ t > 0 \\ u(x,0) = \phi(x), & -\infty < x < \infty, \end{cases} \qquad (2.131)$$

for given function f, and ϕ. The derivation of the solution of (2.131) depends on *Dumamel's principle* and we ask the reader to consult with [1]. The next theorem is stated without proof.

Theorem 2.7 *The heat equation given by* (2.131) *has the solution*

$$u(x,t) = \int_{-\infty}^{\infty} S(x-y,t)\phi(y)dy + \int_0^t \int_{-\infty}^{\infty} S(x-y,t-\tau)f(y,\tau)dyd\tau, \quad for\, t > 0$$
$$(2.132)$$

where $S(x,t)$ *is given by* (2.128).

2.7.2 Heat equation on semi-infinite domain: Dirichlet condition

In the previous subsection, we considered the heat problem on $\mathbb{R}$. Presently, we want to make use of our previous results and obtain an explicit solution for the heat equation on the unbounded semi-infinite domain $[0,\infty]$. Now it makes sense to add a boundary condition at $x = 0$, that is, $u(0,t) = 0$. Thus, we consider the heat problem with Dirichlet boundary condition at the end point $x = 0$,

$$\begin{cases} v_t - kv_{xx} = 0, & 0 < x < \infty, \ t > 0 \\ v(x,0) = \phi(x), & x > 0 \\ v(0,t) = 0, & t > 0. \end{cases} \qquad (2.133)$$

Our aim is to find a solution of (2.133) as we did for the heat problem over the entire real line. There will be no need to start from scratch, but instead we will reintroduce the problem over the entire real line by extending the initial data to the whole line. Whatever method we use to extend to the negative half-line, we should make sure that the boundary condition is automatically satisfied by the solution of the problem on the whole line that arises from the extended data. For heat problems with Dirichlet condition, one would choose the *odd extension* of the initial data $\phi(x)$. If $\psi(x)$ is odd, then $\psi(x) = -\psi(-x)$, from which we get $2\psi(0) = 0$, or $\psi(0) = 0$. This is true for any odd function. We make it formal in the next lemma.

Lemma 3 *Let* $f : (-\infty,\infty) \to \mathbb{R}$ *be an odd function* $(f(x) = -f(-x))$, *that is continuous at* $x = 0$ *then* $f(0) = 0$.

Proof *Since* f *is conitnuous at* $x = 0$ *and odd, we have*

$$f(0) = \lim_{x \to 0^+} f(x) = -\lim_{x \to 0^+} f(x) = \lim_{x \to 0^-} f(x) = -\lim_{x \to 0^-} f(x) = -f(0).$$

this gives $2f(0) = 0$, *or* $f(0) = 0$. *This completes the proof.*

The next lemma assures us that if the initial data is odd, then the solution of the heat equation over the real line is also odd.

Lemma 4 *Let $u(x,t)$ be the solution of the heat equation on $-\infty < x < \infty$. If the initial data $\phi(x) = u(x,0)$ is odd, then for all $t \geq 0$, $u(x,t)$ is an odd function of x.*

Proof *By (2.129), the solution is given by*

$$
\begin{aligned}
u(-x,t) &= \frac{1}{\sqrt{4\pi kt}} \int_{-\infty}^{\infty} e^{-\frac{(-x-y)^2}{4kt}} \phi(y)\,dy \\
&= -\frac{1}{\sqrt{4\pi kt}} \int_{-\infty}^{\infty} e^{-\frac{(-x-y)^2}{4kt}} \phi(-y)\,dy, \quad (y \mapsto -y) \\
&= -\frac{1}{\sqrt{4\pi kt}} \int_{-\infty}^{\infty} e^{-\frac{(-x+y)^2}{4kt}} \phi(y)\,dy \\
&= -\frac{1}{\sqrt{4\pi kt}} \int_{-\infty}^{\infty} e^{-\frac{(x-y)^2}{4kt}} \phi(y)\,dy \\
&= -u(x,t).
\end{aligned}
$$

This completes the proof.

Thus, by Lemmas 3 and 4, we see that if the initial data $\phi(x)$ is odd, then $u(x,t)$ is odd. Since

$$u(x,t) + u(-x,t)$$

solves the heat problem, we have $2u(0,t) = 0$, and hence $u(0,t) = 0$ for any $t > 0$, which is exactly the boundary condition in (2.133) in v. In summary, if one extends the initial data to an odd function on the whole real line, then the solution with the extended initial data automatically satisfies the Dirichlet boundary condition of (2.133). We have the following definition.

Definition 2.6 *The odd extension of a function $f(x)$ denoted by $f_o(x)$ is defined as*

$$
f_o(x) = \begin{cases} f(x), & x > 0 \\ -f(-x), & x < 0 \\ 0, & x = 0. \end{cases} \tag{2.134}
$$

The odd extension f_o is defined for negative x by reflecting the $f(x)$ with respect to the vertical axis, and then with respect to the horizontal axis. This procedure produces a function whose graph is symmetric with respect to the origin, and thus it is odd.

For example. if $f(x) = x$, $x > 0$, then $f_o(x) = x$ for $-\infty < x < \infty$. In light of the above discussion we recast the heat problem in (2.133) with extended data as

$$
\begin{cases} u_t - k u_{xx} = 0, & -\infty < x < \infty,\ t > 0 \\ u(x,0) = \phi_o(x). \end{cases} \tag{2.135}
$$

We already know from (2.127) that the solution of (2.135) is given by

$$u(x,t) = \int_{-\infty}^{\infty} S(x-y,t)\phi_o(y)dy, \quad \text{for } t > 0 \tag{2.136}$$

Since v is a solution for $x \geq 0$, we have $v(x,t) = u(x,t)$ for $x \geq 0$. Notice that

$$v(x,0) = u(x,0)\Big|_{x>0} = \phi_o(x)\Big|_{x>0} = \phi(x)$$

and $v(0,t) = u(0,t) = 0$, since $u(x,t)$ is an odd function of x. Thus, $v(x,t)$ satisfies the boundary condition in (2.135). Substituting $\phi_o(x)$ into the solution given by (2.136), we obtain by splitting the integral over two regions

$$\begin{aligned} u(x,t) &= \int_0^{\infty} S(x-y,t)\phi_o(y)dy + \int_{-\infty}^0 S(x-y,t)\phi_o(y)dy \\ &= \int_0^{\infty} S(x-y,t)\phi(y)dy - \int_{-\infty}^0 S(x-y,t)\phi(-y)dy. \end{aligned}$$

Substituting y for $-y$ in the second integral leads to

$$u(x,t) = \int_0^{\infty} S(x-y,t)\phi(y)dy - \int_0^{\infty} S(x+y,t)\phi(y)dy.$$

Using (2.128)

$$S(x,t) = \frac{1}{\sqrt{4\pi kt}} e^{-\frac{x^2}{4kt}},$$

and $v(x,t)$ in the above expression, we may write the solution formula for (2.133) as follows

$$v(x,t) = \frac{1}{\sqrt{4\pi kt}} \int_0^{\infty} \left[e^{-\frac{(x-y)^2}{4kt}} - e^{-\frac{(x+y)^2}{4kt}} \right] \phi(y)dy, \quad \text{for } t > 0. \tag{2.137}$$

Example 1 *Consider the heat equation in (2.133) with $\phi(x) = u_0$ for constant u_0. Substituting $\phi(y) = u_0$ into the solution $v(x,t)$ in (2.137), we obtain*

$$v(x,t) = \frac{u_0}{\sqrt{4\pi kt}} \int_0^{\infty} \left[e^{-\frac{(x-y)^2}{4kt}} - e^{-\frac{(x+y)^2}{4kt}} \right] dy, \quad \text{for } t > 0.$$

Making the change of variable $s = \frac{x-y}{\sqrt{4kt}}$, we arrive at

$$\begin{aligned} \frac{u_0}{\sqrt{4\pi kt}} \int_0^{\infty} e^{-\frac{(x-y)^2}{4kt}} ds &= -\frac{u_0}{\sqrt{\pi}} \int_{\frac{x}{\sqrt{4kt}}}^{-\infty} e^{-s^2} ds \\ &= \frac{u_0}{\sqrt{\pi}} \int_{-\infty}^{\frac{x}{\sqrt{4kt}}} e^{-s^2} ds. \end{aligned}$$

Similarly, by letting $s = \frac{x+y}{\sqrt{4kt}}$, we arrive at

$$\frac{u_0}{\sqrt{4\pi kt}} \int_0^{\infty} e^{-\frac{(x+y)^2}{4kt}} ds = \frac{u_0}{\sqrt{\pi}} \int_{\frac{x}{\sqrt{4kt}}}^{\infty} e^{-s^2} ds.$$

Thus,

$$
\begin{aligned}
v(x,t) &= \frac{u_0}{\sqrt{\pi}} \left[\int_{-\infty}^{\frac{x}{\sqrt{4kt}}} e^{-s^2} ds - \int_{\frac{x}{\sqrt{4kt}}}^{\infty} e^{-s^2} ds \right] \\
&= \frac{u_0}{\sqrt{\pi}} \left[\int_{-\infty}^{0} e^{-s^2} ds + \int_{0}^{\frac{x}{\sqrt{4kt}}} e^{-s^2} ds \right. \\
&\quad \left. - \left(\int_{0}^{\infty} e^{-s^2} ds - \int_{0}^{\frac{x}{\sqrt{4kt}}} e^{-s^2} ds \right) \right] \\
&= u_0 \frac{2}{\sqrt{\pi}} \int_{0}^{\frac{x}{\sqrt{4kt}}} e^{-s^2} ds \quad (\text{since } \int e^{-s^2} ds \text{ is even}) \\
&= u_0 \, erf\left(\frac{x}{\sqrt{4kt}} \right).
\end{aligned}
$$

$\square$

In the heat problem (2.133), we considered the Dirichlet boundary condition $v(0,t) = 0$, and derived the solution given by (2.137). Currently, we are interested in finding the solution for (2.133), but with a nonzero boundary condition. That is, we consider the heat problem with a nonhomogenous boundary condition

$$
\begin{cases}
v_t - kv_{xx} = 0, & 0 < x < \infty, \ t > 0 \\
v(x,0) = 0, & x > 0 \\
v(0,t) = p(t), & t > 0,
\end{cases}
\tag{2.138}
$$

where the function p is differentiable. Our approach is to make a change of variables and reduce (2.138) to a problem with homogenous boundary condition. To do so, we let

$$
u(x,t) = v(x,t) - p(t).
$$

Then (2.138) is transformed into the heat problem

$$
\begin{cases}
u_t - ku_{xx} = -p'(t), & 0 < x < \infty, \ t > 0 \\
u(x,0) = -p(0), & x > 0 \\
u(0,t) = 0, & t > 0.
\end{cases}
\tag{2.139}
$$

This is an nonhomogenous heat problem with Dirichelet boundary condition. Therefore, the problem can be solved by recasting it over the real line by the concept of odd extensions and then use (2.132) to obtain its solution. To stay compatible with previous notation we let

$$
\phi(x) = -p(0), \quad \text{and} \quad f(x,t) = -p'(t).
$$

Then

$$
\phi_o(x) = \begin{cases}
-p(0), & x > 0 \\
p(0), & x < 0 \\
0, & x = 0
\end{cases}
\quad \text{and} \quad
f_o(x,t) = \begin{cases}
-p'(t), & x > 0 \\
p'(t), & x < 0 \\
0, & x = 0.
\end{cases}
$$

Then, by (2.132), the solution maybe written in terms of the heat kernel

$$u(x,t) = \int_{-\infty}^{\infty} S(x-y,t)\phi_0(y)dy + \int_0^t \int_{-\infty}^{\infty} S(x-y,t-\tau)f_0(y,\tau)dyd\tau, \quad \text{for } t > 0.$$

(2.140)

By substituting ϕ_o and f_o into (2.140), we obtain the explicit solution

$$
\begin{aligned}
u(x,t) &= -\frac{1}{\sqrt{4\pi kt}} \int_0^{\infty} \left(e^{-\frac{(x-y)^2}{4kt}} - e^{-\frac{(x+y)^2}{4kt}} \right) p(0)dy \\
&\quad - \int_0^t \int_0^{\infty} \frac{1}{\sqrt{4\pi k(t-\tau)}} \left(e^{-\frac{(x-y)^2}{4k(t-\tau)}} - e^{-\frac{(x+y)^2}{4k(t-\tau)}} \right) p'(\tau)dyd\tau.
\end{aligned}
$$

(2.141)

Finally, the solution of our original heat problem (2.138) is

$$v(x,t) = u(x,t) + p(t),$$

where $u(x,t)$ is given by (2.141).

2.7.3   Heat equation on semi-infinite domain: Neumann condition

Now we turn our attention to addressing the one-dimensional heat equation on half-line with the Neumann condition, $u_x(0,t) = 0$. We employ a similar procedure as in the previous subsection by extending the initial data to the negative half-line in such a fashion that the boundary condition is automatically satisfied. In the case of the Dirichlet condition, we used the odd extension to satisfy the initial data, whereas here we use the notion of *even extension*. We begin by considering the heat problem under Neumann condition at the end point $x = 0$,

$$
\begin{cases}
v_t - kv_{xx} = 0, & 0 < x < \infty, \, t > 0 \\
v(x,0) = \phi(x), & x > 0 \\
v_x(0,t) = 0, & t > 0.
\end{cases}
$$

(2.142)

Our aim is to find a solution to (2.142) as we did for the heat problem over the entire real line. There will be no need to start from scratch, but instead we will try to recast the problem over the entire real line by extending the initial data to the whole line. For heat problems with Neumann conditions, one would choose the *even extension* of the initial data $\phi(x)$. Note that if $\psi(x)$ is even, then $\psi(x) = \psi(-x)$, from which we get $\psi'(x) = -\psi'(-x)$, and hence $2\psi'(0) = 0$, or $\psi'(0) = 0$. This is true for any even function. We make it formal in the next lemma.

Lemma 5 *Let $f : (-\infty, \infty) \to \mathbb{R}$ be an even function $(f(x) = f(-x))$, that is differentiable at $x = 0$ then $f'(0) = 0$.*

Proof *Recall from Chapter 1 that the derivative of a function g is given by*

$$g'(x) = \lim_{h \to 0} \frac{g(x+h) - g(x)}{h}.$$

Thus, since f is differentiable and $f(h) = f(-h)$, we have

$$
\begin{aligned}
f'(0) &= \lim_{h\to 0}\frac{f(h)-f(0)}{h} = \lim_{h\to 0}\frac{f(-h)-f(0)}{h} \\
&= -\lim_{h\to 0}\frac{f(0)-f(-h)}{h} = -f'(0).
\end{aligned}
$$

This implies $2f'(0) = 0$, or $f'(0) = 0$. This completes the proof.

Now we state a similar lemma to Lemma 4 for even initial data.

Lemma 6 *Let $u(x,t)$ be the solution of the heat equation on $-\infty < x < \infty$. If the initial data $\phi(x) = u(x,0)$ is even, then for all $t \geq 0$, $u(x,t)$ is an even function of x.*

As before, by Lemma 6, we have that if the initial data $\phi(x)$ is even, then $u(x,t)$ is even. Thus, by Lemma 5, $u_x(0,t) = 0$ for any $t > 0$. Hence u automatically satisfies the Neumann boundary condition at $t = 0$. We have the following definition.

Definition 2.7 *The even extension of a function $f(x)$ denoted by $f_e(x)$ is defined as*

$$
f_e(x) = \begin{cases} f(x), & x > 0 \\ f(-x), & x < 0 \\ 0, & x = 0. \end{cases} \tag{2.143}
$$

In light of the above discussion, we recast the heat problem in (2.142) with extended data as

$$
\begin{cases} u_t - ku_{xx} = 0, & -\infty < x < \infty,\ t > 0 \\ u(x,0) = \phi_e(x) \\ u_x(0,t) = 0, & t > 0. \end{cases} \tag{2.144}
$$

We already know from (2.127) that the solution of (2.144) is given by

$$
u(x,t) = \int_{-\infty}^{\infty} S(x-y,t)\phi_e(y)dy, \quad \text{for } t > 0. \tag{2.145}
$$

Since v is a solution for $x \geq 0$, we have $v(x,t) = u(x,t)$ for $x \geq 0$. Notice that

$$
v(x,0) = u(x,0)\Big|_{x>0} = \phi_e(x)\Big|_{x>0} = \phi(x)
$$

and $v_x(0,t) = u_x(0,t) = 0$, since $u(x,t)$ is an even function of x. Thus, $v(x,t)$ satisfies the Neumann condition in (2.144). Substituting $\phi_e(x)$ into the solution given by (2.145), we obtain by splitting the integral over two regions

$$
\begin{aligned}
u(x,t) &= \int_0^{\infty} S(x-y,t)\phi_e(y)dy + \int_{-\infty}^0 S(x-y,t)\phi_e(y)dy \\
&= \int_0^{\infty} S(x-y,t)\phi(y)dy + \int_{-\infty}^0 S(x-y,t)\phi(-y)dy.
\end{aligned}
$$

Making the change of variables $y \mapsto -y$ in the second integral leads to

$$u(x,t) = \int_0^\infty S(x-y,t)\phi(y)dy - \int_\infty^0 S(x+y,t)\phi(y)dy.$$

Using (2.128)

$$S(x,t) = \frac{1}{\sqrt{4\pi kt}} e^{-\frac{x^2}{4kt}},$$

and $v(x,t)$ in the above expression, we may write the solution formula for (2.142) as follows

$$v(x,t) = \frac{1}{\sqrt{4\pi kt}} \int_0^\infty \left[e^{-\frac{(x-y)^2}{4kt}} + e^{-\frac{(x+y)^2}{4kt}} \right] \phi(y)dy, \quad \text{for } t > 0. \qquad (2.146)$$

We have the following example.

Example 2.28 Consider the heat equation given by (2.142) with $\phi(x) = u_0$ for constant u_0. Substituting $\phi(y) = u_0$ into the solution $v(x,t)$ in (2.146) we obtain

$$v(x,t) = \frac{u_0}{\sqrt{4\pi kt}} \int_0^\infty \left[e^{-\frac{(x-y)^2}{4kt}} + e^{-\frac{(x+y)^2}{4kt}} \right] dy, \quad \text{for } t > 0.$$

Making the change of variable $s = \frac{x-y}{\sqrt{4kt}}$, and $s = \frac{x+y}{\sqrt{4kt}}$, in the first integral and the second integral, respectively, we arrive at

$$
\begin{aligned}
v(x,t) &= \frac{u_0}{\sqrt{\pi}} \left[\int_{-\infty}^{\frac{x}{\sqrt{4kt}}} e^{-s^2} ds + \int_{\frac{x}{\sqrt{4kt}}}^\infty e^{-s^2} ds \right] \\
&= \frac{u_0}{\sqrt{\pi}} \left[\int_{-\infty}^0 e^{-s^2} ds + \int_0^{\frac{x}{\sqrt{4kt}}} e^{-s^2} ds \right. \\
&\quad + \left. \left(\int_0^\infty e^{-s^2} ds - \int_0^{\frac{x}{\sqrt{4kt}}} e^{-s^2} ds \right) \right] \\
&= u_0 \frac{2}{\sqrt{\pi}} \int_0^\infty e^{-s^2} ds \quad (\text{since } \int e^{-s^2} ds \text{ is even}) \\
&= u_0 \frac{2}{\sqrt{\pi}} (\frac{\sqrt{\pi}}{2}), \quad (\text{by Lemma 1}) \\
&= u_0.
\end{aligned}
$$

$\square$

2.7.4 Exercises

Exercise 2.71 *Prove parts (a)–(c) of Lemma 2.*

Exercise 2.72 *For constant u_0, write the solution in terms of the error function for the heat problem*

$$
\begin{cases}
u_t - ku_{xx} = 0, & -\infty < x < \infty, \ t > 0 \\
u(x,0) = \phi(x), & -\infty < x < \infty
\end{cases}
$$

where

$$\phi(x) = \begin{cases} u_0, & |x| < l \\ 0, & |x| > l \end{cases}$$

Exercise 2.73 *Consider the heat problem*

$$\begin{cases} v_t - kv_{xx} = 0, & 0 < x < \infty, \ t > 0 \\ v(x,0) = \phi(x), & x > 0 \\ v(0,t) = 0, & t > 0 \end{cases}$$

with initial data

$$\phi(x) = \begin{cases} 0, & 0 < x < l \\ 1, & x > l. \end{cases}$$

Show its solution $u(x,t)$ can be written as

$$v(x,t) = \frac{1}{2} erf\left(\frac{l+x}{\sqrt{4kt}}\right) - \frac{1}{2} erf\left(\frac{l-x}{\sqrt{4kt}}\right).$$

Exercise 2.74 *Solve the heat problem*

$$\begin{cases} u_t - ku_{xx} = 0, & -\infty < x < \infty, \ t > 0 \\ u(x,0) = e^{-2|x|}, & -\infty < x < \infty. \end{cases}$$

Exercise 2.75 *Solve*

$$\begin{cases} u_t - ku_{xx} = 0, & 0 < x < \infty, \ t > 0 \\ u(x,0) = \phi(x), & x > 0 \\ u(0,t) = 0, & t > 0 \end{cases}$$

(a)

$$\phi(x) = \begin{cases} 1-x, & 0 < x < 1 \\ 0, & x \geq 1 \end{cases}$$

(b)

$$\phi(x) = \begin{cases} xe^{-x^2}, & 0 < x < 1 \\ 0, & x \geq 1. \end{cases}$$

Exercise 2.76 *Provide all the details in obtaining (2.141).*

Exercise 2.77 *Consider the nonlinear heat equation*

$$u_t - ku_{xx} + bu_x^2 = 0 \quad for \ -\infty < x < \infty \ and \ t > 0,$$

subject to an initial condition $u(x,0) = f(x)$. This form of PDE makes its presence in stochastic optimal control theory. In this question you will derive a representation formula for the solution $u(x,t)$.

(a) Define the Cole-Hopf transformation $w(x,t) = e^{-\frac{b}{k}u(x,t)}$. Show that w is a solution of the linear heat equation $w_t - kw_{xx} = 0$.

(b) Use the fundamental solution of the heat equation to solve for w(x,t).

(c) Invert the Cole-Hopf transformation to find a formula for u.

Exercise 2.78 *Solve the heat equation $u_t - u_{xx} = 0$ on the entire real line $-\infty < x < \infty$ with initial condition $u(x,0) = \cos(x)$ without using the fundamental solution of the heat equation. Use your answer to deduce the value of the integral*

$$\int_{-\infty}^{\infty} \cos(x) e^{-\frac{x^2}{4t}} dx.$$

Hint: Search for a solution of the form $u(x,t) = h(t)\cos(x)$.

Exercise 2.79 *Solve*

$$\begin{cases} u_t - ku_{xx} = 0, & 0 < x < \infty, \ t > 0 \\ u(x,0) = 0, & x > 0 \\ u(0,t) = p(t), & t > 0 \end{cases}$$

(a)

$$p(t) = \begin{cases} 1, & 0 < t < 1 \\ 0, & t \geq 1. \end{cases}$$

(b)

$$p(t) = 1, \quad t > 0.$$

Exercise 2.80 *Provide all the details in obtaining (2.130).*

Exercise 2.81 *For positive constant c, consider the heat equation with convection term*

$$u_t + cu_x - ku_{xx} = 0, \tag{2.147}$$

(a) Determine the values of α and β so that the transformation

$$u(x,t) = v(x,t)e^{\alpha x + \beta t}$$

transforms (2.147) to the heat equation $v_t - kv_{xx} = 0$.

(b) Find the solution of (2.147) on $-\infty < x < \infty$, with initial condition $u(x,0) = \phi(x)$.

(c) Find the solution of (2.147) on $x > 0$, with initial and boundary conditions $u(x,0) = \phi(x)$, and $u(x,0) = 0, t > 0$.

Exercise 2.82 *Let $f : (-\infty, \infty) \to \mathbb{R}$ be an even function $(f(z) = f(-z))$, that is differentiable. Use the definition of the derivative and show that $f'(-x) + f'(x) = 0$.*

Exercise 2.83 *Solve*

$$\begin{cases} u_t - ku_{xx} = 0, & 0 < x < \infty, \ t > 0 \\ u(x,0) = \phi(x), & x > 0 \\ u_x(0,t) = 0, & t > 0 \end{cases}$$

(a)
$$\phi(x) = \begin{cases} 1, & 0 < x < 1 \\ 0, & x \geq 1. \end{cases} \quad , \quad (b)\ \phi(x) = xe^{-ax}.$$

(c)
$$\phi(x) = \begin{cases} 1-x^2, & 0 < x < 1 \\ 0, & x \geq 1. \end{cases}$$

Exercise 2.84 *Consider the heat equation over* $\mathbb{R}$ *along with its solution given by* (2.129). *Show that if the initial function* $\phi(x)$ *is uniformly bounded on* $\mathbb{R}$, *then the solution* $u(x,t)$ *satisfies*

$$|u(x,t)| \leq \max_{-\infty < x < \infty} |\phi(x)|.$$

Hint: Make use of the substitution $s = \frac{x-y}{4kt}$.

2.8   Wave Equation on Semi-Infinite Domain

We take advantage of the development of solutions of the heat equation on semi-infinite domain with Dirichlet conditions or Neumann conditions, and establish the solution of the wave equation on semi unbounded domain with Dirichlet condition or Neumann condition. We begin by considering the wave equation of semi-infinite string with a fixed end or Dirichlet condition

$$\begin{cases} u_{tt} - c^2 u_{xx} = 0, & 0 < x < \infty,\ t > 0 \\ u(x,0) = f(x), & x > 0 \\ u_t(x,0) = g(x), & x > 0 \\ u(0,t) = 0, & t > 0. \end{cases} \tag{2.148}$$

Using the same concept as we did for the heat equation with Dirichlet condition, we use the odd extensions $f_o(x)$ and $g_o(x)$ of $f(x)$ and $g(x)$, respectively and solve the wave equation on the whole real line with initial conditions $u(x,0) = f_o(x)$ and $u_t(x,0) = g_o(x)$. In other words, we consider

$$\begin{cases} u_{tt} - c^2 u_{xx} = 0, & -\infty < x < \infty,\ t > 0 \\ u(x,0) = f_o(x), \quad u_t(x,0) = g_o(x), & x > 0. \end{cases} \tag{2.149}$$

D'Alembert formula gives

$$u(x,t) = \frac{1}{2}[f_o(x+ct) + f_o(x-ct)] + \frac{1}{2c}\int_{x-ct}^{x+ct} g_o(s)ds. \tag{2.150}$$

By Exercise 2.89, $u(x,t)$ given in (2.150) is odd and so $u(0,t) = 0$. Now we try to make some sense out of the solution in (2.150).

Remember that $x > 0$. We do this in two cases.

(a) First, suppose that $x > ct$. Then $x + ct \geq 0$ and $x - ct \geq 0$, and so

$$f_0(x + ct) = f(x + ct) \quad \text{and} \quad f_0(x - ct) = f(x - ct).$$

In addition, on $[x - ct, x + ct]$ we have $g_0(s) = g(s)$. Thus if $x > ct$, we have from (2.150) that

$$u(x,t) = \frac{1}{2}[f(x + ct) + f(x - ct)] + \frac{1}{2c}\int_{x-ct}^{x+ct} g(s)ds.$$

(b) Second, if $0 < x < ct$. Then $x + ct \geq 0$ and $x - ct < 0$, and so

$$f_0(x + ct) = f(x + ct) \quad \text{and} \quad f_0(x - ct) = -f(ct - x).$$

Moreover, on $[x - ct, 0]$ we have $g_0(s) = -g(-s)$ and on $[0, x + ct]$ we have $g_0(s) = g(s)$. Therefore if $0 < x < ct$ we have from (2.150) that

$$
\begin{aligned}
u(x,t) &= \frac{1}{2}[f(x + ct) - f(ct - x)] + \frac{1}{2c}\int_{x-ct}^{0}(-g(-s))ds + \frac{1}{2c}\int_{0}^{x+ct} g(s)ds \\
&= \frac{1}{2}[f(x + ct) - f(ct - x)] - \frac{1}{2c}\int_{0}^{ct-x} g(s)ds \\
&\quad + \frac{1}{2c}\int_{0}^{x+ct} g(s)ds \ (s \mapsto -s) \\
&= \frac{1}{2}[f(x + ct) - f(ct - x)] + \frac{1}{2c}\int_{ct-x}^{ct+x} g(s)ds.
\end{aligned}
$$

In summary, the solution of the half-line wave equation with Dirichlet boundary condition is

$$u(x,t) = \begin{cases} \frac{1}{2}[f(x + ct) + f(x - ct)] + \frac{1}{2c}\int_{x-ct}^{x+ct} g(s)ds, & x \geq ct \\ \frac{1}{2}[f(x + ct) - f(ct - x)] + \frac{1}{2c}\int_{ct-x}^{ct+x} g(s)ds, & 0 < x < ct. \end{cases} \tag{2.151}$$

Fig. 2.20 shows the two regions of the existence of the solution.

Now we turn our attention to establishing the solution of the wave equation on semi unbounded domain with Neumann condition. We begin by considering the wave equation of semi-infinite string with a Neumann condition

$$\begin{cases} u_{tt} - c^2 u_{xx} = 0, & 0 < x < \infty, \, t > 0 \\ u(x,0) = f(x), & x > 0 \\ u_t(x,0) = g(x), & x > 0 \\ u_x(0,t) = 0, & t > 0. \end{cases} \tag{2.152}$$

Using the same concept as we did for the wave equation with Dirichlet condition, we use the even extensions $f_e(x)$ and $g_e(x)$ of $f(x)$ and $g(x)$, respectively, and solve

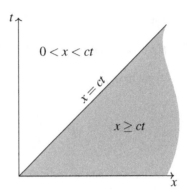

FIGURE 2.20
Regions of existence of the solution.

the wave equation on the whole real line with initial conditions $u(x,0) = f_e(x)$ and $u_t(x,0) = g_e(x)$. In other words, we consider

$$\begin{cases} u_{tt} - c^2 u_{xx} = 0, & -\infty < x < \infty, \ t > 0 \\ u(x,0) = f_e(x), \quad u_t(x,0) = g_e(x), & x > 0. \end{cases} \tag{2.153}$$

D'Alembert formula gives

$$u(x,t) = \frac{1}{2}[f_e(x+ct) + f_e(x-ct)] + \frac{1}{2c}\int_{x-ct}^{x+ct} g_e(s)ds. \tag{2.154}$$

By Exercise 2.88, $u(x,t)$ given by (2.154) is even and since the derivative of an even function is odd, and so u_x will be odd in x, and hence $u_x(0,t) = 0$. As before we can simplify (2.150). We do this in two cases and recall that $x > 0$.

(a) First, suppose that $x > ct$. Then $x + ct \geq 0$ and $x - ct \geq 0$, and so

$$f_e(x+ct) = f(x+ct) \quad \text{and} \quad f_e(x-ct) = f(x-ct).$$

Moreover, on $[x - ct, x + ct]$ we have $g_e(s) = g(s)$. Thus if $x > ct$, we have from (2.154) that

$$u(x,t) = \frac{1}{2}[f(x+ct) + f(x-ct)] + \frac{1}{2c}\int_{x-ct}^{x+ct} g(s)ds.$$

(b) Second, if $0 < x < ct$. Then $x + ct \geq 0$ and $x - ct < 0$, and so

$$f_e(x+ct) = f(x+ct) \quad \text{and} \quad f_e(x-ct) = f(ct-x).$$

Moreover, on $[x - ct, 0]$ we have $g_e(s) = g(-s)$ and on $[0, x+ct]$ we have $g_e(s) = g(s)$. Therefore if $0 < x < ct$ we have from (2.154) that

$$u(x,t) = \frac{1}{2}[f(x+ct) + f(ct-x)] + \frac{1}{2c}\int_{x-ct}^{0} g(-s)ds + \frac{1}{2c}\int_{0}^{x+ct} g(s)ds$$

$$= \frac{1}{2}[f(x+ct)+f(ct-x)]+\frac{1}{2c}\int_0^{ct-x} g(s)ds$$
$$+\frac{1}{2c}\int_0^{x+ct} g(s)ds \; (s\mapsto -s)$$

In summary, the solution of the half-line wave equation with Neumann boundary condition is

$$u(x,t) = \begin{cases} \frac{1}{2}[f(x+ct)+f(x-ct)]+\frac{1}{2c}\int_{x-ct}^{x+ct} g(s)ds, & x\geq ct \\ \frac{1}{2}[f(x+ct)+f(ct-x)]+\frac{1}{2c}\left[\int_0^{ct-x} g(s)ds+\int_0^{x+ct} g(s)ds\right], & 0<x<ct. \end{cases}$$
(2.155)

Example 2.29 Consider the wave equation (2.148) with $c=1$ and with initial data $f(x)=\sin(x)$, $g(x)=0$, and the Dirichlet condition $u(0,t)=0$. Then using (2.151), we have for $x>t$, that

$$u(x,t) = \frac{1}{2}[\sin(x+t)+\sin(x-t)] = \sin(x)\cos(t).$$

On the other hand, for $0<x<t$, we have

$$u(x,t) = \frac{1}{2}[\sin(x+t)-\sin(t-x)] = \sin(x)\cos(t).$$

Thus,

$$u(x,t) = \sin(x)\cos(t), \; x>0.$$

$\square$

Example 2.30 Consider the wave equation (2.148) with $c=1$ and with initial data $f(x)=\sin(x)$, $g(x)=0$, and the Neumann condition $u_x(0,t)=0$. Then using (2.155), One can easily verify that

$$u(x,t) = \sin(t)\cos(x), \quad x>0.$$

$\square$

2.8.1 Exercises

Exercise 2.85 *Solve*

$$\begin{cases} u_{tt}-c^2 u_{xx}=0, & 0<x<\infty, \; t>0 \\ u(x,0)=0, & x>0 \\ u_t(x,0)=\cos(x), & x>0 \\ u(0,t)=0, & t>0. \end{cases}$$

Exercise 2.86 *Solve*

$$\begin{cases} u_{tt}-c^2 u_{xx}=0, & 0<x<\infty, \; t>0 \\ u(x,0)=0, & x>0 \\ u_t(x,0)=\sin(x), & x>0 \\ u(0,t)=0, & t>0. \end{cases}$$

Exercise 2.87 *Solve*

$$\begin{cases} u_{tt} - c^2 u_{xx} = 0, & 0 < x < \infty, \, t > 0 \\ u(x,0) = 0, & x > 0 \\ u_t(x,0) = 1, & x > 0 \\ u(0,t) = 0, & t > 0. \end{cases}$$

Exercise 2.88 *Let $u(x,t)$ be the D'alembert's solution on $-\infty < x < \infty$ of the wave equation given by 2.82. If the initial data given by $f(x) = u(x,0)$ and $g(x) = u_x(x,0)$ is even, then for all $t \geq 0$, $u(x,t)$ is an even function of x.*

Exercise 2.89 *Let $u(x,t)$ be the D'alembert's solution on $-\infty < x < \infty$ of the wave equation given by 2.82. If the initial data given by $f(x) = u(x,0)$ and $g(x) = u_x(x,0)$ is odd, then for all $t \geq 0$, $u(x,t)$ is an odd function of x.*

Exercise 2.90 *Solve*

$$\begin{cases} u_{tt} - c^2 u_{xx} = 0 & 0 < x < \infty, \, t > 0 \\ u(x,0) = 0, & x > 0 \\ u_t(x,0) = \sin(x), & x > 0 \\ u_x(0,t) = 0, & > 0. \end{cases}$$

Exercise 2.91 *Solve*

$$\begin{cases} u_{tt} - c^2 u_{xx} = 0, & 0 < x < \infty, \, t > 0 \\ u(x,0) = 0, & x > 0 \\ u_t(x,0) = \cos(x), & > 0 \\ u_x(0,t) = 0, & t > 0. \end{cases}$$

Exercise 2.92 *Solve*

$$\begin{cases} u_{tt} - c^2 u_{xx} = 0, & 0 < x < \infty, \, t > 0 \\ u(x,0) = 1, & x > 0 \\ u_t(x,0) = 0, & x > 0 \\ u_x(0,t) = 0, & t > 0. \end{cases}$$

Exercise 2.93 *Solve*

$$\begin{cases} u_{tt} - c^2 u_{xx} = 0, & 0 < x < \infty, \, t > 0 \\ u(x,0) = 0, & x > 0 \\ u_t(x,0) = 1, & x > 0 \\ u_x(0,t) = 0, & t > 0. \end{cases}$$

Exercise 2.94 *Consider the wave problem with nonhomogenous Dirichlet boundary condition*

$$\begin{cases} v_{tt} - c^2 v_{xx} = 0, & 0 < x < \infty, \, t > 0 \\ v(x,0) = f(x), & x > 0 \\ v_t(x,0) = g(x), & x > 0 \\ v(0,t) = p(t), & t > 0, \end{cases}$$

where the function p is twice differentiable. Make the change of variables

$$u(x,t) = v(x,t) - p(t)$$

and reduce the wave problem to a problem with homogenous boundary condition and then use the D'alembert solution given by (2.85) to find the solution u and then find the solution v of the original wave problem.

3

Matrices and Systems of Linear Equations

In this chapter, we look at systems of equations, matrix algebra, and applications to linear algebra by solving linear systems of differential equations. In addition, we will study quadratic forms and their applications and functions in symmetric matrices.

3.1 Systems of Equations and Gaussian Elimination

We are interested in solving a nonhomogeneous system of m linear equations with n unknown variables $x_1, x_2, \ldots, x_n$ of the form

$$
\begin{aligned}
a_{11}x_1 + a_{12}x_2 + a_{13}x_3 + \ldots + a_{1n}x_n &= b_1 \\
a_{21}x_1 + a_{22}x_2 + a_{23}x_3 + \ldots + a_{2n}x_n &= b_2 \\
a_{31}x_1 + a_{32}x_2 + a_{33}x_3 + \ldots + a_{3n}x_n &= b_3 \\
&\vdots \\
a_{m1}x_1 + a_{m2}x_2 + a_{m3}x_3 + \ldots + a_{mn}x_n &= b_m
\end{aligned}
\tag{3.1}
$$

where the a_{ij}, b_i are constants for $1 \le i \le m, 1 \le j \le n$. By a solution of the system (3.1), we mean an n-tuple of real numbers of the form $(p_1, p_2, \ldots, p_n)$ that when plugged into (3.1) produces a true statement. For example, $(3, -1, 0, 2)$ is a solution of the system

$$
\begin{aligned}
x_1 + x_2 - x_3 + 4x_4 &= 10 \\
-x_1 - x_2 + 2x_3 + x_4 &= 0 \\
10x_1 + 3x_2 + x_4 &= 29
\end{aligned}
$$

The set of all solutions is called *the solution set*. Under the assumption that the system (3.1) has a solution, we use the Gaussian elimination method to find the solution set. We now describe the method in few steps.

DOI: 10.1201/9781003449881-3

Step 1. In this step, we try to eliminate x_1 from the second , third, $\ldots, mth$ equation. Suppose that $a_{11} \neq 0$. If not, renumber the equations or variables so that this is the case. We may achieve the elimination of x_1 by multiplying the first equation with $\dfrac{a_{21}}{a_{11}}$ and then subtracting the resulting equation from the second equation and by multiplying the first equation with $\dfrac{a_{31}}{a_{11}}$ and then subtracting the resulting equation from the third equation, and so forth. This will result of the new system of equations

$$a_{11}x_1 + a_{12}x_2 + l_{13}x_3 + \ldots + al_{1n}x_n = b_1$$
$$l_{22}x_2 + l_{23}x_3 + \ldots + l_{2n}x_n = b_2' \qquad (3.2)$$
$$\vdots$$
$$l_{m2}x_2 + l_{m3}x_3 + \ldots + l_{mn}x_n = b_m'.$$

Since any solution of (3.1) is a solution of (3.2) and conversely, because steps are reversible, we may obtain (3.1) from (3.2).

Step 2. In this step we try to eliminate x_2 from the third, $\ldots, mth$ equation in (3.2). Suppose that $l_{22} \neq 0$. (Otherwise, renumber the equations or variables so that this is so.) We do this by multiplying the second equation with $\dfrac{l_{32}}{l_{22}}$ and then subtracting the resulting equation from the third equation. Similarly, we multiply the second equation with $\dfrac{l_{42}}{l_{22}}$ and then subtracting the resulting equation from the fourth equation, and so forth. The further steps are now obvious. For example, in the third step, we eliminate x_3 and in the fourth step, we eliminate x_4, etc. The process will only stop when no equations are left or when the coefficients of all the unknowns in the remaining equations are all zero. This leads to the system of equations

$$a_{11}x_1 + a_{12}x_2 + a_{13}x_3 + \ldots + a_{1n}x_n = b_1$$
$$l_{22}x_2 + l_{23}x_3 + \ldots + l_{2n}x_n = b_2' \qquad (3.3)$$
$$\vdots$$
$$d_{rr}x_r + \ldots + d_{rn}x_n = b_m''$$

where either $r = m$, or $r < m$. If $r < m$, the remaining equations have the form

$$0 = b_{r+1}'', 0 = b_{r+2}'', \ldots, 0 = b_m''$$

and the system has no solution, unless

$$b_{r+1}'' = 0, \ldots, b_m'' = 0.$$

If the system has a solution we may obtain it by assigning arbitrary values for the unknown $x_{r+1}, \ldots, x_n$, solving the last equation in (3.3) for x_r, the next to the last for x_{r-1}, and so on up to the line. When $m = n = r$, the system (3.3) has triangular form and there is one, and only one , solution. We will illustrate the method in a series of examples.

Example 3.1 Consider the system

$$x_1 + 2x_2 - 2x_3 + 4x_4 = 11$$
$$2x_1 + x_2 - x_3 + 3x_4 = 9$$
$$x_1 - x_2 + x_3 - x_4 = -2.$$

In the first step, we eliminate x_1 from the last two equations. This is done by multiplying the first equation by 2 and then subtracting the resulting equation from the second equation. Similarly, we subtract the third equation from the first equation. This leads us to the new system of equations

$$x_1 + 2x_2 - 2x_3 + 4x_4 = 11$$
$$-3x_2 + 3x_3 - 5x_4 = -13$$
$$-3x_2 + 3x_3 - 5x_4 = -13.$$

Note that the last two equations are identical. In the second step we eliminate x_2 from the third equation by subtracting the third equation from the second equation. This results into the new system of equations

$$x_1 + 2x_2 - 2x_3 + 4x_4 = 11$$
$$-3x_2 + 3x_3 - 5x_4 = -13$$
$$0x_2 + 0x_3 + 0x_4 = 0.$$

The third equation is satisfied for any value for x_2, x_3, and x_4. Thus the third equation puts no constraint on the solution. However, the first and second equations represent four unknowns with two constraints and hence there are two arbitrary unknowns ($4 - 2 = 2$). We may choose x_3 and x_4 arbitrarily. Thus, we let $x_3 = s$ and $x_4 = t$, where s and t are arbitrary. Then the second equation gives

$$x_2 = x_3 - \frac{5}{3}x_4 + \frac{13}{3} = s - \frac{5}{3}t + \frac{13}{3}.$$

Using the first equation, we solve for x_1 and obtain

$$x_1 = -2x_2 + 2x_3 - 4x_4 + 11 = -\frac{2}{3}t + \frac{7}{3}.$$

☐

Example 3.2 Consider the system

$$x_1 + x_2 - 3x_3 = 4$$
$$2x_1 + x_2 - x_3 = 2$$
$$3x_1 + 2x_2 - 4x_3 = 7.$$

We start by using the first equation. We perform two steps simultaneously. Multiply the first equation by -2 and add it to the second equation. Then multiply the first equation by -3 and add it to the third equation to obtain

$$x_1 + x_2 - 3x_3 = 4$$
$$-x_2 + 5x_3 = -6$$
$$-x_2 + 5x_3 = -5.$$

Next, -1 times the second equation is added to the third equation produces

$$x_1 + x_2 - 3x_3 = 4$$
$$-x_2 + 5x_3 = -6$$
$$0x_1 + 0x_2 + 0x_3 = 1.$$

The third equation can not be satisfied for any values x_1, x_2, and x_3 and hence the process stops and the system has no solution. $\qquad\square$

Example 3.3 Consider the system

$$2x_1 - 2x_2 = -6$$
$$x_1 - x_2 + x_3 = 1$$
$$3x_2 - 2x_3 = -5.$$

We start by multiplying the first equation by $1/2$

$$x_1 - x_2 = -3$$
$$x_1 - x_2 + x_3 = 1$$
$$3x_2 - 2x_3 = -5.$$

Now, adding -1 times the first equation to the second equation yields

$$x_1 - x_2 = -3$$
$$x_3 = 4$$
$$3x_2 - 2x_3 = -5.$$

From the second equation we immediately have $x_3 = 4$. Substituting this value into the third equation gives $3x_2 - 2(4) = -5$, and hence $x_2 = 1$. Similarly, substituting into the first equation gives $x_1 = -2$. Thus the system has the solution $(x_1, x_2, x_3) = (-2, 1, 4)$. $\qquad\square$

Example 3.4 Consider the system

$$x_1 + x_2 - 3x_3 = 4$$
$$2x_1 + x_2 - x_3 = 2$$
$$3x_1 + 2x_2 - 4x_3 = 6.$$

We will perform two steps at once. First, -2 times the first equation and then add it to the second equation; second, -3 times the first equation and then add it to the third equation yield the following equivalent system

$$x_1 + x_2 - 3x_3 = 4$$
$$-x_2 + 5x_3 = -6$$
$$-x_2 + 5x_3 = -6.$$

Now, adding -1 times the second equation to the third equation yields

$$x_1 + x_2 - 3x_3 = 4$$
$$-x_2 + 5x_3 = -6$$
$$0x_1 + 0x_2 + 0x_3 = 0.$$

The third equation is satisfied for any value for x_1, x_2 and x_3. Thus the third equation puts no constraint on the solution. However, the first and second equations represent three unknowns with two constraints and hence there is one arbitrary unknown ($3 - 2 = 1$.) To simplify calculation, it is more convenient to let $x_3 = s$ where s is arbitrary. Then the second equation gives $x_2 = 5s + 6$. Similarly, the first equation yields $x_1 = -2 - 2s$. Since s is arbitrary, the system has infinitely many solutions. □

3.2   Homogeneous Systems

We begin by considering the linear homogeneous system of equations

$$a_{11}x_1 + a_{12}x_2 + a_{13}x_3 + \ldots + a_{1n}x_n = 0$$
$$a_{21}x_1 + a_{22}x_2 + a_{23}x_3 + \ldots + a_{2n}x_n = 0$$
$$\cdots$$
$$a_{m1}x_1 + a_{m2}x_2 + a_{m3}x_3 + \ldots + a_{mn}x_n = 0$$

where the a_{ij}, $1 \leq i \leq m, 1 \leq j \leq n$, are constants. Note that the above homogeneous system will always have the *trivial* solution $(x_1 = x_2, \ldots, x_n = 0)$ as a solution. Any other solution is *nontrivial*. In the case $m < n$, that is, there are more unknowns than equations, then the above homogeneous system (3.4) has a nontrivial solution, as seen in the next theorem.

Theorem 3.1 *In the homogeneous system (3.4), if $m < n$, then the system has a nontrivial solution.*

Proof *The proof is based on the Gauss elimination method. We may assume the coefficient a_{11} of x_1 is not zero. This is a fair assumption since if all the coefficients of x_1 are zero; that is $a_{11} = a_{21} = \ldots = a_{m1} = 0$, then, $x_1 = 1, x_2 = x_3 = \ldots = x_n = 0$*

is a nontrivial solution. Thus, we may assume $a_{11} \neq 0$. Divide the first equation in (3.4) *by a_{11} to obtain the equation*

$$x_1 + b_{12}x_2 + b_{13}x_3 + \ldots b_{1n}x_n = 0. \tag{3.4}$$

Multiply (3.4) *successively by $a_{21}, a_{31}, \ldots, a_{m1}$, and subtract the respective resultant equations from the second, third, $\ldots, m$th equations of* (3.4), *to reduce* (3.4) *to the form*

$$\begin{aligned} x_1 + b_{12}x_2 + b_{13}x_3 + \ldots + b_{1n}x_n &= 0 \\ b_{22}x_2 + b_{23}x_3 + \ldots + b_{2n}x_n &= 0 \\ &\cdots \\ b_{m1}x_1 + b_{m2}x_2 + b_{m3}x_3 + \ldots + b_{mn}x_n &= 0. \end{aligned}$$

Now we repeat the same process but now we assume the coefficient b_{22} of x_2 is not zero. Hence, by applying the Gaussian procedure again produces the third system

$$\begin{aligned} x_1 + c_{13}x_3 + \ldots + c_{1n}x_n &= 0 \\ x_2 + c_{23}x_3 + \ldots + c_{2n}x_n &= 0 \\ c_{33}x_3 + \ldots + c_{3n}x_n &= 0 \\ &\cdots \\ c_{m3}x_3 + \ldots + c_{mn}x_n &= 0. \end{aligned}$$

By continuing this process and in particular at the r stage and by using the fact that the numbers of variables is less than the number of equations, we ultimately, arrive a system of m equations of the form

$$\begin{aligned} x_1 + d_{1r}x_r + \ldots + d_{1n}x_n &= 0 \\ x_2 + d_{2r}x_r + \ldots + d_{2n}x_n &= 0 \\ \vdots \\ x_{r-1} + d_{r-1}x_r + \ldots + d_{r-1,n}x_n &= 0 \\ 0 &= 0. \end{aligned}$$

If we let $x_r = 1, x_{r+1} = \cdots = x_n = 0$, and $x_1 = -d_{1r}, x_2 = -d_{2r}, \ldots, x_{r-1} = -d_{r-1,r}$, we obtain a nontrivial solution. The proof is done since the systems are equivalent.

Remark 8 *In fact, the homogeneous system* (3.4) *has infinitely many solutions since the choice of x_r is arbitrary.*

3.2.1 Exercises

Exercise 3.1 *Solve the given system*

$$\begin{aligned} x_1 - 2x_2 - x_3 + 3x_4 &= 1 \\ 2x_1 - 4x_2 + x_3 &= 5 \\ x_1 - 2x_2 + 2x_3 - 3x_4 &= 4. \end{aligned}$$

Exercise 3.2 *Solve the given system*

$$3x_1 + 7x_2 - x_3 = -1$$
$$x_1 + 3x_2 + x_3 = 1$$
$$-x_1 - 2x_2 + x_3 = 1.$$

Exercise 3.3 *Determine all solutions of the system*

$$x_1 - 3x_2 + 4x_3 = 1$$
$$-2x_1 + x_2 + 2x_4 = -1$$
$$-x_1 - 2x_2 + 4x_3 + 2x_4 = 0.$$
$$4x_1 + 3x_2 - 8x_3 - 6x_4 = 1.$$

Exercise 3.4 *Solve the given system*

$$-2x_1 + x_2 + x_3 = -3$$
$$x_1 + 3x_2 - x_3 = 5$$
$$3x_1 + 2x_2 - 2x_3 = 8.$$

Exercise 3.5 *Solve the given system*

$$-2x_1 + x_2 + x_3 = 0$$
$$x_1 + 3x_2 - x_3 = 0$$
$$3x_1 + 2x_2 - 2x_3 = 0.$$

Exercise 3.6 *Find necessary ans sufficient conditions on a, b and c so the system has a solution*

$$x_1 + 7x_2 + 4x_3 = a$$
$$2x_1 + 8x_2 + 5x_3 = b$$
$$3x_1 + 9x_2 + 6x_3 = c.$$

Exercise 3.7 *Find q so that the following system has a nontrivial solution*

$$qx_1 + x_2 - (4 - q)x_3 = 0$$
$$2qx_1 + (2 - q)x_2 - x_3 = 0$$
$$3x_1 + (1 + q)x_2 - (5 - q)x_3 = 0.$$

Exercise 3.8 *Solve*

$$x_1 + x_2 + x_3 = 6$$
$$x_1 + 2x_2 - 3x_3 = -4$$
$$-x_1 - 4x_2 + 9x_3 = 18.$$

Exercise 3.9 *Solve*

$$x_1 + 2x_2 - 3x_3 = -1$$
$$3x_1 - x_2 + 2x_3 = 7$$
$$5x_1 + 3x_2 - 4x_3 = 2.$$

3.3 Matrices

In this section we look at matrix algebra and related issues. We begin with the definition of a matrix. A *matrix A* is a rectangular array of real or complex numbers of the form

$$A = \begin{pmatrix} a_{11} & a_{12} & \cdots & a_{1n} \\ a_{21} & a_{22} & \cdots & a_{2n} \\ \vdots & \vdots & \ddots & \vdots \\ a_{m1} & a_{m2} & \cdots & a_{mn} \end{pmatrix}. \tag{3.5}$$

The element in the ith row and jth column of the matrix A is denoted by a_{ij}. Sometimes we use the more compact notation

$$A = (a_{ij}), \quad i = 1, 2, \ldots, m, \quad j = 1, 2, \ldots n.$$

The matrix A has m rows and n columns and we say it is an $m \times n$ matrix and we may write it as $A_{m \times n}$. When $m = n$, then A is said to be a *square* matrix. Two matrices are said to be equal when and only when they have the same size (that is same numbers of rows and columns) and have the same entry in each position. In other words, if $B_{p \times q}$ is another matrix with $B = (b_{ij})$, $i = 1, 2, \ldots, p$, $j = 1, 2, \ldots q$, then $A = B$ if and only if $m = p$ and $n = q$, and $a_{ij} = b_{ij}$ for all i and j. As for addition, if A and B are two matrices with the same size, then

$$A + B = (a_{ij} + b_{ij})_{m \times n}.$$

The product of two matrices $A_{m \times n}$ and $B_{n \times p}$ is another matrix $C_{m \times p}$, where the matrix

$$C = \left(\sum_{j=1}^{n} a_{ij} b_{jk} \right)_{m \times p}.$$

To be more explicit, the product of the two matrices A and B has the general formula

$$\begin{aligned} AB &= (a_{ij})_{m \times n} (b_{ij})_{n \times p} \\ &= \begin{pmatrix} a_{11} & a_{12} & \cdots & a_{1n} \\ a_{21} & a_{22} & \cdots & a_{2n} \\ \vdots & \vdots & \ddots & \vdots \\ a_{m1} & a_{m2} & \cdots & a_{mn} \end{pmatrix} \begin{pmatrix} b_{11} & b_{12} & \cdots & b_{1p} \\ b_{21} & b_{22} & \cdots & b_{2p} \\ \vdots & \vdots & \ddots & \vdots \\ b_{n1} & b_{n2} & \cdots & b_{np} \end{pmatrix} \\ &= \begin{pmatrix} \sum_{k=1}^{n} a_{1k} b_{k1} & \cdots & \sum_{k=1}^{n} a_{1k} b_{kp} \\ \vdots & \ddots & \vdots \\ \sum_{k=1}^{n} a_{mk} b_{k1} & \cdots & \sum_{k=1}^{n} a_{mk} b_{kp} \end{pmatrix}. \end{aligned}$$

Notice that the resulting matrix from the product AB has the same number of rows as A and the same number of columns as B. Thus, if AB is defined, BA may not be defined. Moreover, the multiplication of two matrices, is, in general, not commutative.

Associative law If A is an $m \times n$ matrix, B is an $n \times p$ matrix, and C is a $p \times q$ matrix, then

$$(AB)C = A(BC). \tag{3.6}$$

Moreover, if α is a constant, then clearly

$$\alpha A = (\alpha a_{ij})_{m \times n}.$$

As for an application, the system (3.1) in matrix notation may be written as

$$\begin{pmatrix} a_{11} & a_{12} & \cdots & a_{1n} \\ a_{21} & a_{22} & \cdots & a_{2n} \\ \vdots & \vdots & \ddots & \vdots \\ a_{m1} & a_{m2} & \cdots & a_{mn} \end{pmatrix} \begin{pmatrix} x_1 \\ x_2 \\ \vdots \\ x_n \end{pmatrix} = \begin{pmatrix} b_1 \\ b_2 \\ \vdots \\ b_m \end{pmatrix},$$

or, $Ax = b$, where $x = \begin{pmatrix} x_1 \\ x_2 \\ \vdots \\ x_n \end{pmatrix}$, $b = \begin{pmatrix} b_1 \\ b_2 \\ \vdots \\ b_m \end{pmatrix}$, and A is given by (3.5).

Definition 3.1 *The transpose of the $m \times n$ matrix (3.5) is the $n \times m$ matrix A^T defined by*

$$A^T = \begin{pmatrix} a_{11} & a_{21} & \cdots & a_{m1} \\ a_{12} & a_{22} & \cdots & a_{m2} \\ \vdots & \vdots & \ddots & \vdots \\ a_{1n} & a_{2n} & \cdots & a_{mn} \end{pmatrix}.$$

Let the matrix $A_{n \times n}$ be a square matrix. That is

$$A = \begin{pmatrix} a_{11} & a_{12} & \cdots & a_{1n} \\ a_{21} & a_{22} & \cdots & a_{2n} \\ \vdots & \vdots & \ddots & \vdots \\ a_{n1} & a_{n2} & \cdots & a_{nn} \end{pmatrix}. \tag{3.7}$$

Then the diagonal containing the entries

$$a_{11}, a_{22}, \ldots, a_{nn}$$

is called the *principal diagonal*. The square *identity matrix*, denoted by I has 1s for its principal diagonal entries and has 0s elsewhere. In other words,

$$I = \begin{pmatrix} 1 & 0 & \cdots & 0 \\ 0 & 1 & \cdots & 0 \\ \vdots & \vdots & \ddots & \vdots \\ 0 & 0 & \cdots & 1 \end{pmatrix}.$$

Thus, if A is given by (3.7), then

$$AI = IA = A.$$

Also, it is readily verified that if X is an $n \times 1$ matrix, then

$$IX = X.$$

The *Kronecker delta* is defined as follows:

$$\delta_{ij} = \begin{cases} 1, & i = j \\ 0, & i \neq j \end{cases}$$

Using the Kronecker delta we may write the identity matrix as

$$I = (\delta_{ij})_{n \times n}.$$

Throughout this chapter, we denote the identity $n \times n$ matrix by I.

Definition 3.2 *a) A real square matrix $A = (a_{ij})$ is said to be symmetric if it is equal to its transpose, that is,*

$$A = A^T.$$

b) A real square matrix $A = (a_{ij})$ is said to be skew-symmetric if

$$A^T = -A.$$

As a consequence of the above definition, we know that any square matrix may be written as the sum of a symmetric matrix **R**, and a skew-symmetric matrix **S**, where

$$\mathbf{R} = \frac{1}{2}(A + A^T) \text{ and } \mathbf{S} = \frac{1}{2}(A - A^T). \tag{3.8}$$

Definition 3.3 *Let A be an $n \times n$ matrix.*

a) If all the elements above the principal diagonal (or below the principal diagonal) are zero, then the matrix A is called a triangular matrix

b) If all the elements above and below the principal diagonal are zero, then the matrix A is called a diagonal matrix.

c) A matrix whose entries are all zero is called a zero matrix or null matrix.

3.3.1 Exercises

Exercise 3.10 *Prove (3.6).*

Exercise 3.11 *a) Prove Associative law for matrix addition:*

$$(A + B) + C = A + (B + C).$$

b) *Prove the distributive law for matrix multiplication:*

$$(B+C)A = BA + CA.$$

Exercise 3.12 *Suppose A is an $m \times n$ matrix such that $Ax = b$, and $Ax = c$, where b and c are $m \times 1$ matrices and x is an $n \times 1$ matrix have solutions. Show that $Ax = b + c$ has a solution.*

Exercise 3.13 *Suppose A and B are two matrices such that AB and BA are defined and $AB = BA$. Then show that A and B are square matrices with the same numbers of rows and columns.*

Exercise 3.14 *Let A and B be 2×2 matrices that each one of them commutes with the matrix $\begin{pmatrix} 0 & 1 \\ -1 & 0 \end{pmatrix}$. Show that*

$$AB = BA.$$

Exercise 3.15 *Give an example of two matrices A and B such that $AB = 0$, but neither $A = 0$ nor $B = 0$.*

Exercise 3.16 *Show that $(AB)^T = B^T A^T$.*

Exercise 3.17 *Let A be a square matrix given by $A = (a_{ij})$, $i, j = 1, 2, \ldots, n$. Suppose A is skew-symmetric matrix. Show that if $i = j$, then all the entries in its principle diagonal are zero.*

Exercise 3.18 *Give an example of a 3×3 matrix that is skew-symmetric.*

Exercise 3.19 *Write the matrix $A = \begin{pmatrix} 2 & 3 \\ 5 & -1 \end{pmatrix}$ as the sum of $\mathbf{R}$ and $\mathbf{S}$ as given in* (3.8).

Exercise 3.20 *Show that the transpose of a triangular matrix is triangular.*

Exercise 3.21 *Write $A = \begin{pmatrix} 1 & 2 & 6 \\ 3 & 4 & 7 \\ 5 & 8 & 9 \end{pmatrix}$ as a sum of two triangular matrices. In this sum unique?*

3.4   Determinants and Inverse of Matrices

Determinants of square matrices naturally arise when solving linear equations. For example, consider the 2 equations with 2 unknowns

$$ax + by = k_1$$

$$cx + dy = k_2.$$

Using the Gauss elimination process we find the solution to be

$$x = \frac{k_1 a - k_2 b}{ad - bc}, \quad y = \frac{k_2 a - k_2 c}{ad - bc},$$

provided $ad - bc \neq 0$.

In matrix form, the above system is written

$AX = k$ where

$$A = \begin{pmatrix} a & b \\ c & d \end{pmatrix}, \quad X = \begin{pmatrix} x \\ y \end{pmatrix} \quad \text{and} \quad k = \begin{pmatrix} k_1 \\ k_2 \end{pmatrix}.$$

Define the determinant of the square matrix $A_{2 \times 2}$ by

$$\det(A) = |A| = \begin{vmatrix} a & b \\ c & d \end{vmatrix} = ad - bc.$$

Then we may write the solution of the above system using determinants notations

$$x = \frac{k_1 a - k_2 b}{|A|}, \quad \text{and} \quad y = \frac{k_2 a - k_2 c}{|A|}.$$

The determinant of order n of the square matrix $A_{n \times n}$, denoted by $|A|$ is symbolically defined by

$$\det(A) = |A| = \begin{vmatrix} a_{11} & a_{12} & \cdots & a_{1n} \\ a_{21} & a_{22} & \cdots & a_{2n} \\ \vdots & \vdots & \ddots & \vdots \\ a_{n1} & a_{n2} & \cdots & a_{nn} \end{vmatrix} \tag{3.9}$$

and how to compute its value we explain now. Let M_{ik} denote the resulting determinant after the ith row and kth column from $|A|$ are deleted. We are left with an $(n-1)$th order determinant. The *cofactor* of a_{ik} is denoted by C_{ik} and defined by

$$C_{ik} = (-1)^{i+k} M_{ik}. \tag{3.10}$$

Example 3.5 Consider the third order determinant

$$|D| = \begin{vmatrix} a_{11} & a_{12} & a_{13} \\ a_{21} & a_{22} & a_{23} \\ a_{31} & a_{32} & a_{33} \end{vmatrix}.$$

Then, we have

$$C_{11} = M_{11} = \begin{vmatrix} a_{22} & a_{23} \\ a_{32} & a_{33} \end{vmatrix}, \quad C_{12} = -M_{12} = -\begin{vmatrix} a_{21} & a_{23} \\ a_{31} & a_{33} \end{vmatrix},$$

$$C_{32} = -M_{32} = -\begin{vmatrix} a_{11} & a_{13} \\ a_{21} & a_{23} \end{vmatrix}, \quad \text{etc.}$$

□

We are ready to state the following definition.

Definition 3.4 *The determinant of the $n \times n$ matrix A in (3.9) is the sum of the products of the elements of any row or column and their respective cofactors. That is*

$$|A| = a_{i1}C_{i1} + a_{i2}C_{i2} + \ldots + a_{in}C_{in}, \quad i = 1, 2, \ldots, n,$$

or

$$|A| = a_{1k}C_{1k} + a_{2k}C_{2k} + \ldots + a_{nk}C_{nk}, \quad k = 1, 2, \ldots, n.$$

If all the entries of the matrix A are real constants, then the value of the determinant is a real constant.

Remark 9 *The determinant of an $n \times n$ matrix is the same regardless of which row or column is chosen.*

Example 3.6 Find A of

$$A = \begin{vmatrix} 1 & 2 & -1 \\ 3 & 6 & 0 \\ 0 & 4 & 2 \end{vmatrix}.$$

We make use of the first row

$$|A| = 1 \begin{vmatrix} 6 & 0 \\ 4 & 2 \end{vmatrix} - 2 \begin{vmatrix} 3 & 0 \\ 0 & 2 \end{vmatrix} - 1 \begin{vmatrix} 3 & 6 \\ 0 & 4 \end{vmatrix} = -12.$$

□

Below we state a theorem that contains certain facts concerning determinants. We leave the proofs to you.

Theorem 3.2 *1) If all elements of one row or one column of an $n \times n$ matrix are multiplied by a constant k, then the determinant is k times the determinant of the original matrix.*

2) If all entries of a row or a column of an $n \times n$ matrix are zero, then the determinant is zero.

3) If any two rows or columns of an $n \times n$ matrix are interchanged, then the determinant is -1 times the original determinant.

4) If any two rows (or two columns) of an $n \times n$ matrix are constant multiples of each other, then the determinant is zero.

5) If the entries of any row (or column) of an $n \times n$ matrix are altered by adding to them any constant multiple of the corresponding elements in any other row (or column) then the determinant does not change.

Theorem 3.3 *Let A and B be two $n \times n$ matrices. Then,*

$$\det(AB) = \det(A)\det(B).$$

Theorem 3.4 *Let A be an $n \times n$ matrix. Then*

$$\det(A) = \det(A^T).$$

Now we transition to the concept on the inverse of a matrix. We have the following definition.

Definition 3.5 *Let A be an $n \times n$ matrix. If there exists an $n \times n$ matrix B such that*

$$AB = BA = I,$$

then A is said to be invertible and B is said to be the inverse of A. We denote the inverse matrix of A by A^{-1}. Matrices that do not have inverses are said to be noninvertible or singular.

Theorem 3.5 *A square matrix A can not have more than one inverse.*

Proof *Suppose the matrix A has two inverses B and C. That is, $AB = BA = I$, and $AC = CA = I$. Then,*

$$B = BI = B(AC) = (BA)C = IC = C.$$

In the next example we show how to find the inverse of a 2×2 matrix by solving systems of equations.

Example 2 Find the inverse of $A = \begin{pmatrix} 1 & 2 \\ 3 & 4 \end{pmatrix}$. Suppose the matrix $B = \begin{pmatrix} a & b \\ c & d \end{pmatrix}$ is the inverse matrix of A. Then, it must satisfy $AB = BA = I$. In Other words:

$$\begin{pmatrix} 1 & 2 \\ 3 & 4 \end{pmatrix} \begin{pmatrix} a & b \\ c & d \end{pmatrix} = \begin{pmatrix} 1 & 0 \\ 0 & 1 \end{pmatrix}.$$

We obtain the following system of linear equations,

$$a + 2c = 1, \ b + 2d = 0, \ 3a + 4c = 0, \ \text{and} \ 3b + 4d = 1.$$

Solving for a in the third equation and substituting into the first equation gives $a = -2$ and $c = 3/2$. Similarly, solving for b in the second equation and substituting it into the fourth equation gives $d = -1/2$ and $d = 1$. Hence $B = \begin{pmatrix} -2 & 1 \\ \frac{3}{2} & -\frac{1}{2} \end{pmatrix}$. We can easily check that $AB = BA = I$. We conclude the matrix B is the inverse matrix of A. $\square$

Theorem 3.6 *Let A be an $n \times n$ matrix. Then A is invertible if and only if*

$$\det(A) \neq 0.$$

Proof *Since A is invertible, we have that* $AA^{-1} = I$ *It follows that* $\det(AA^{-1}) = \det(I) = 1$. *By Theorem 3.3, we see that*

$$\det(A)\det(A^{-1}) = 1,$$

from which it follows that $\det(A) \neq 0$. *The second part of the proof is left as an exercise. This completes the proof.*

Now we are ready to give a formula for the inverse of a square matrix, but first we make the following definition.

Definition 3.6 *Let A be an* $n \times n$ *matrix. The adjoint of A, Adj A is*

$$Adj\, A = (C_{ik})^T,$$

where C_{ik} *is given by* (3.10).

The next definition provides an alternative for finding the inverse of a matrix it term of its adjoint.

Definition 3.7 *Let A be an* $n \times n$ *matrix with* $\det(A) \neq 0$. *Then*

$$A^{-1} = \frac{1}{\det(A)}Adj\, A. \tag{3.11}$$

Let $A = \begin{pmatrix} a & b \\ c & d \end{pmatrix}$. Then from (3.11) we have that

$$A^{-1} = \frac{1}{\det(A)}\begin{pmatrix} d & -b \\ -c & a \end{pmatrix}. \tag{3.12}$$

In the case of an $n \times n$ diagonal matrix

$$A = \begin{pmatrix} a_{11} & 0 & \cdots & 0 \\ 0 & a_{22} & \cdots & 0 \\ \vdots & \vdots & \ddots & \vdots \\ 0 & 0 & \cdots & a_{nn} \end{pmatrix},$$

we have that

$$A^{-1} = \begin{pmatrix} 1/a_{11} & 0 & \cdots & 0 \\ 0 & 1/a_{22} & \cdots & 0 \\ \vdots & \vdots & \ddots & \vdots \\ 0 & 0 & \cdots & 1/a_{nn} \end{pmatrix}.$$

Let A be an $n \times n$ matrix that is invertible. Then

$$AA^{-1} = I.$$

Replace A by the matrix C and get

$$CA^{-1} = I.$$

If we now take for C the inverse A^{-1}, this becomes

$$A^{-1}(A^{-1})^{-1} = I.$$

By multiplying both sides from the left by A we obtain

$$(A^{-1})^{-1} = A.$$

This shows that the inverse of the inverse of an invertible matrix A is the matrix A.

Theorem 3.7 *Let A and B be two $n \times n$ invertible matrices. Then*

$$(AB)^{-1} = B^{-1}A^{-1}.$$

Proof *We begin with $AA^{-1} = I$. Next we replace A by AB and obtain*

$$AB(AB)^{-1} = I.$$

Multiply both sides of the preceding expression from the left with A^{-1} we arrive at

$$B(AB)^{-1} = A^{-1}.$$

By multiplying both sides from the left by B^{-1} the results follows. This completes the proof.

Of course the Theorem 3.7 can be easily generalized to products of more than two matrices. Hence, by induction one might have

$$\left(ABC\ldots PQ\right)^{-1} = Q^{-1}P^{-1}\ldots B^{-1}A^{-1}. \tag{3.13}$$

Example 3.7 Consider the third order matrix

$$A = \begin{pmatrix} 1 & 2 & 0 \\ -1 & 1 & 1 \\ 1 & 2 & 3 \end{pmatrix}.$$

Then, $\det(A) = 9$. We are interested in finding A^{-1}. We begin by computing the cofactors using (3.10). Then we have

$$C_{11} = M_{11} = \begin{vmatrix} 1 & 2 \\ 1 & 3 \end{vmatrix} = 1, \; C_{12} = -M_{12} = -\begin{vmatrix} -1 & 1 \\ 1 & 3 \end{vmatrix} = 4, \; C_{13} = M_{13} = \begin{vmatrix} -1 & 1 \\ 1 & 2 \end{vmatrix} = -3,$$

$$C_{21} = -M_{21} = -\begin{vmatrix} 2 & 0 \\ 2 & 3 \end{vmatrix} = -6, \; C_{22} = M_{22} = \begin{vmatrix} 1 & 0 \\ 1 & 3 \end{vmatrix} = 3, \; C_{23} = -M_{23} = -\begin{vmatrix} 1 & 2 \\ 1 & 2 \end{vmatrix} = 0,$$

$$C_{31} = M_{21} = \begin{vmatrix} 2 & 0 \\ 1 & 1 \end{vmatrix} = 2, \; C_{32} = -M_{22} = -\begin{vmatrix} 1 & 0 \\ -1 & 1 \end{vmatrix} = -1, \; C_{33} = M_{33} = \begin{vmatrix} 1 & 2 \\ -1 & 1 \end{vmatrix} = 3.$$

Thus,

$$C = (C_{ij}) = \begin{pmatrix} 1 & 4 & -3 \\ -6 & 3 & 0 \\ 2 & -1 & 3 \end{pmatrix}.$$

Moreover,

$$\text{Adj } A = C^T = \begin{pmatrix} 1 & -6 & 2 \\ 4 & 3 & -1 \\ -3 & 0 & 3 \end{pmatrix}.$$

Finally, using (3.11) we obtain

$$A^{-1} = \frac{1}{9} \begin{pmatrix} 1 & -6 & 2 \\ 4 & 3 & -1 \\ -3 & 0 & 3 \end{pmatrix} = \begin{pmatrix} 1/9 & -2/3 & 2/9 \\ 4/9 & 1/3 & -1/3 \\ -1/3 & 0 & 1/3 \end{pmatrix}.$$

□

For application we consider the nonhomogenous system of n equations in n unknowns given by

$$a_{11}x_1 + a_{12}x_2 + a_{13}x_3 + \ldots + a_{1n}x_n = b_1$$
$$a_{21}x_1 + a_{22}x_2 + a_{23}x_3 + \ldots + a_{2n}x_n = b_2$$
$$a_{31}x_1 + a_{32}x_2 + a_{33}x_3 + \ldots + a_{3n}x_n = b_3 \qquad (3.14)$$
$$\vdots$$
$$a_{n1}x_1 + a_{n2}x_2 + a_{n3}x_3 + \ldots + a_{nn}x_n = b_n$$

where the a_{ij}, b_i are constants for $1 \le i \le n, 1 \le j \le n$. In matrix form the system may be written as

$$AX = b, \qquad (3.15)$$

where

$$A = \begin{pmatrix} a_{11} & a_{12} & \cdots & a_{1n} \\ a_{21} & a_{22} & \cdots & a_{2n} \\ \vdots & \vdots & \ddots & \vdots \\ a_{n1} & a_{n2} & \cdots & a_{nn} \end{pmatrix}, \quad X = \begin{pmatrix} x_1 \\ x_2 \\ \vdots \\ x_n \end{pmatrix}, \quad \text{and} \quad b = \begin{pmatrix} b_1 \\ b_2 \\ \vdots \\ b_n \end{pmatrix}.$$

If $\det(A) \ne 0$, then by multiplying (3.15) from the left by A^{-1} we get

$$X = A^{-1}b. \qquad (3.16)$$

Clearly, (3.16) is a solution of (3.15). To see this, substitute X into (3.15) and get

$$A(A^{-1}b) = (AA^{-1})b = Ib = b.$$

As for uniqueness, suppose there is another solution Y such that

$$AY = b.$$

Then multiplying from the left by A^{-1} we arrive at

$$X = A^{-1}b, \ Y = A^{-1}b,$$

from which we conclude that $X = Y$. Now that we have established the system has a unique solution, we try to give an explicit formula for such solution. Using (3.11) along with (3.16) and Definition 3.6 we have

$$X = \frac{1}{\det(A)}(\operatorname{Adj} A)b.$$

Or,

$$
\begin{pmatrix} x_1 \\ x_2 \\ \vdots \\ x_n \end{pmatrix} = \frac{1}{\det(A)}
\begin{pmatrix}
C_{11} & C_{21} & \cdots & C_{n1} \\
C_{12} & C_{22} & \cdots & C_{n2} \\
\vdots & \vdots & \ddots & \vdots \\
C_{1n} & a_{2n} & \cdots & C_{nn}
\end{pmatrix}
\begin{pmatrix} b_1 \\ b_2 \\ \vdots \\ b_n \end{pmatrix}
$$

$$
= \frac{1}{\det(A)}
\begin{pmatrix}
b_1 C_{11} + b_2 C_{21} + \cdots + b_n C_{n1} \\
b_1 C_{12} + b_2 C_{22} + \cdots + b_n C_{n2} \\
\vdots \\
b_1 C_{1n} + b_2 C_{2n} + \cdots b_n C_{nn}
\end{pmatrix}.
$$

It follows from the above calculation that the components of the solutions are given by

$$x_i = \frac{b_1 C_{1i} + b_2 C_{2i} + \cdots + b_n C_{ni}}{\det(A)}, \quad i = 1, 2, \ldots, n \tag{3.17}$$

We summarize the results in the following theorem.

Theorem 3.8 *Consider the nonhomogeneous system (3.14) of n linear equations with n unknowns. If its coefficient matrix A has $\det(A) \neq 0$, then the system has a unique solution $x_1, x_2, \ldots, x_n$ given by (3.17).*

As a direct consequence of (3.17) we have the following corollary.

Corollary 3 *A homogeneous system of n linear equations with n unknowns and a coefficients matrix A with $\det(A) \neq 0$ has just the trivial solution.*

Proof *Formula (3.17) is valid since $\det(A) \neq 0$. The results follow since each $b_i = 0$, $i = 1, 2, \ldots, n$.*

Example 3.8 Solve the system

$$x_1 + 2x_2 = 5$$
$$-x_1 + x_2 + x_3 = 4$$
$$x_1 + 2x_2 + 3x_3 = 14.$$

In matrix notation, we have

$$A = \begin{pmatrix} 1 & 2 & 0 \\ -1 & 1 & 1 \\ 1 & 2 & 3 \end{pmatrix}, \quad X = \begin{pmatrix} x_1 \\ x_2 \\ x_3 \end{pmatrix}, \quad b = \begin{pmatrix} 5 \\ 4 \\ 14 \end{pmatrix}.$$

Notice that the matrix A is the same as the one in example (3.7). Thus, the C_{ij} are readily available and using (3.17), we obtain

$$x_1 = \frac{b_1 C_{11} + b_2 C_{21} + b_3 C_{31}}{\det(A)} = \frac{5(1) + 4(-6) + 14(2)}{9} = 1,$$

$$x_2 = \frac{b_1 C_{12} + b_2 C_{22} + b_3 C_{32}}{\det(A)} = \frac{5(4) + 4(3) + 14(-1)}{9} = 2,$$

and

$$x_3 = \frac{b_1 C_{13} + b_2 C_{23} + b_3 C_{33}}{\det(A)} = \frac{5(-3) + 4(0) + 14(3)}{9} = 3.$$

$\square$

We have the following theorem regarding the homogenous system (3.4).

Theorem 3.9 *If*

$$X_{(1)} = \begin{pmatrix} x_1 \\ x_2 \\ \vdots \\ x_n \end{pmatrix}, \quad X_{(2)} = \begin{pmatrix} y_1 \\ y_2 \\ \vdots \\ y_n \end{pmatrix}$$

are solutions of the homogenous system (3.4), then

$$X = c_1 X_{(1)} + c_2 X_{(2)},$$

where c_1, c_2 are constants, is also a solution of the homogenous system (3.4).

Proof *The homogenous system can be written as $AX = 0$. By assumption, $AX_{(1)} = AX_{(1)} = 0$. Hence,*

$$AX = A(c_1 X_{(1)} + c_2 X_{(2)}) = c_1 AX_{(1)} + c_2 AX_{(2)} = 0.$$

We note that the theorem does not hold for nonhomogenous systems.

Definition 3.8 *Any matrix obtained by omitting some rows or columns from a given $A_{m \times n}$ matrix is said to be a submatrix of A. We note a submatrix includes the matrix A itself.*

Example 3.9 The matrix

$$A = \begin{pmatrix} a_{11} & a_{12} & a_{13} \\ a_{21} & a_{22} & a_{23} \end{pmatrix}$$

contains the following submatrices: A itself; the three 2×2 submatrices

$$\begin{pmatrix} a_{11} & a_{12} \\ a_{21} & a_{22} \end{pmatrix}, \quad \begin{pmatrix} a_{11} & a_{13} \\ a_{21} & a_{23} \end{pmatrix}, \quad \begin{pmatrix} a_{12} & a_{13} \\ a_{22} & a_{23} \end{pmatrix};$$

the two 1×3 submatrices

$$\begin{pmatrix} a_{11} & a_{12} & a_{13} \end{pmatrix}, \begin{pmatrix} a_{21} & a_{22} & a_{23} \end{pmatrix};$$

the three 1×2 submatices

$$\begin{pmatrix} a_{11} \\ a_{21} \end{pmatrix}, \quad \begin{pmatrix} a_{12} \\ a_{22} \end{pmatrix}, \quad \begin{pmatrix} a_{13} \\ a_{23} \end{pmatrix};$$

the six 1 submatrices,

$$\begin{pmatrix} a_{11} & a_{12} \end{pmatrix}, \quad \begin{pmatrix} a_{11} & a_{13} \end{pmatrix}, \quad \begin{pmatrix} a_{12} & a_{13} \end{pmatrix},$$

$$\begin{pmatrix} a_{21} & a_{22} \end{pmatrix}, \quad \begin{pmatrix} a_{21} & a_{23} \end{pmatrix}, \quad \begin{pmatrix} a_{22} & a_{23} \end{pmatrix},$$

and the six 1×1 submatrices,

$$(a_{11}), \quad (a_{12}), \quad (a_{13}), \quad (a_{21}), \quad (a_{22}), \quad (a_{23}).$$

□

Definition 3.9 *The rank of a matrix A is the order of the largest square submatrix with a nonzero determinant.*

To expand on the above definition, a matrix A is said to be of rank r if it contains at least one r-rowed square submatrix with nonvanishing determinant, while the determinant of any square submatrix having $r+1$ or more rows, possibly contained in A, is zero.

Example 3.10 *Let*

$$A = \begin{pmatrix} -3 & 3 & 0 \\ 1 & -2 & -1 \\ 2 & 2 & 4 \end{pmatrix}.$$

Then $\det(A) = 0$. *Now the 2×2 submatrix* $A = \begin{pmatrix} 1 & -2 \\ 2 & 2 \end{pmatrix}$ *has nonzero determinant and we conclude the rank of A is 2.* □

Example 3.11 Let

$$A = \begin{pmatrix} 6 & 3 & 4 & 7 \\ 4 & 2 & 1 & 3 \\ 2 & 1 & 0 & 1 \end{pmatrix}.$$

Then all of its 3×3 submatrices

$$\begin{pmatrix} 6 & 3 & 4 \\ 4 & 2 & 1 \\ 2 & 1 & 0 \end{pmatrix}, \quad \begin{pmatrix} 3 & 4 & 7 \\ 2 & 1 & 3 \\ 1 & 0 & 1 \end{pmatrix}, \quad \begin{pmatrix} 6 & 4 & 7 \\ 4 & 1 & 3 \\ 2 & 0 & 1 \end{pmatrix}, \quad \begin{pmatrix} 6 & 3 & 7 \\ 4 & 2 & 3 \\ 2 & 1 & 1 \end{pmatrix}$$

have determinant zero. However, the 2×2 submatrix

$$\begin{pmatrix} 6 & 4 \\ 4 & 1 \end{pmatrix}$$

has nonzero determinant and the matrix A has rank 2.

$\square$

Theorem 3.10 *Suppose the rank of the $A_{m \times n}$ matrix is r. Then then rank of its transpose, A^T is r too.*

Proof *Let R be an r-rowed submatrix of A with $\det(R) \neq 0$. It is obvious that R^T is a submatrix of A^T. Then by Theorem 3.4, $\det(R^T) = \det(R)$. This implies that rank of $A^T \geq r$. On the other hand, if A contains an $(r+1)$-rowed square matrix M, then, by the rank definition, $\det(M) = 0$. Since M corresponds to M^T in A^T, and $\det(M^T) = 0$, it follows that A^T can not contain an $(r+1)$-rowed square matrix with nonzero determinant. Thus, rank of A^T is r. This completes the proof.*

Theorem 3.11 *If a matrix A is of rank r, and a set of r rows (or columns) containing a non-singular submatrix of order r is selected, then any other row (column) in the matrix A is a linear combination of these r rows (or columns).*

Proof *To simplify notation, we suppose that the submatrix R of order r is the upper left corner of the matrix A has a non-vanishing determinant, and consider the submatrix of A*

$$M = \begin{pmatrix} a_{11} & a_{12} & \cdots & a_{1r} & a_{1s} \\ a_{21} & a_{22} & \cdots & a_{2r} & a_{2s} \\ \vdots & \vdots & \ddots & \vdots & \vdots \\ a_{r1} & a_{r2} & \cdots & a_{rr} & a_{rs} \\ a_{q1} & a_{q2} & \cdots & a_{qr} & a_{qs} \end{pmatrix} = \begin{pmatrix} & & & \vdots & a_{1s} \\ & R & & \vdots & a_{2s} \\ & & & \vdots & \vdots \\ & & & \vdots & a_{rs} \\ a_{q1} & a_{q2} & \cdots & a_{qr} & a_{qs} \end{pmatrix}$$

where $s > r$ and $q > r$. Since A is of rank r, $|M| = 0$ for all such q and s. Now the system,

$$\alpha_1 a_{11} + \alpha_2 a_{21} + \ldots + \alpha_r a_{r1} = a_{q1}$$

$$\alpha_1 a_{12} + \alpha_2 a_{22} + \ldots + \alpha_r a_{r2} = a_{q2}$$

$$\vdots$$

$$\alpha_1 a_{1r} + \alpha_2 a_{2r} + \ldots + \alpha_r a_{rr} = a_{qr}, \; or$$

$$R^T \alpha = b, \quad where \quad \alpha = \begin{pmatrix} \alpha_1 \\ \vdots \\ \alpha_r \end{pmatrix}, \quad b = \begin{pmatrix} a_{q1} \\ \vdots \\ a_{qr} \end{pmatrix},$$

has a unique solution $\alpha = (R^T)^{-1} b$, *since* $|R^T| = |R| \neq 0$. *Rewrite this system in the following form:*

$$\alpha_1 b_1 + \alpha_2 b_2 + \ldots + \alpha_r b_r,$$

where b_j *is the jth row of R. We see that we can determine a row of elements which indeed is a linear combination of the first row of M, and which we will have its first elements identical with the first r elements of the last row of M. Let the last elements of that combination be* a'_{qs}. *In evaluating the determinant* $|M|$, *we may subtract this linear combination of the first r rows from the last row without changing the value of the determinant. Thus,*

$$|M| = \begin{pmatrix} a_{11} & a_{12} & \cdots & a_{1r} & a_{1s} \\ a_{21} & a_{22} & \cdots & a_{2r} & a_{2s} \\ \vdots & \vdots & \ddots & \vdots & \vdots \\ a_{r1} & a_{r2} & \cdots & a_{rr} & a_{rs} \\ 0 & 0 & \cdots & 0 & a_{qs} - a'_{qs} \end{pmatrix},$$

where

$$a'_{qs} = \alpha_1 a_{1s} + \alpha_2 a_{2s} + \ldots + \alpha_r a_{rs},$$

and

$$|M| = \pm (a_{qs} - a'_{qs}) |R| = 0.$$

Hence the last row of M is a linear combination of the first r rows. Since this is true for any q and s, the result follows.

As a consequence of Theorem 3.10, we have the following two corollaries.

Corollary 4 *A homogeneous system of n linear equations with n unknowns has a nontrivial solution if and only if, its coefficients matrix A satisfies* $\det(A) = 0$.

Corollary 5 *A square matrix is singular if and only if one of its rows (or columns) is a linear combination of the others.*

3.4.1 Application to least square fitting

Least square approximation is a mathematical process that determines the curve that best fits a set of points by minimizing the sum of the squares of the offsets (also known as "the residuals") of the points from the curve. Finding the best-fitting straight line through a group of points is a problem that can be solved using the linear least squares fitting technique, which is the most straightforward and widely used type of linear regression.

We begin with the simple problem by trying to find the straight line $y = ax + b$, that best fits an n-observations given by (x_n, y_n), for $n = 1, 2, \ldots, N$. From the linear equation, it is intuitive to define the error by

$$E(a,b) = \sum_{n=1}^{N} \left(y_n - (ax_n + b) \right)^2. \tag{3.18}$$

This is simply N times the variance of the data collection

$$\{ y_1 - (ax_1 + b), \ldots, y_N - (ax_N + b) \}.$$

For best fitting, we must minimize the error given by (3.18). That is, we must find the values of (a, b) such that

$$\frac{\partial E}{\partial a} = 0, \quad \frac{\partial E}{\partial b} = 0. \tag{3.19}$$

Using (3.19), we obtain

$$\frac{\partial E}{\partial a} = 2 \sum_{n=1}^{N} \left(y_n - (ax + b) \right)(-x_n),$$

$$\frac{\partial E}{\partial b} = 2 \sum_{n=1}^{N} \left(y_n - (ax + b) \right)(-1). \tag{3.20}$$

Setting $\frac{\partial E}{\partial a} = \frac{\partial E}{\partial b} = 0$, we arrive at the system of equations with two unknown

$$(\sum_{n=1}^{N} x_n^2)a + (\sum_{n=1}^{N} x_n)b = \sum_{n=1}^{N} x_n y_n,$$

$$(\sum_{n=1}^{N} x_n)a + (\sum_{n=1}^{N} 1)b = \sum_{n=1}^{N} y_n, \tag{3.21}$$

In Matrix from,

$$\begin{pmatrix} \sum_{n=1}^{N} x_n^2 & \sum_{n=1}^{N} x_n \\ \sum_{n=1}^{N} x_n & N \end{pmatrix} \begin{pmatrix} a \\ b \end{pmatrix} = \begin{pmatrix} \sum_{n=1}^{N} x_n y_n \\ \sum_{n=1}^{N} y_n \end{pmatrix}.$$

This implies that

$$\begin{pmatrix} a \\ b \end{pmatrix} = \begin{pmatrix} \sum_{n=1}^{N} x_n^2 & \sum_{n=1}^{N} x_n \\ \sum_{n=1}^{N} x_n & N \end{pmatrix}^{-1} \begin{pmatrix} \sum_{n=1}^{N} x_n y_n \\ \sum_{n=1}^{N} y_n \end{pmatrix}.$$

Consequently,

$$\begin{pmatrix} a \\ b \end{pmatrix} = \frac{1}{N\sum_{n=1}^{N} x_n^2 - \sum_{n=1}^{N} x_n \sum_{n=1}^{N} x_n} \begin{pmatrix} N & -\sum_{n=1}^{N} x_n \\ -\sum_{n=1}^{N} x_n & \sum_{n=1}^{N} x_n^2 \end{pmatrix} \begin{pmatrix} \sum_{n=1}^{N} x_n y_n \\ \sum_{n=1}^{N} y_n \end{pmatrix}.$$

The above concept can be easily generalized to functions that are not straight lines. Given functions $f_1, \ldots, f_k$, find the values of coefficients $a_1, \ldots, a_k$, such that the linear combination

$$y = a_1 f_1(x) + \ldots + a_k f_k(x)$$

is the best approximation to the data. Staying with the same set up, we define the error by

$$E(a_1, \ldots, a_k) = \sum_{n=1}^{N} \left(y_n - (a_1 f_1(x_n) + \ldots + a_k f_k(x_n)) \right)^2. \tag{3.22}$$

To find the values of $(a_1, \ldots, a_k)$ we set

$$\frac{\partial E}{\partial a_1} = 0, \ldots, \frac{\partial E}{\partial a_k} = 0. \tag{3.23}$$

To be more specific, we consider fitting the parabola $y = a + bx + cx^2$. Then

$$E(a, b, c) = \sum_{n=1}^{N} \left(y_n - (a + bx_n + cx_n^2) \right)^2.$$

Then

$$\frac{\partial E}{\partial a} = \frac{\partial E}{\partial b} = \frac{\partial E}{\partial c} = 0,$$

implies that

$$2 \sum_{n=1}^{N} \left(y_n - (a + bx_n + cx_n^2) \right)(-1) = 0,$$

$$2 \sum_{n=1}^{N} \left(y_n - (a + bx_n + cx_n^2) \right)(-x_n) = 0$$

and

$$2 \sum_{n=1}^{N} \left(y_n - (a + bx_n + cx_n^2) \right)(-x_n^2) = 0 \tag{3.24}$$

The system (3.24) reduces to

$$\sum_{n=1}^{N} y_n = Na + b \sum_{n=1}^{N} x_n + c \sum_{n=1}^{N} x_n^2$$

$$\sum_{n=1}^{N} x_n y_n = a \sum_{n=1}^{N} x_n + b \sum_{n=1}^{N} x_n^2 + c \sum_{n=1}^{N} x_n^3$$

and

$$\sum_{n=1}^{N} x_n^2 y_n = a \sum_{n=1}^{N} x_n^2 + b \sum_{n=1}^{N} x_n^3 + c \sum_{n=1}^{N} x_n^4. \tag{3.25}$$

Example 3.12 Suppose we want to find the values a, b, and c so that $y = a + bx + cx^2$ is the best approximation to the data

$$(0, 1), \quad (1, 1.8), \quad (2, 1.3), \quad (3, 2.5), \quad (4, 6.3).$$

Then, we have $N = 5$. Using the given data, one can easily compute

$$\sum_{n=1}^{5} x_n = 10, \quad \sum_{n=1}^{5} y_n = 12.9, \quad \sum_{n=1}^{5} x_n^2 = 30, \quad \sum_{n=1}^{5} x_n^3 = 100, \quad \sum_{n=1}^{5} x_n^4 = 354,$$

$$\sum_{n=1}^{5} x_n y_n = 37.1, \quad \sum_{n=1}^{5} x_n^2 y_n = 130.3.$$

Thus, system (3.25) reduces to

$$5a + 10b + 30c = 12.9$$
$$10a + 30b + 100c = 37.1$$
$$30a + 100b + 354c = 130.3.$$

Using matrix notation, we arrive at the solution

$$\begin{pmatrix} a \\ b \\ c \end{pmatrix} = \begin{pmatrix} 5 & 10 & 30 \\ 10 & 30 & 100 \\ 30 & 100 & 354 \end{pmatrix}^{-1} \begin{pmatrix} 12.9 \\ 37.1 \\ 130.3 \end{pmatrix}.$$

This gives $a = 1.42, b = -1.07, c = 0.55$, and the desired parabola is

$$y = 1.42 - 1.07x + 0.55x^2.$$

$\square$

3.4.2 Exercises

Exercise 3.22 *Use the method of Example 2 to find the inverse matrix of each of the following matrices.*

(a) $A = \begin{pmatrix} 3 & 5 \\ 1 & 2 \end{pmatrix}$, *(b)* $A = \begin{pmatrix} 5 & 2 \\ -7 & -3 \end{pmatrix}$, *(c)* $A = \begin{pmatrix} 1 & 3 & 3 \\ 1 & 3 & 4 \\ 1 & 5 & 3 \end{pmatrix}$.

Exercise 3.23 *Prove Theorem 3.3.*

Exercise 3.24 *Prove Theorem 3.4.*

Exercise 3.25 *Let A be an $n \times n$ matrix with $\det(A) \neq 0$. Show that*

$$\det(A^{-1}) = \frac{1}{\det(A)}.$$

Exercise 3.26 *Find the inverse of each of the following matrices.*

(a) $A = \begin{pmatrix} 3 & 1 \\ 2 & 4 \end{pmatrix}$, (b) $A = \begin{pmatrix} -3 & 6 & -11 \\ 3 & -4 & 6 \\ 4 & -8 & 13 \end{pmatrix}$

(c) $A = \begin{pmatrix} 3 & 5 & 7 \\ 1 & 2 & 3 \\ 2 & 3 & 5 \end{pmatrix}$ (d) $A = \begin{pmatrix} 1 & 2 & 3 & 4 \\ 2 & 3 & 4 & 1 \\ 3 & 4 & 1 & 2 \\ 4 & 1 & 2 & 3 \end{pmatrix}$.

Exercise 3.27 *Show that if $AB = AC$, then $B = C$, provided that $|A| \neq 0$.*

Exercise 3.28 *Suppose A and B are symmetric and invertible. Show that if $AB = BA$, then $AB, A^{-1}B, AB^{-1}$, and $A^{-1}B^{-1}$ are symmetric.*

Exercise 3.29 *Find 2×2 matrices A and B that satisfy*

$$\begin{pmatrix} 3 & 1 \\ 2 & 1 \end{pmatrix} A + \begin{pmatrix} 3 & 4 \\ 2 & 3 \end{pmatrix} B = \begin{pmatrix} 1 & 1 \\ 1 & 1 \end{pmatrix}$$

$$\begin{pmatrix} 3 & 1 \\ 2 & 1 \end{pmatrix} A - \begin{pmatrix} 3 & 4 \\ 2 & 3 \end{pmatrix} B = \begin{pmatrix} 1 & 0 \\ 0 & 1 \end{pmatrix}.$$

Exercise 3.30 *Solve*

$$2x_1 + 2x_2 - x_3 = 4$$
$$3x_1 + x_2 + 4x_3 = -9$$
$$x_1 + 2x_2 + x_3 = 1.$$

Exercise 3.31 *Solve*

$$x_1 - x_2 - x_3 + x_4 = -2$$
$$2x_1 + x_2 + x_3 + x_4 = 3$$
$$-x_1 + x_2 + x_3 = 1$$
$$2x_1 + x_3 - 3x_4 = 7.$$

Exercise 3.32 *Find the rank of the following matrices.*

a) $\begin{pmatrix} 2 & 5 \\ 3 & 8 \end{pmatrix}$, b) $\begin{pmatrix} 0 & 5 & 8 \\ 3 & 1 & 4 \\ -3 & 4 & 4 \end{pmatrix}$, c) $\begin{pmatrix} 3 & 0 & 4 \\ 1 & 7 & -1 \end{pmatrix}$,

d) $\begin{pmatrix} 8 & 1 & 3 & 6 \\ -8 & -1 & -3 & 4 \\ 0 & 3 & 2 & 2 \end{pmatrix}$.

Exercise 3.33 *Let a, b, be real numbers and consider the matrix $A = \begin{pmatrix} 1 & a & a^2 \\ 1 & b & b^2 \\ 1 & c & c^2 \end{pmatrix}$.*

Show that

$$\det(A) = (b - a)(c - a)(c - b).$$

Exercise 3.34 *Let A be an $n \times n$ time-dependent matrix. That is, the entries of A depend on the independent variable t. Suppose all its entries are continuously differentiable on some interval I.*

Show that

$$\frac{d}{dt}\big(\det(A)\big) = tr\Big(adj(A)\frac{d(A)}{dt}\Big).$$

Exercise 3.35 *Let A, and B be two $n \times n$ nonzero matrices such that $AB = 0$. Then show that both A and B are singular.*

Exercise 3.36 *Find the line $y = a + bx$ to best fit the data of Example 3.12.*

Exercise 3.37 *Find the line $y = a + bx$ to best fit the data*

$$(1,2), \quad (2,5), \quad (3,3), \quad (4,8), \quad (5,7).$$

Exercise 3.38 *Generalize the method of least squares to find the function*

$$y = a_m x^m + a_{m-1} x^{m-1} + \ldots + a_0,$$

that best fits a given data.

3.5   Vector Spaces

Vector spaces play a fundamental role in the development of linear algebra. We will state some definitions and go over a few concepts, such as bases, dimensions, and span.

Definition 3.10 *A triple $(V, +, \cdot)$ is said to be a linear (or vector) space over a field F if V is a set and the following are true.*

1. Properties of $+$

   (a) $+$ is a function from $V \times V$ to V. Outputs are denoted $x + y$.

   (b) for all $x, y \in V$, $x + y = y + x$. ($+$ is commutative)

   (c) for all $x, y, w \in V$, $x + (y + w) = (x + y) + w$. ($+$ is associative)

   (d) there is a unique element of V which we denote 0 such that for all $x \in V$, $0 + x = x + 0 = x$. (additive identity)

   (e) for each $x \in V$ there is a unique element of V which we denote $-x$ such that $x + (-x) = -x + x = 0$. (additive inverse)

2. Properties of $\cdot$

   (a) $\cdot$ is a function from $F \times V$ to V. Outputs are denoted $\alpha \cdot x$, or αx.

   (b) for all $\alpha, \beta \in F$ and $x \in V$, $\alpha(\beta x) = (\alpha\beta)x$.

(c) for all $x \in V$, $1 \cdot x = x$.

(d) for all $\alpha, \beta \in F$ and $x \in V$, $(\alpha + \beta)x = \alpha x + \beta x$.

(e) for all $\alpha \in F$ and $x, y \in V$, $\alpha(x + y) = \alpha x + \alpha y$.

Commonly, the real numbers or complex numbers are the field in the above definition.

Example 3.13 The set of real numbers $\mathbb{R}$ is a vector space over the field $F = \mathbb{R}$ under the usual addition and multiplication. $\square$

Example 3.14 The set $\mathbb{R}^n = \{x = (x_1, x_2, \ldots, x_n)^T\}$ is a vector space over the field $F = \mathbb{R}$ under the usual addition and multiplication. That is for $x, y \in \mathbb{R}^n$ and $\alpha \in F$ the addition and multiplication are defined as

$$x + y = (x_1 + y_1, x_2 + y_2, \ldots, x_n + y_n)^T$$

and

$$\alpha x = (\alpha x_1, \alpha x_2, \ldots, \alpha x_n)^T$$

is a vector space. $\square$

Example 3.15 Let $a_n \neq 0$, and define the set $V = \{p(x) = a_n x^n + a_{n-1} x^{n-1} + \ldots + a_1 x + a_0 : a_i \in \mathbb{R}, i = 0, 1, \ldots, n\}$. Then V is a vector space over the field $F = \mathbb{R}$. We define the addition of two polynomial and multiplication as follows: if $p, q \in V$, $q(x) = b_n x^n + b_{n-1} x^{n-1} + \ldots + b_1 x + b_0 : b_i \in \mathbb{R}, i = 0, 1, \ldots, n$ then

$$(p+q)(x) = (a_n + b_n)x^n + (a_{n-1} + b_{n-1})x^{n-1} + \ldots + (a_1 + b_1)x + a_0 + b_0 = p(x) + q(x),$$

and

$$(\alpha p)(x) = \alpha a_n x^n + \alpha a_{n-1} x^{n-1} + \ldots + \alpha a_1 x + \alpha a_0 = \alpha p(x),$$

for $\alpha \in \mathbb{R}$, we have V is a vector space. For example, if $p(x) = 3x^2 + 4x + 6, q(x) = x^3 - 2x^2 + 2x + 5$, then $(p+q)(x) = x^3 + (3-2)x^2 + (2+4)x + 5 + 6$, and $(3p)(x) = 9x^2 + 12x + 18$. The additive inverse of $p(x)$ is $-p(x) = -a_n x^n - a_{n-1} x^{n-1} - \ldots - a_1 x - a_0$. $\square$

Example 3.16 Let the set $C(D)$ be the set of all continuous functions $f : D \to \mathbb{R}$. For $f, g \in C(D)$, we define addition and multiplication pointwise as follows:

$$(f+g)(x) = f(x) + g(x), \text{ for all } x \in D,$$

and

$$(cf)(x) = cf(x), \ c \in \mathbb{R}$$

then $C(D)$ is a vector space. $\square$

Definition 3.11 *The $n \times 1$ vectors $v_1, v_2, \ldots, v_n$ in V are said to be linearly dependent if there exists constants $c_1, c_2, \ldots, c_n$ not all zero, such that*

$$c_1 v_1 + c_2 v_2 + \ldots + c_n v_n = 0$$

If the vectors are not linearly dependent, then they are called linearly independent.

Example 3.17 The vectors

$$e_1 = (1,0,\ldots,0), \ e_2 = (0,1,0,\ldots,0),\ldots, \ e_n = (0,0,\ldots,1)$$

are linearly independent since if

$$(0,\ldots,0) = c_1(1,0,\ldots,0) + c_2(0,1,0,\ldots,0) + c_n(0,0,\ldots,1) = (c_1,c_2,\ldots,c_n),$$

then,

$$(0,\ldots,0) = (c_1,c_2,\ldots,c_n).$$

This has the only solution $c_1 = c_2 = \ldots = c_n = 0$.   □

Definition 3.12 *Let V be a vector space over $\mathbb{R}$. Let $v_1, v_2, \ldots, v_n$ in V. A vector $v \in V$ is a linear combination of $\{v_1, v_2, \ldots, v_n\}$ if there exists scalars $b_1, b_2, \ldots, b_n \in \mathbb{R}$ such that*

$$v = b_1v_1 + b_2v_2 + \ldots + b_nv_n.$$

Definition 3.13 *(Span) The span of $\{v_1, v_2, \ldots, v_n\}$ is defined as*

$$span(v_1, v_2, \ldots, v_n) := \{b_1v_1 + b_2v_2 + \ldots + b_nv_n \mid b_1, b_2, \ldots, b_n \in \mathbb{R}\}.$$

Example 3.18 Consider the vectors

$$v_1 = (3,2,1), \quad v_2 = (2,-3,2), \quad \text{and} \quad v_3 = (-12,5,-8).$$

Then,

$$a_1v_1 + a_2v_2 + a_3v_3 = (0,0,0)$$

implies that

$$(3a_1 + 2a_2 - 12a_3, 2a_1 - 3a_2 + 5a_3, a_1 + 2a_2 - 8a_3) = (0,0,0).$$

So we have the three equations with three unknowns

$$3a_1 + 2a_2 - 12a_3 = 0$$
$$2a_1 - 3a_2 + 5a_3 = 0$$
$$a_1 + 2a_2 - 8a_3 = 0.$$

Using Gauss-elimination method we see the system has infinitely many solutions; namely,

$$a_1 = 2a_3, \ a_2 = 3a_3.$$

Setting $a_3 = 1$, we obtain a solution $(2,3,1)$. In general, the set of all solutions (or solutions space) is given by

$$S = \{(a_1, a_2, a_3) \mid a_1 = 2a_3, \quad a_2 = 3a_3\} = span\big((2,3,1)\big).$$

□

We have the following important Lemma concerning linear independence.

Lemma 7 *(Linear Independence Lemma) The set of vectors $\{v_1, v_2, \ldots, v_p\}$ is linearly independent if and only if every vector $v \in span\{v_1, v_2, \ldots, v_p\}$ can be uniquely written as a linear combination of $\{v_1, v_2, \ldots, v_p\}$.*

Proof *Suppose $v_1, v_2, \ldots, v_p$ are linearly independent and suppose any vector $v \in span\{v_1, v_2, \ldots, v_p\}$ can be written in two different ways.*

That is there are constants $b_i, \ c_i, i = 1, 2, \ldots p$ such that

$$v = b_1 v_1 + b_2 v_2 + \ldots + b_p v_p$$

and

$$v = c_1 v_1 + c_2 v_2 + \ldots + c_p v_p.$$

By subtracting the two equations we arrive at

$$0 = (b_1 - c_1)v_1 + (b_2 - c_2)v_2 + \ldots + (b_p - c_p)v_p. \tag{3.26}$$

Since the set of vectors $\{v_1, v_2, \ldots, v_p\}$ is linearly independent, the only solution to equation (3.26) is

$$b_1 - c_1 = 0, \quad b_2 - c_2 = 0, \quad \ldots, \quad b_p - c_p = 0.$$

Thus,

$$b_1 = c_1, \ b_2 = c_2, \ldots, \ b_p = c_p.$$

This proves the necessary part of the lemma. For the proof of the sufficient condition, for every $v \in span\{v_1, v_2, \ldots, v_p\}$, there are unique $b_i, i = 1, 2, \ldots p$ such that $v = b_1 v_1 + b_2 v_2 + \ldots + c_p v_p$. This implies that the zero vector $v = 0$ can be written as a linear combination of $v_1, v_2, \ldots, v_p$, only when

$$b_1 = b_2 = \ldots = b_p = 0.$$

This shows the set of vectors $\{v_1, v_2, \ldots, v_p\}$ is linearly independent. This completes the proof.

Definition 3.14 *(subspace) Let V be a vector space over F. Then U is a subspace of V if and only if the following properties are satisfied:*

1. *$0 \in U$; additive identity*

2. *If $u_1, u_2 \in U$, then $u_1 + u_2 \in U$; (closure under addition)*

3. *For scalar $a \in F, u \in U$, then $au \in U$; (closure under scalar multiplication).*

Example 3.19 The set
$$U = \{(a, 0) \mid a \in \mathbb{R}\}$$
is a subspace of $\mathbb{R}^2$. $\qquad \square$

Example 3.20 The set

$$U = \{(a,b,c) \in \mathbb{R}^3 \mid b + 4c = 0\}$$

is a subspace of $\mathbb{R}^3$. To see this , we make sure the requirements of (3.14) are met. As for 1., we easily see that $(0,0,0) \in U$, since $b + 4c = 0$ is satisfied. To verify 2., we let $u = (u_1, u_2, u_3)$ and $v = (v_1, v_2, v_3)$. Then we have

$$u_2 + 4u_3 = 0, \text{ and } v_2 + 4v_3 = 0$$

by adding the two equations we easily arrive at

$$(u_2 + v_2) + 4(u_3 + v_3) = 0.$$

Let $K = (u_1 + v_1, u_2 + v_2, u_3 + v_3) \in U$. Then it must satisfy $(u_2 + v_2) + 4(u_3 + v_3) = 0$. This shows that $K := u + v \in U$. It remains to be shown that 3. holds. Let $\alpha \in \mathbb{R}$, and $u = (u_1, u_2, u_3) \in U$. Then, $\alpha u = (\alpha u_1, \alpha u_2, \alpha u_3)$ satisfies the equation $\alpha u_2 + 4\alpha u_3 = \alpha(u_2 + 4u_3) = 0$, and so $\alpha u \in U$. $\square$

Lemma 8 *Let $v_1, v_2, \ldots, v_n$ be vectors in the vector space V. Then*

1. *$v_j \in span(v_1, v_2, \ldots, v_n)$,*

2. *$span(v_1, v_2, \ldots, v_n)$ is a subspace of V.*

3. *If $v_1, v_2, \ldots, v_n$ are vectors in the vector space V and $U \subset V$ is a subspace such that $v_1, v_2, \ldots, v_n \in U$, then $span(v_1, v_2, \ldots, v_n) \subset U$.*

Proof *As for 1. let v_j be any vector of $v_1, v_2, \ldots, v_n \in V$. Since V is a vector space, the result follows from parts b) of addition and e) of multiplication. To prove 2. we observe that $0 \in span(v_1, v_2, \ldots, v_n)$ and that $span(v_1, v_2, \ldots, v_n)$ is closed under addition and scalar multiplication.*

Remark 10 *Lemma 8 implies that $span(v_1, v_2, \ldots, v_n)$ is the smallest subspace of V containing $v_1, v_2, \ldots, v_n$.*

Example 3.21 The vectors $v_1 = (2,2,0)$, $v_2 = (2,-2,0)$ spans a subspace of $\mathbb{R}^3$. Actually, if $v = (x_1, x_2, x_3) \in \mathbb{R}^3$, then $span(v_1, v_2)$ is $\mathbb{R}^2$ as a subset of R^3. To see this, we write v as a combination of v_1, and v_2.

$$(x_1, x_2, x_3) = a_1(2,2,0) + a_2(2,-2,0),$$

implies that

$$(x_1, x_2, x_3) = (2a_1 + 2a_2, 2a_1 - 2a_2, 0).$$

Clearly, $a_1 = \frac{x_1 + x_2}{4}$, and $a_1 = \frac{x_1 - x_2}{4}$ form a solution for any $x_1, x_2 \in \mathbb{R}$ and $x_3 = 0$. $\square$

Definition 3.15 *If $span(v_1, v_2, \ldots, v_n) = V$, then we say that $(v_1, v_2, \ldots, v_n)$ spans V. In this case the vector space V is finite-dimensional . A vector space that is not finite-dimensional is called infinite-dimensional .*

Example 3.22 The vectors

$$e_1 = (1,0,\ldots,0), \quad e_2 = (0,1,0,\ldots,0), \quad \ldots, \quad e_n = (0,0,\ldots,1)$$

span $\mathbb{R}^n$ since any vector $u = (u_1,u_2,\ldots,u_n) \in \mathbb{R}^n$ can be written as a combination of $e_1,e_2,\ldots,e_n$. That is

$$u = u_1 e_1 + u_2 e_2 + \ldots + u_n e_n.$$

Thus, the vector space $\mathbb{R}^n$ is finite-dimensional with dimension n. □

Example 3.23 Consider the polynomial $p(x) = a_n x^n + a_{n-1}x^{n-1} + \ldots + a_1 x + a_0 \in V[x]$ with coefficients in $\mathbb{R}$ such that $a_n \neq 0$. Thus, the polynomial has degree n and we write $\deg(p(x)) = n$. By convention the degree of the zero polynomial $p(x) = 0$, is $-\infty$. Let $V_n[x] = \{p(x) \in V[x] \mid \deg(p(x)) \leq n\}$. Then by Example 3.15, $V_n[x]$ is a vector space. Then $V_n[x]$ is a subspace of $V[x]$. That is

$$V_n[x] \subset V[x].$$

This is the case since the zero polynomial is in $V_n[x]$. Moreover, $V_n[x]$ is closed under vector addition and scalar multiplication. Since

$$V_n[x] = \operatorname{span}(1,x,x^2,\ldots,x^n),$$

the subspace $V_n[x]$ is of finite dimension. On the other hand, we assert that $V[x]$ is infinite-dimensional. Assume the contrary; that is

$$V[x] = \operatorname{span}(p_1(x),p_2(x),\ldots,p_k(x))$$

for finite index k. Let

$$n = \max\Big(\deg(p_1(x)),\deg(p_2(x)),\ldots,\deg(p_k(x))\Big).$$

Then, $x^{n+1} \in V[x]$, but

$$x^{n+1} \notin \operatorname{span}(p_1(x),p_2(x),\ldots,p_k(x)).$$

Hence, $V[x]$ is infinite-dimensional. □

Definition 3.16 *(Bases) The set of vectors $\{v_1,v_2,\ldots,v_n\}$ is a basis for the finite-dimensional vector space V if $\{v_1,v_2,\ldots,v_p\}$ is linearly independent and $V = \operatorname{span}(v_1,v_2,\ldots,v_n)$.*

Example 3.24 The vectors $e_i, i = 1,2,\ldots,n$ of Example 3.22 form a basis for $\mathbb{R}^n$. □

Example 3.25 The vectors $e_i, i = 1,2,\ldots,n$ of Example 3.22 form a basis for $\mathbb{R}^n$. Along the same lines of thinking, the vectors $(1,2),(1,1)$ forms a basis for $\mathbb{R}^2$. □

Example 3.26 The set $\{1,x,x^2,\ldots,x^n\}$ forms a basis for $V_n[x]$, where $V_n[x]$ is defined in Example 3.23. □

Observe that the set $\{1, x, x^2\}$ is a basis for the vector space of polynomials in x with real coefficients having degree at most 2. Note that $V_2[x]$ has infinitely many polynomial with degree at most 2, yet we managed to have a description of all them using the set $\{1, x, x^2\}$.

Recall that a vector space V is called finite-dimensional if V has a basis consisting of a finite numbers of vectors; otherwise, V is infinite-dimensional.

Remark 11 *The dimension of a vector space is the number of vectors in a basis. It can be shown that in an n-dimensional vector space, any set of $n+1$ vectors is linearly dependent. Thus, the dimension of a vector space could be defined as the number of vectors in a maximal linearly independent set.*

Remark 12 *By Lemma 7, If $\{v_1, v_2, \ldots, v_n\}$ forms a basis of V, then every vector $v \in V$ can be uniquely written as a linear combination of $v_1, v_2, \ldots, v_n$.*

To see the difference between basis and span, we consider the vectors

$$v_1 = (1,0), \quad v_2 = (0,1), \quad \text{and} \quad v_3 = (2,0).$$

Clearly, $\text{span}(v_1, v_2) = \mathbb{R}^2$. Moreover, $\text{span}(v_1, v_2, v_3) = \mathbb{R}^2$. (We ask you to verify this). However, only this set $\{v_1, v_2\}$ is a basis of $\mathbb{R}^2$, because the vector v_3 makes the set $\{v_1, v_2, v_3\}$ linearly dependent.

3.5.1   Exercises

Exercise 3.39 *Show the set*

$$U = \{(a,b,c) \in \mathbb{R}^3 \mid a+b+4c = 0\}$$

is a vector space under the usual operations vector addition and scalar multiplication on $\mathbb{R}^3$.

Exercise 3.40 *Show the set*

$$U = \{(a,b,c) \in \mathbb{R}^3 \mid a+2b = 0\}$$

is a subspace under the usual operations vector addition and scalar multiplication on $\mathbb{R}^3$.

Exercise 3.41 *Show the set*

$$U = \{(a,0) \in \mathbb{R}^2 \mid a \in \mathbb{R}\}$$

is a subspace under the usual operations addition and multiplication on $\mathbb{R}^2$.

Exercise 3.42 *Show that the vectors $v_1 = (1,1,1)$, $v_2 = (2,1,3)$, and $v_3 = (-1,2,1)$ are linearly independent in $\mathbb{R}^3$. Write $v = (2,4,3)$ as a linear combination of the vectors $v_1, v_2,$ and v_3.*

Exercise 3.43 *Redo Example 3.18 for the following set of vectors.*

(a) $v_1 = (1,1,1)$, $v_2 = (1,2,0)$, *and* $v_3 = (0,-1,1)$.

(b) $v_1 = (1,1,1)$, $v_2 = (1,2,0)$, *and* $v_3 = (0,-1,2)$.

Exercise 3.44 *Explain why the set of vectors given by*

$$v_1 = (3,4,5), \quad v_2 = (-3,0,5), \quad v_3 = (4,4,4) \quad and \quad v_4 = (3,4,0)$$

is linearly dependent.

Exercise 3.45 *Prove 3. of Lemma 8.*

Exercise 3.46 *Either show the set is a vector space or explain why it is not. All functions are assumed to be continuous.*

(a) $U = \{(a,2) \in \mathbb{R}^2 | a \in \mathbb{R}\}$ *under the usual operations of addition and multiplication on* $\mathbb{R}^2$.

(b) $U = \{(a,b) \in \mathbb{R}^2 \mid a,b \geq 0\}$ *under the usual operations of addition and multiplication on* $\mathbb{R}^2$.

(c) $U = \{f : \mathbb{R} \to \mathbb{R} \mid \frac{d}{dx}f \text{ exists}\}$ *under the usual operations of addition and multiplication on functions.*

(d) $U = \{f : \mathbb{R} \to \mathbb{R} \mid f(x) \neq 0 \text{ for any } x \in \mathbb{R}\}$ *under the usual operations of addition and multiplication on functions.*

(e) The solution set to a linear nonhomogeneous equations.

(f) $U = \{A_{2 \times 2} | \det(A) = 0\}$ *under the usual operations of addition and multiplication for matrices.*

(g) $U = \{f : [-1,1] \to [-1,\infty)\}$ *under the usual operations of addition and multiplication on functions.*

(h) $U = \{f : \mathbb{R} \to \mathbb{R} \mid f(0) = 0\}$ *under the usual operations of addition and multiplication on functions.*

(i) $U = \{f : \mathbb{R} \to \mathbb{R} \mid f(x) \leq 0, \text{ for all } x \in \mathbb{R}\}$ *under the usual operations of addition and multiplication on functions.*

Exercise 3.47 *Show that any set of vectors* $\{v_1, v_2, \ldots, v_n\}$, *which spans a vector space V contains a linearly independent subset which also spans V.*

Exercise 3.48 *Show the vectors* $v_1 = (1,1)$, $v_2 = (1,2)$, *and* $v_3 = (1,0)$ *span* $\mathbb{R}^2$.

Exercise 3.49 *Give a basis of*

$$M_{2 \times 2} = \{ \begin{pmatrix} a & b \\ c & d \end{pmatrix} \mid a,b,c, \text{ and } d \in \mathbb{R} \}.$$

There are many possible answers.

Exercise 3.50 *Let $v_1 = (1,0,0)$, $v_2 = (0,1,0)$, $v_3 = (0,0,1)$,
and $v_4 = (1,1,1)$.*

(a) Show that $\{v_1, v_2, v_3, v_4\}$ spans $\mathbb{R}^3$.

(b) Does the set of vectors in part a) form a basis for $\mathbb{R}^3$? Explain.

3.6   Eigenvalues-Eigenvectors

For motivational purpose we begin the the following example.

Example 3.27 *(Lotka–Volterra Predator–Prey Model)*   We consider the Lotka–
Volterra Predator–Prey model. Let $x = x(t)$ and $y = y(t)$ be the number of preys
and predators at time t, respectively. To keep the model simple, we will make the
following assumptions:

- the predator species is dependent on a single prey species as its only food supply,

- the prey species has an unlimited food supply, and

- there is no threat to the prey other than the specific predator.

We observe that, in the absence of predation, the prey population would grow at a
natural rate

$$\frac{dx}{dt} = ax,\, a > 0.$$

On the other hand, in the absence of prey, the predator population would decline at a
natural rate

$$\frac{dy}{dt} = -cy,\, c > 0.$$

The effects of predators eating prey is an interaction rate of decline $(-bxy,\, b > 0)$
in the prey population x, and an interaction rate of growth $(dxy,\, d > 0)$ of predator
population y. Hence, one obtains the predator-prey model

$$\frac{dx}{dt} = ax - bxy$$
$$\frac{dy}{dt} = -cy + dxy. \tag{3.27}$$

The Lotka-Volterra model consists of a system of linked differential equations that
cannot be separated from each other and that cannot be solved in closed form. Since,
$(0,0)$ is a solution of the system, we linearize around it and rewrite the systems as

$$X' = AX + g(x,y), \tag{3.28}$$

where

$$X = \begin{pmatrix} x \\ y \end{pmatrix}, \quad A = \begin{pmatrix} a & 0 \\ 0 & -c \end{pmatrix}, \quad g = \begin{pmatrix} -bxy \\ dxy \end{pmatrix}.$$

Since the function g is continuously differentiable in both variables near the origin, the stability of the nonlinear system (3.28) is heavily influenced by the stability of linear system

$$X' = AX. \tag{3.29}$$

We search for solutions to (3.29) of the form

$$X = z e^{\lambda t},$$

where $z = \begin{pmatrix} z_1 \\ z_2 \end{pmatrix}$, for a parameter λ. Substituting into (3.29) we arrive at the relation

$$Az = \lambda z. \tag{3.30}$$

It is evident that the zero vector $z = \begin{pmatrix} 0 \\ 0 \end{pmatrix}$ is a solution of (3.30) for any value of λ. We are interested in the values of λ for which (3.30) has a nonzero solution. Such values are called eigenvalues and the corresponding vector solutions given by z are called eigenvectors. We have this important definition below. □

Definition 3.17 *Let A be an $n \times n$ constant matrix, in short "matrix." A number λ is said to be an eigenvalue of A if there exists a nonzero vector v such that*

$$Av = \lambda v. \tag{3.31}$$

The vector v is said to be an eigenvector corresponding to the eigenvalue λ. We may refer to λ and v as an eigenpair.

Theorem 3.12 *If λ_0, v_0 is an eigenpair of A, then*

$$X(t) = e^{\lambda_0 t} v_0 = v_0 e^{\lambda_0 t}$$

is a solution of

$$X' = AX. \tag{3.32}$$

Proof *Let A be an $n \times n$ matrix. Let $X(t) = e^{\lambda_0 t} v_0$. Then*

$$X'(t) = \lambda_0 e^{\lambda_0 t} v_0 = \lambda_0 v_0 e^{\lambda_0 t} = A v_0 e^{\lambda_0 t} = A e^{\lambda_0 t} v_0 = AX,$$

as desired. This completes the proof.

Consider the $n \times n$ matrix $A = (a_{ij})$ such that

$$Ax = \lambda x, \tag{3.33}$$

where

$$A = \begin{pmatrix} a_{11} & a_{12} & \cdots & a_{1n} \\ a_{21} & a_{22} & \cdots & a_{2n} \\ \vdots & \vdots & \ddots & \vdots \\ a_{n1} & a_{n2} & \cdots & a_{nn} \end{pmatrix}, \quad x = \begin{pmatrix} x_1 \\ x_2 \\ \vdots \\ x_n \end{pmatrix}.$$

For the purpose of finding the eigenvalues and corresponding eigenvectors, we rewrite (3.33) as

$$a_{11}x_1 + a_{12}x_2 + a_{13}x_3 + \ldots + a_{1n}x_n = \lambda x_1$$
$$a_{21}x_1 + a_{22}x_2 + a_{23}x_3 + \ldots + a_{2n}x_n = \lambda x_2$$
$$a_{31}x_1 + a_{32}x_2 + a_{33}x_3 + \ldots + a_{3n}x_n = \lambda x_3$$
$$\vdots$$
$$a_{n1}x_1 + a_{n2}x_2 + a_{n3}x_3 + \ldots + a_{nn}x_n = \lambda x_n$$

By transferring the terms on the right-hand side to the left-hand side, we arrive at

$$(a_{11} - \lambda)x_1 + a_{12}x_2 + a_{13}x_3 + \ldots + a_{1n}x_n = 0$$
$$a_{21}x_1 + (a_{22} - \lambda)x_2 + a_{23}x_3 + \ldots + a_{2n}x_n = 0$$
$$a_{31}x_1 + a_{32}x_2 + (a_{33} - \lambda)x_3 + \ldots + a_{3n}x_n = 0$$
$$\vdots$$
$$a_{n1}x_1 + a_{n2}x_2 + a_{n3}x_3 + \ldots + (a_{nn} - \lambda)x_n = 0.$$

By Corollary 4 this homogeneous system has a nontrivial solution if and only if the corresponding determinant of the coefficients is zero. That is

$$D(\lambda) = \det(A - \lambda I) = \begin{vmatrix} a_{11} - \lambda & a_{12} & \cdots & a_{1n} \\ a_{21} & a_{22} - \lambda & \cdots & a_{2n} \\ \vdots & \vdots & \ddots & \vdots \\ a_{n1} & a_{n2} & \cdots & a_{nn} - \lambda \end{vmatrix} = 0. \qquad (3.34)$$

Equation (3.34) is called the *characteristic equation* corresponding to the matrix A. By expanding $D(\lambda)$ we obtain a polynomial of nth degree in λ. This is called the *characteristic polynomial* corresponding to the matrix A. Thus, we have proved the following theorem.

Theorem 3.13 *The eigenvalues of an $n \times n$ matrix A are the roots of its corresponding characteristic equation* (3.34).

In general, if $D(\lambda)$ is an nth degree polynomial then it can be factored into linear terms over $\mathbb{C}$. of the form

$$D(\lambda) = (\lambda - \lambda_1)^{k_1}(\lambda - \lambda_2)^{k_2} \ldots (\lambda - \lambda_p)^{k_p},$$

where $k_1 + k_2 + \ldots + k_p = n$, and k_i is called the *multiplicity* of the eigenvalue λ_i. For example, if $A_{5 \times 5}$ is a matrix with characteristic polynomial

$$D(\lambda) = \lambda^5 - 3\lambda^2 + 6\lambda^3 - 4\lambda^2,$$

then

$$D(\lambda) = \lambda^2(\lambda - 1)(\lambda - 2)^2.$$

So, $\lambda_1 = 0$ has multiplicity $k_1 = 2$, $\lambda_2 = 1$ has multiplicity $k_2 = 1$, and $\lambda_3 = 2$ has multiplicity $k_3 = 2$. Once the eigenvalues are obtained, we use (3.33) to find the corresponding eigenvectors. We will apply the concept of eigenvalues and eigenvectors to solving linear systems of differential equations, As a result, we need to undertake some initial work.

Theorem 3.14 (*Independent eigenvectors*) *Let* $v_1, v_2, \ldots, v_p$ *be the corresponding eigenvectors to the distinct eigenvalues* $\lambda_1, \lambda_2, \ldots, \lambda_p$ *of a matrix A. Then* $v_1, v_2, \ldots, v_p$ *are linearly independent.*

Proof *Suppose* $v_1, v_2, \ldots, v_j$ *are linearly independent for positive integer* j, *where* j *is maximal. If* $j < p$, *then* v_{j+1} *can be written as a linear combination of the vectors,* $v_1, v_2, \ldots, v_j$. *That is there are constants* $c_1, c_2, \ldots, c_j$ *such that*

$$v_{j+1} = c_1 v_1 + c_2 v_2 + \ldots + c_j v_j.$$

Multiply from the left by the matrix A and apply the fact that $Av_i = \lambda_i v_i$ *for* $i = 1, 2, \ldots j$ *to arrive at*

$$\begin{aligned} Av_{j+1} &= \lambda_{j+1} v_{j+1} \\ &= \lambda_{j+1}(c_1 v_1 + c_2 v_2 + \ldots + c_j v_j) \\ &= c_1 \lambda_{j+1} v_1 + c_2 \lambda_{j+1} v_2 + \ldots + c_j \lambda_{j+1} v_j. \end{aligned}$$

On the other hand

$$\begin{aligned} Av_{j+1} &= A(c_1 v_1 + c_2 v_2 + \ldots + c_j v_j) \\ &= c_1 Av_1 + c_2 Av_2 + \ldots + c_j Av_j \\ &= c_1 \lambda_1 v_1 + c_2 \lambda_2 v_2 + \ldots + c_j \lambda_j v_j. \end{aligned}$$

Subtracting the two equations gives

$$c_1(\lambda_{j+1} - \lambda_1)v_1 + c_2(\lambda_{j+1} - \lambda_2)v_2 + \ldots + c_j(\lambda_{j+1} - \lambda_j)v_j = 0.$$

Since $v_1, v_2, \ldots, v_j$ *are linearly independent, we must have that*

$$c_1(\lambda_{j+1} - \lambda_1) = 0, \quad c_2(\lambda_{j+1} - \lambda_2) = 0, \quad \ldots \quad , c_j(\lambda_{j+1} - \lambda_j) = 0.$$

But then

$$\lambda_{j+1} - \lambda_j \neq 0, \quad for\ all \quad j = 1, 2, \ldots, n$$

which can only hold when

$$c_1 = c_2, \ldots, c_j = 0.$$

This implies that the vector

$$v_{j+1} = 0,\ (zero\ vector)$$

which is a contradiction, since v_{j+1} *is the eigenvector corresponding to* λ_{j+1}. *This completes the proof.*

The solution of a given linear system of differential equations is the focus of the following theorem.

Theorem 3.15 *(Distinct eigenvalues) Let λ_1, λ_2, ..., λ_n be n distinct real eigenvalues of the matrix A of (3.32) and let $K_1, K_2, \ldots, K_n$ be the corresponding eigenvectors. Then the general solution of (3.32) on the interval $I = (-\infty, \infty)$ is given by*

$$X(t) = c_1 K_1 e^{\lambda_1 t} + c_2 K_2 e^{\lambda_2 t}, \ldots, c_n K_n e^{\lambda_n t}$$

for constants c_i, $i = 1, 2, \ldots, n$.

Example 3.28 Consider the linear homogeneous system of differential equations

$$
\begin{aligned}
x_1' &= 5x_1 + 2x_2 + 3x_3 \\
x_2' &= 8x_2 + 3x_3 \\
x_3' &= 4x_3.
\end{aligned}
$$

In matrix form we have

$$x' = Ax,$$

where

$$x = \begin{pmatrix} x_1 \\ x_2 \\ x_3 \end{pmatrix}, \quad A = \begin{pmatrix} 5 & 2 & 3 \\ 0 & 8 & 3 \\ 0 & 0 & 4 \end{pmatrix}.$$

Then

$$D(\lambda) = \det(A - \lambda I) = \begin{vmatrix} 5 - \lambda & 2 & 3 \\ 0 & 8 - \lambda & 3 \\ 0 & 0 & 4 - \lambda \end{vmatrix}$$

and the system has a nontrivial solution if and only if $D(\lambda) = 0$. Expanding the determinant along the first row we obtain the characteristic equation

$$(5 - \lambda)(8 - \lambda)(4 - \lambda) = 0,$$

which has the three distinct eigenvalues

$$\lambda_1 = 5, \quad \lambda_2 = 8, \quad \text{and} \quad \lambda_3 = 4.$$

To compute the corresponding eigenvectors, we let $K_1 = \begin{pmatrix} k_1 \\ k_2 \\ k_3 \end{pmatrix}$. Then using (3.31) we have $(A - \lambda I)K_1 = 0$, or

$$
\begin{aligned}
(5 - \lambda)k_1 + 2k_2 + 3k_3 &= 0 \\
(8 - \lambda)k_2 + 3k_3 &= 0 \\
(4 - \lambda)k_3 &= 0.
\end{aligned}
\tag{3.35}
$$

From the third and second equations, it is obvious that, with $\lambda = 5$, that $k_3 = k_2 = 0$. The first equation implies that $0k_1 + 0 + 0 = 0$, from which we conclude that k_1 is arbitrary. So, if we set $k_1 = 1$, then the corresponding eigenvector is given by

$K_1 = \begin{pmatrix} 1 \\ 0 \\ 0 \end{pmatrix}$. Similarly, if we substitute $\lambda = 8$ in (3.35), we arrive at the corresponding

eigenvector $K_2 = \begin{pmatrix} 2 \\ 3 \\ 0 \end{pmatrix}$. Finally, the third eigenvector corresponding to $\lambda = 4$ is

$K_3 = \begin{pmatrix} 6 \\ 3 \\ -4 \end{pmatrix}$. Using Theorem 3.15, we arrive at the solution

$$x(t) = c_1 \begin{pmatrix} 1 \\ 0 \\ 0 \end{pmatrix} e^{5t} + c_2 \begin{pmatrix} 2 \\ 3 \\ 0 \end{pmatrix} e^{8t} + c_3 \begin{pmatrix} 6 \\ 3 \\ -4 \end{pmatrix} e^{4t}.$$

□

In some cases a repeated eigenvalue gives one independent eigenvector and the others must be found using the following method as the next example demonstrates.

We consider the system

$$x' = \begin{pmatrix} 3 & -18 \\ 2 & -19 \end{pmatrix} \begin{pmatrix} x_1 \\ x_2 \end{pmatrix}. \tag{3.36}$$

Then the coefficient matrix has the repeated eigenvalue $\lambda_1 = \lambda_2 = -3$. If $K_1 = \begin{pmatrix} k_1 \\ k_2 \end{pmatrix}$ is the corresponding eigenvector, then we have the two equations $6k_1 - 18k_2 = 0$, $2k_1 - 6k_2 = 0$, which are both equivalent to $k_1 = 3k_2$. By setting $k_2 = 1$, we obtain the single eigenvector $K_1 = \begin{pmatrix} 3 \\ 1 \end{pmatrix}$ and it follows that the corresponding solution is given by

$$\phi_1 = \begin{pmatrix} 3 \\ 1 \end{pmatrix} e^{-3t}.$$

But since we are interested in finding the general solution, we need to examine the question of finding another solution.

In general, if m is a positive integer and $(\lambda - \lambda_1)^m$ is a factor of the characteristic equation $\det(A - \lambda I) = 0$, while $(\lambda - \lambda_1)^{m+1}$ is not a factor, then λ_1 is said to be an eigenvalue of multiplicity m. Below, we discuss two such scenarios:

(a) For some $n \times n$ matrice A it may be possible to find m linearly independent eigenvectors $K_1, K_2, \ldots, K_n$ corresponding to an eigenvalue λ_1 of multiplicity $m \le n$. In this case the general solution of the system contains the linear combination

$$c_1 K_1 e^{\lambda_1 t} + c_2 K_2 e^{\lambda_2 t} + \ldots + c_n K_n e^{\lambda_n t}.$$

(b) If there is one eigenvector corresponding to an eigenvalue of multiplicity m, then m linearly independent solutions of the form

$$\phi_1 = K_{11}e^{\lambda_1 t}$$
$$\phi_2 = K_{21}te^{\lambda_1 t} + K_{22}e^{\lambda_1 t}$$
$$\vdots$$
$$\phi_m = K_{m1}\frac{t^{m-1}}{(m-1)!}e^{\lambda_1 t} + K_{m2}\frac{t^{m-2}}{(m-2)!}e^{\lambda_1 t} + \ldots + K_{mm}e^{\lambda_1 t},$$

where K_{ij} are columns vectors that can always be found, and they are known as *generalized eigenvectors*. For an illustration of case (b), we suppose λ_1 is an eigenvalue of multiplicity two with only one corresponding eigenvector K_1. To find the second eigenvector, we assume a second solution of

$$X' = AX$$

of the form

$$\phi_2(t) = K_1 te^{\lambda_1 t} + Pe^{\lambda_1 t}, \tag{3.37}$$

where

$$P = \begin{pmatrix} p_1 \\ p_2 \\ \vdots \\ p_n \end{pmatrix} \quad \text{and} \quad K = \begin{pmatrix} k_1 \\ k_2 \\ \vdots \\ k_n \end{pmatrix}$$

are to be found. Differentiate $\phi_2(t)$ and substitute back into $x' = Ax$ to get

$$(AK_1 - \lambda_1 K_1)te^{\lambda_1 t} + (AP - \lambda_1 P - K_1)e^{\lambda_1 t} = 0.$$

Since the above equation must hold for all t, it follows that

$$(A - \lambda_1 I)K_1 = 0, \tag{3.38}$$

and

$$(A - \lambda_1 I)P = K_1. \tag{3.39}$$

Equation (3.38) reaffirm that K_1 is the eigenvector of A associated with the eigenvalue λ_1. Thus, we obtained one solution $\phi_1(t) = K_1 e^{\lambda_1 t}$. To find the second solution given by (3.37) we must solve for the vector P in (3.39). To find a second solution for (3.36), we let $P = \begin{pmatrix} p_1 \\ p_2 \end{pmatrix}$. Then from equation (3.39), we have $(A + 3I)P = K_1$, which implies that $6p_1 - 18p_2 = 3$, or $2p_1 - 6p_2 = 1$. Since these two equations are equivalent, we may chose $p_1 = 1$ and find $p_2 = 1/6$. However, for simplicity, we shall choose $p_1 = 1/2$ so that $p_2 = 0$. Using (3.37) we find that

$$\phi_2(t) = (31)te^{-3t} + \begin{pmatrix} 1/2 \\ 0 \end{pmatrix}e^{-3t}.$$

Finally, the general solution is

$$X = c_1\phi_1(t) + c_2\phi_2(t).$$

3.6.1 Exercises

Exercise 3.51 *Find the eigenvalues and the corresponding eigenvectors.*

(a) $\begin{pmatrix} 2 & 5 \\ 3 & 8 \end{pmatrix}$, (b) $\begin{pmatrix} 3 & -1 & 0 \\ 4 & 0 & 0 \\ 2 & 5 & -3 \end{pmatrix}$, (c) $\begin{pmatrix} 13 & -3 & 5 \\ 0 & 4 & 0 \\ -15 & 9 & -7 \end{pmatrix}$.

Exercise 3.52 *Show that if A is an $n \times n$ matrix with $\det(A) \neq 0$, then all of its eigenvalues are different from zero.*

Exercise 3.53 *Show that if A is an $n \times n$ matrix with eigenvalues λ_i, $i = 1, 2, \ldots, n$, then the eigenvalues of A^2 are λ_i^2, $i = 1, 2, \ldots, n$.*

Exercise 3.54 *Solve thew following systems of differential equations.*

(a) $x' = \begin{pmatrix} 5 & -1 & 0 \\ 0 & -5 & 9 \\ 5 & -1 & 0 \end{pmatrix} \begin{pmatrix} x_1 \\ x_2 \\ x_3 \end{pmatrix}$, (b) $x' = \begin{pmatrix} 1 & 2 \\ 4 & 3 \end{pmatrix} \begin{pmatrix} x_1 \\ x_2 \end{pmatrix}$,

(c) $x' = \begin{pmatrix} 3 & -1 & -1 \\ 1 & 1 & -1 \\ 1 & -1 & 1 \end{pmatrix} \begin{pmatrix} x_1 \\ x_2 \\ x_3 \end{pmatrix}$, (d) $x' = \begin{pmatrix} -4 & 2 \\ -\frac{5}{2} & 2 \end{pmatrix} \begin{pmatrix} x_1 \\ x_2 \end{pmatrix}$.

Exercise 3.55 *Let A and B be two square matrices with $AB = BA$. Let λ be an eigenvalue of A with corresponding eigenvector k. If $Bk \neq 0$, show that Bk is an eigenvector of A, with eigenvalue λ.*

Exercise 3.56 *Let A be an $n \times n$ matrix. Show that if the sum of all entries of each column is r, then r is an eigenvalue of A.*

Exercise 3.57 *Let A be a non-zero $n \times n$ matrix. Show that if $A^T = \lambda A$, then $\lambda = \pm 1$.*

Exercise 3.58 *Solve*

(a)

$$x' = \begin{pmatrix} 1 & -2 & 2 \\ -2 & 1 & -2 \\ 2 & -2 & 1 \end{pmatrix} \begin{pmatrix} x_1 \\ x_2 \\ x_3 \end{pmatrix},$$

(b)

$$x' = \begin{pmatrix} 3 & -18 \\ 2 & -9 \end{pmatrix} \begin{pmatrix} x_1 \\ x_2 \end{pmatrix}.$$

3.7 Inner Product Spaces

In this section we introduce inner product spaces, normed vector spaces and orthogonality of vectors. We begin with the following definition.

Definition 3.18 *(Inner product) If for any vectors u, v, and w in a vector space V and a scalar $a \in \mathbb{R}$ we can define an inner (or scalar) product (u, v) such that*

1. $(u, v) = (v, u)$,

2. $(u, v + w) = (u, v) + (u, w)$,

3. $(au, v) = a(u, v)$,

4. $(u, u) \geq 0$, and $(u, u) = 0$ if and only if $u = 0$,

then V is called an inner product space.

Example 3.29 Let V be a finite dimensional vector space. For

$$u = (u_1, u_2, \ldots, u_n) \in V, \quad v = (v_1, v_2, \ldots, v_n) \in V,$$

we define

$$
\begin{aligned}
(u, v) &= u \cdot v = u_1 v_1 + u_2 v_2 + \ldots + u_n v_n \\
&= \sum_{i=1}^{n} u_i v_i. \tag{3.40}
\end{aligned}
$$

Clearly, (3.40) satisfies 1. $-$ 4. For the purpose of illustration, we quickly go over the verifications. Now

$$(u, v) = \sum_{i=1}^{n} u_i v_i = \sum_{i=1}^{n} v_i u_i = (v, u).$$

This verifies 1. As for 2. we let $w = (w_1, w_2, \ldots, w_n) \in V$. Then

$$
\begin{aligned}
(u, v + w) &= (u_1, u_2, \ldots, u_n) \cdot (v_1 + w_1, v_2 + w_2, \ldots, v_n + w_n) \\
&= u_1(v_1 + w_1) + u_2(v_2 + w_2) + \ldots + u_n(v_n + w_n) \\
&= u_1 v_1 + u_2 v_2 + \ldots u_n v_n + u_1 w_1 + u_2 w_2 + \ldots + u_n w_n \\
&= (u, v) + (u, w).
\end{aligned}
$$

On the other hand,

$$
\begin{aligned}
(au, v) &= (au_1, au_2, \ldots, au_n) \cdot (v_1, v_2, \ldots, v_n) \\
&= au_1 v_1 + au_2 v_2 + \ldots + au_n v_n \\
&= a(u_1 v_1 + u_2 v_2 + \ldots + u_n v_n) \\
&= a(u, v).
\end{aligned}
$$

This verifies 3. For verifying 4. we see that

$$(u, u) = u_1^2 + u_2^2 + \ldots + u_n^2 = 0$$

if and only if $u_1 = u_2 = \ldots = u_n = 0$. Thus, $(u, u) > 0$, if and only if, $u \neq 0$. We conclude that (3.40) defines an inner product. We note that if u and v are two vectors in $\mathbb{R}^n$, then

$$(u, v) = uv^T = vu^T = u_1 v_1 + u_2 v_2 + \ldots + u_n v_n.$$

$\square$

For the next example we define the space $C_0[a,b]$ to be the set of all continuous functions $f : [a,b] \to \mathbb{R}$.

Example 3.30 Consider the vector space $C_0[a,b]$. Let $f,g \in C_0[a,b]$. If we define

$$(f,g) = \int_a^b f(x)g(x)dx, \qquad (3.41)$$

then $C_0[a,b]$ is an inner product space. ☐

Definition 3.19 (Norm) *A linear space is said to have a norm on it if there is a rule that uniquely determines the size of a given element in the space. In particular, if V is a linear space, then a norm on V is a mapping that associates to each $y \in V$ a nonnegative real number denoted by*

$$||y|| = \sqrt{(y,y)}$$

called the norm of y, and that satisfies the following conditions:

1. $||y|| > 0$ *and* $||y|| = 0$ *if and only if* $y = 0$.

2. $||\alpha y|| = |\alpha| ||y||$ *for all* $y \in V, \quad \alpha \in \mathbb{R}$.

3. $||y+z|| \le ||y|| + ||z||$ *for* $y,z \in V$ *(triangle inequality).*

Definition 3.20 (Normed space) *A normed linear space is a linear space V on which there is defined a norm $|| \cdot ||$.*

Remark 13 *The number $||y||$ is interpreted as the magnitude on the size of y. Thus a norm puts geometric structure on V.*

Lemma 9 (Schwartz inequality) *Let V be a linear normed space. Then for $y,z \in V$ we have*

$$||(y,z)|| \le ||y|| \, ||z||.$$

Proof *Let $\lambda \in \mathbb{R}$ and consider*

$$
\begin{aligned}
0 \le ||y+\lambda z||^2 &= (y+\lambda z, y+\lambda z) \\
&= (y,y) + 2\lambda(y,z) + \lambda^2(z,z) \\
&= ||y||^2 + 2(y,z)\lambda + ||z||^2\lambda^2,
\end{aligned}
$$

which is quadratic in λ. Thus, we must have

$$4(y,z)^2 - 4||y||^2 \, ||z||^2 \le 0.$$

By remarking that $(y,z)^2 = ||(y,z)||^2$, the above inequality gives

$$||(y,z)||^2 \le ||y||^2 \, ||z||^2.$$

Taking the square root implies the result.

Example 3.31 Let $V = \mathbb{R}^n$. For $v = (v_1, v_2, \ldots, v_n) \in V$, we claim

$$||v|| = \sqrt{(v,v)} = \sqrt{\sum_{i=1}^{n} v_i^2} \tag{3.42}$$

defines a norm on $\mathbb{R}^n$. The verifications of 1. and 2. are similar to those in Example 3.29. We verify 3. Let $y, z \in V$. Then by (3.42) we have $||y||^2 = (y,y)$. Thus,

$$
\begin{aligned}
0 \leq ||y+z||^2 &= ||y+z|| \, ||y+z|| \\
&= (y+z, y+z) \\
&= (y,y) + (y,z) + (z,y) + (z,z) \\
&= ||y||^2 + 2(y,z) + ||z||^2 \\
&\leq ||y||^2 + 2||y|| \, ||z|| + ||z||^2 \text{ (Using Lemma 9)} \\
&= \left(||z|| + ||z|| \right)^2.
\end{aligned}
$$

Taking the square root on both sides we arrive at

$$||y+z|| \leq ||y|| + ||z||.$$

Thus, $\mathbb{R}^n$ is a normed space. □

Definition 3.21 *i) A set of vectors $\{v_i\}_{i=1}^{n}$ is called orthogonal if*

$$(v_i, v_j) = \delta_{ij} = \begin{cases} 1, & i = j \\ 0, & i \neq j \end{cases}$$

ii) A matrix A is called orthogonal if

$$A^T A = I, \quad or \quad A^{-1} = A^T.$$

iii) A set of vectors $S = \{v_i\}_{i=1}^{n}$ is called orthonormal if every vector in S has magnitude 1 and the set of vectors are mutually orthogonal.

For example the matrix

$$A = \frac{1}{3} \begin{pmatrix} 2 & -2 & 1 \\ 1 & 2 & 2 \\ 2 & 1 & -2 \end{pmatrix}$$

is orthogonal since $AA^T = I$, or $A^{-1} = A^T$. Now we present an elementary review of complex numbers.

Definition 3.22 *A complex number z is any relation of the form*

$$z = a + ib = (a,b), \quad where \quad i^2 = -1.$$

The real numbers a and b are called the real and imaginary parts of z, respectively. The set of all complex numbers is denoted by $\mathbb{C}$. The number $\bar{z} = a - ib$ is called the conjugate of z.

By placing a on the x-axis and b on the y-axis, we can interpret $z = a + ib$ as a vector from the origin terminating at (a, b). The length of the vector is called the *modulus* or *magnitude* of z and is denoted by

$$||z|| = \sqrt{a^2 + b^2}.$$

Notice that

$$||z||^2 = a^2 + b^2 = z\bar{z}.$$

Next, we state two of the most important characteristics of symmetric matrices.

Theorem 3.16 *Suppose A is an $n \times n$ real symmetric matrix.*

a) *If λ_1 and λ_2 are two distinct eigenvalues of A, then their corresponding eigenvectors y_1 and y_2 are orthogonal. That is*

$$(y_1, y_2) = 0.$$

b) *All the eigenvalues of A are real.*

Proof *a) Let λ_1 and λ_2 be two distinct eigenvalues of A, with corresponding eigenvectors y_1 and y_2. This implies that*

$$Ay_1 = \lambda_1 y_1, \quad Ay_2 = \lambda_2 y_2. \tag{3.43}$$

Multiplying the transpose of the first equation in (3.43) from the right by y_2 we get

$$(Ay_1)^T y_2 = \lambda_1 y_1^T y_2,$$

or

$$y_1^T A^T y_2 = \lambda_1 y_1^T y_2.$$

Multiplying the second equation in (3.43) from the left by y_1^T, we obtain

$$y_1^T A y_2 = \lambda_2 y_1^T y_2.$$

Subtracting the last two expressions yields

$$y_1^T A y_2 - y_1^T A^T y_2 = \lambda_2 y_1^T y_2 - \lambda_1 y_1^T y_2.$$

This results into,

$$0 = (\lambda_2 - \lambda_1) y_1^T y_2.$$

Since $\lambda_1 \neq \lambda_2$, we must have

$$y_1^T y_2 = (y_1, y_2) = 0.$$

This proves the first part. As for the second part, suppose λ is a complex eigenvalue of the symmetric matrix A with the possibility of a complex eigenvector V such that $Av = \lambda v$. We take the complex conjugate on both sides of the preceding equation and obtain $\bar{A}v = \bar{\lambda}\bar{v}$. This implies that $A\bar{v} = \bar{\lambda}\bar{v}$. Using $A^T = A$, we have the following manipulation:

$$\bar{v}^T A v = \bar{v}^T (Av) = \bar{v}^T (\lambda v) = \lambda (\bar{v}^T, v).$$

Similarly,

$$\bar{v}^T A v = (A\bar{v})^T v = (\bar{\lambda}\bar{v})^T v = \bar{\lambda}(\bar{v}, v).$$

Subtracting the above two expressions, we obtain

$$0 = (\bar{\lambda} - \lambda)(\bar{v}, v).$$

Since $(\bar{v}, v) > 0$, we must have $(\bar{\lambda} - \lambda) = 0$, or $\bar{\lambda} = \lambda$. This completes the proof.

Next, we define the *Gram-Schmidt process* which is a procedure that converts a set of linearly independent vectors into a set of orthonormal vectors that spans the same space as the original set.

Theorem 3.17 (Gram-Schmidt Process) *Suppose the vectors $u_1, u_2, \ldots, u_n$ form a basis for a vector space V. Then, from the vectors $u_i, i = 1, 2, \ldots, n$ we can form an orthonornal basis $x_1, x_2, \ldots, x_n$ for V.*

Proof *We first let $v_1 = u_1$ and take*

$$x_1 = \frac{v_1}{||v_1||}, \quad \left(= \frac{u_1}{||u_1||}\right).$$

Then

$$(x_1, x_1) = \left(\frac{u_1}{||u_1||}, \frac{u_1}{||u_1||}\right) = (u_1, u_1)\frac{1}{||u_1||}\frac{||u_1||}{||u_1||} = 1.$$

Next we seek x_2 such that $||x_2|| = 1$, and $(x_1, x_2) = 0$. For a constant c to be determined, we set

$$v_2 = u_2 - cx_1.$$

Then,

$$
\begin{aligned}
(x_1, v_2) &= (x_1, u_2 - cx_1) \\
&= (x_1, u_2) - c(x_1, x_1) \\
&= (x_1, u_2) - c \overset{\text{want}}{=} 0.
\end{aligned}
$$

This implies $c = (x_1, u_2)$, which is scalar component of u_2 in the direction of x_1. Thus,

$$v_2 = u_2 - (x_1, u_2)x_1.$$

So we take

$$x_2 = \frac{v_2}{||v_2||}.$$

In a similar fashion, we set

$$v_3 = u_3 - c_1 x_1 - c_2 x_2.$$

Then

$$(x_1, v_3) = (x_1, u_3) - c_1(x_1, x_1) - c_2(x_1, x_2) \overset{\text{want}}{=} 0.$$

Since $(x_1, x_2) = 0$, the above expression implies that $c_1 = (x_1, u_3)$. Also,

$$(x_2, v_3) = (x_2, u_3) - c_1(x_2, x_1) - c_2(x_2, x_2) \overset{\text{want}}{=} 0,$$

implies $c_2 = (x_2, u_3)$. Thus,

$$v_3 = u_3 - (x_1, u_3)x_1 - (x_2, u_3)x_2.$$

So we take

$$x_3 = \frac{v_3}{||v_3||}.$$

Continuing in this process, we obtain a general formula for all vectors given by

$$x_j = \frac{v_j}{||v_j||}, \quad \text{where} \quad v_j = u_j - \sum_{i=1}^{j-1}(x_i, u_j)x_i. \tag{3.44}$$

Left to show that $v_j \neq 0$, for all $j = 1, 2, \ldots n$. If not, then u_j is a linear combination of $u_1, u_2, \ldots, u_n$. This is impossible since $u_1, u_2, \ldots, u_n$ are linearly independent. This completes the proof.

Example 3.32 Find an orthonormal basis for the subspace of $\mathbb{R}^4$ spanned by the vectors

$$u_1 = \begin{pmatrix} 1 \\ 1 \\ 1 \\ 1 \end{pmatrix}, \quad u_2 = \begin{pmatrix} 1 \\ 1 \\ -1 \\ -1 \end{pmatrix}, \quad u_3 = \begin{pmatrix} 0 \\ -1 \\ 2 \\ 1 \end{pmatrix}.$$

According to (3.44), we have

$$x_1 = \frac{u_1}{||u_1||} = \frac{1}{2}\begin{pmatrix} 1 \\ 1 \\ 1 \\ 1 \end{pmatrix}$$

and $x_2 = \frac{v_2}{||v_2||}$, where $v_2 = u_2 - (x_1, u_2)x_1$. Now,

$$v_2 = \begin{pmatrix} 1 \\ 1 \\ -1 \\ -1 \end{pmatrix} - \frac{1}{2}\begin{pmatrix} 1 \\ 1 \\ 1 \\ 1 \end{pmatrix} \cdot \begin{pmatrix} 1 \\ 1 \\ -1 \\ -1 \end{pmatrix} \frac{1}{2}\begin{pmatrix} 1 \\ 1 \\ 1 \\ 1 \end{pmatrix},$$

or

$$v_2 = \begin{pmatrix} 1 \\ 1 \\ -1 \\ -1 \end{pmatrix} - \frac{1}{4}(0) \begin{pmatrix} 1 \\ 1 \\ 1 \\ 1 \end{pmatrix} = \begin{pmatrix} 1 \\ 1 \\ -1 \\ -1 \end{pmatrix}.$$

Thus, $x_2 = \frac{v_2}{\|v_2\|} = \frac{1}{2} \begin{pmatrix} 1 \\ 1 \\ -1 \\ -1 \end{pmatrix}$. Similarly, $x_3 = \frac{v_3}{\|v_3\|}$, where

$$v_3 = u_3 - (x_1, u_3)x_1 - (x_2, u_3)x_2.$$

From this we get $v_3 = \begin{pmatrix} 1/2 \\ -1/2 \\ 1/2 \\ -1/2 \end{pmatrix}$. Hence $x_3 = \frac{v_3}{\|v_3\|} = \begin{pmatrix} 1/2 \\ -1/2 \\ 1/2 \\ -1/2 \end{pmatrix}$.   □

In the next section, we define the diagonalization of matrices. To do so, we must first define similarity, a crucial concept in linear algebra. We would like to begin classifying matrices at this point in our work. How can we determine whether matrices A and B are similar in type, or, to put it another way, whether they are comparable? We begin with the following definition.

Definition 3.23 *Let A and B be two $n \times n$ matrices. We say A is similar to B if there exists an invertible matrix P such that $A = P^{-1}BP$.*

Note that if A is similar to B then B is similar to A (see Exercise 3.72). Thus, we may say A and B are similar matrices, or simply, similar. We have the following theorem regarding similar matrices.

Theorem 3.18 *Suppose A and B are similar. Then the following is true.*

(a) $\det(A) = \det(B)$.

(b) $\text{rank}(A) = \text{rank}(B)$.

(c) A and B have the same eigenvalues.

Proof *We only prove part (c). For parts (a) and (b), see Exercise 3.74. Since A and B are similar, there exists a nonsingular matrix P such that $A = PBP^{-1}$. Using $I = PP^{-1}$, we have*

$$\begin{aligned} \det(A - \lambda I) &= \det(A - \lambda PP^{-1}) \\ &= \det(PBP^{-1} - \lambda PP^{-1}) \\ &= \det\left(P(B - \lambda I)P^{-1}\right) \\ &= \det(P)\det(B - \lambda I)\det(P^{-1}) \end{aligned}$$

$$= \det(P)\frac{1}{\det(P)}\det(B - \lambda I)$$
$$= \det(B - \lambda I),$$

given that A and B share the same characteristic polynomial. This concludes the proof.

3.7.1 Exercises

Exercise 3.59 *Verify $C_0[a,b]$ of Example 3.30 is an inner product space.*

Exercise 3.60 *Let $V = \mathbb{R}^n$ and for $y \in V$ show that*

$$||y|| = \max_{1 \le i \le n}\{|y_i|\}$$

defines a norm.

Exercise 3.61 *For $f \in C_0([a,b])$, we define*

$$||f||_M = \max_{a \le x \le b}\{|f(x)|\}, \quad (maximum\ norm)$$

and

$$||f||_1 = \int_a^b |f(x)|dx.$$

Show that $||f||_M$ and $||f||_1$, define norms on $C_0([a,b])$.

Exercise 3.62 *Show that every finite-dimensional inner product space has an orthonormal basis.*

Exercise 3.63 *Every orthonormal list of vectors in V can be extended to an orthonormal basis of V.*

Exercise 3.64 *Use the inner product defined by (3.41) to find all values of a so that the two functions*

$$f(x) = ax, \ g(x) = x^2 - ax + 2$$

are orthogonal on $[0,1]$.

Exercise 3.65 *Show the two vectors*

$$u_1 = \begin{pmatrix} 1 \\ -2 \\ 1 \end{pmatrix}, \quad u_2 = \begin{pmatrix} 2 \\ 3 \\ 4 \end{pmatrix},$$

are orthogonal but not orthonormal. Use u_1, u_2 to form two vectors v_1, and v_2 that are orthonormal and span the same space.

Exercise 3.66 *Suppose the vectors $u_1, u_2, \ldots, u_n$ are orthogonal to the vector y. Then show that any vector in the $span(u_1, u_2, \ldots, u_n)$ is orthogonal to y.*

Exercise 3.67 *For any two vectors u and v in a vector space V, show that*

$$\frac{1}{2}\left(||u+v||^2 + ||u-v||^2\right) = ||u||^2 + ||v||^2.$$

Exercise 3.68 *Let* $w: \mathbb{R} \to (0,\infty)$ *be continuous. Show that for any two polynomials f and g of degree n,*

$$(f,g) = \int_a^b f(x)g(x)w(x)dx$$

defines an inner product. Actually this is called "weighted inner product."

Exercise 3.69 *Consider the vector space* $\mathbb{C}$ *over* $\mathbb{R}$. *Show that if* $z, w \in \mathbb{C}$, *then*

$$(z,w) = \frac{1}{2}(z\bar{w} + w\bar{z}),$$

is an inner product.

Exercise 3.70 *Use the inner product defined by (3.41) to show that the set of functions*

$$\{f_n(x)\}_{n=1}^\infty = \{\frac{\sin(n\pi \ln(x))}{\sqrt{x}}\}, \quad x \in [1,e]$$

is orthogonal.

Exercise 3.71 *Apply the Gram-Schmidt process to the following vectors*

(a)

$$u_1 = \begin{pmatrix} 5 \\ -2 \\ 4 \end{pmatrix}, \quad u_2 = \begin{pmatrix} 3 \\ -1 \\ 7 \end{pmatrix}, \quad u_3 = \begin{pmatrix} 3 \\ -3 \\ 6 \end{pmatrix}.$$

(b)

$$u_1 = \begin{pmatrix} 1 \\ 1 \\ 0 \\ 1 \end{pmatrix}, \quad u_2 = \begin{pmatrix} 1 \\ -2 \\ 0 \\ 0 \end{pmatrix}, \quad u_3 = \begin{pmatrix} 1 \\ 0 \\ -1 \\ 2 \end{pmatrix}.$$

(c)

$$u_1 = \begin{pmatrix} 1 \\ 1 \\ 0 \end{pmatrix}, \quad u_2 = \begin{pmatrix} 2 \\ 1 \\ 1 \end{pmatrix}.$$

Exercise 3.72 *Consider the three* $n \times n$ *matrices A, B, and C. Show that*

(a) A is similar to A. (Reflexive)

(b) If A is similar to B, then B is similar to A. (Symmetric)

(c) If A is similar to B, and B is similar to C, then A is similar to C. (Transitive)

Exercise 3.73 *Find the matrix A that is similar to the matrix B given that*

$$B = \begin{pmatrix} -13 & -8 & -4 \\ 12 & 7 & 4 \\ 24 & 16 & 7 \end{pmatrix}, \quad P = \begin{pmatrix} 1 & 1 & 2 \\ -2 & -1 & -3 \\ 1 & -2 & 0 \end{pmatrix}.$$

Exercise 3.74 *Prove parts (a) and (b) of Theorem 3.18.*

3.8 Diagonalization

In this section, we look at the concept of matrix diagonalization, which is the process of transformation on a matrix in order to recover a similar matrix that is diagonal. Once a matrix is diagonalized, it becomes very easy to raise it to integer powers. We begin with the following definition.

Definition 3.24 *Let A be an $n \times n$ matrix. We say that A is diagonalizable if there exists an invertible matrix P such that*

$$D = P^{-1}AP$$

where D is a diagonal matrix.

Theorem 3.19 *Let A be an $n \times n$ matrix. The following are equivalent.*

(a) The matrix A is diagonalizable.

(b) The matrix A has n linearly independent eigenvectors.

Proof *Assume (a) and consider the invertible matrix*

$$P = \begin{pmatrix} p_{11} & p_{12} & \cdots & p_{1n} \\ p_{21} & p_{22} & \cdots & p_{2n} \\ \vdots & \vdots & \ddots & \vdots \\ p_{n1} & p_{n2} & \cdots & p_{nn} \end{pmatrix}.$$

Then the relation $D = P^{-1}AP$ implies that $AP = PD$, where

$$D = \begin{pmatrix} k_1 & 0 & \cdots & 0 \\ 0 & k_2 & \ddots & \vdots \\ \vdots & \ddots & \ddots & \vdots \\ 0 & \cdots & 0 & k_n \end{pmatrix}.$$

Thus, if we denote the columns of the matrix P with $\mathbf{p_1}, \mathbf{p_2}, \ldots, \mathbf{p_n}$, *then*

$$AP = PD = \begin{pmatrix} k_1 p_{11} & k_2 p_{12} & \cdots & k_n p_{1n} \\ k_1 p_{21} & k_2 p_{22} & \ddots & k_n p_{2n} \\ \vdots & \ddots & \ddots & \vdots \\ k_1 p_{n1} & k_2 p_{n2} & \cdots & k_n p_{nn} \end{pmatrix}.$$

yields

$$A\mathbf{p_1} = k_1 \mathbf{p_1}, \; A\mathbf{p_2} = k_1 \mathbf{p_2}, \; \ldots, \; A\mathbf{p_n} = k_1 \mathbf{p_n},$$

where $A\mathbf{p_i} = k_i \mathbf{p_i}$, $i = 1, 2, \ldots, n$ *are the successive columns of AP. Since P is invertible, each of its column vector is nonzero. Thus the above relation implies that* $k_1, k_2, \ldots, k_n$ *are eigenvalues of A with correponding eigenvectors* $\mathbf{p_1}, \mathbf{p_2}, \ldots, \mathbf{p_n}$. *Since P is invertible, it follows from Corollary 5 that* $\mathbf{p_1}, \mathbf{p_2}, \ldots, \mathbf{p_n}$ *are linearly independent eigenvectors. As for (b) implying (a), we assume* $\mathbf{p_1}, \mathbf{p_2}, \ldots, \mathbf{p_n}$ *are linearly independent eigenvectors with corresponding eigenvectors,* $k_1, k_2, \ldots, k_n$. *Let the matrix P be given as in the proof of part (a). Then the product of the two matrices AP has the columns* $A\mathbf{p_i}$, $i = 1, 2, \ldots, n$. *But* $A\mathbf{p_i} =$

$k_i \mathbf{p_i}$, $i = 1, 2, \ldots, n$, *and this translates into* $AP = \begin{pmatrix} k_1 p_{11} & k_2 p_{12} & \cdots & k_n p_{1n} \\ k_1 p_{21} & k_2 p_{22} & \ddots & k_n p_{2n} \\ \vdots & \ddots & \ddots & \vdots \\ k_1 p_{n1} & k_2 p_{n2} & \cdots & k_n p_{nn} \end{pmatrix} =$

$$\begin{pmatrix} p_{11} & p_{12} & \cdots & p_{1n} \\ p_{21} & p_{22} & \cdots & p_{2n} \\ \vdots & \vdots & \ddots & \vdots \\ p_{n1} & p_{n2} & \cdots & p_{nn} \end{pmatrix} \begin{pmatrix} k_1 & 0 & \cdots & 0 \\ 0 & k_2 & \ddots & \vdots \\ \vdots & \ddots & \ddots & \vdots \\ 0 & \cdots & 0 & k_n \end{pmatrix} = PD, \textit{ where D is the diagonal ma-}$$

trix having its diagonal entries the eigenvalues $k_1, k_2, \ldots, k_n$. *The matrix P is invertible, since its column vectors are linearly independent. Thus, the relation* $AP = PD$ *implies that* $D = P^{-1} AP$. *This completes the proof.*

Note that not every matrix is diagonalizable. To see this we consider the matrix $A = \begin{pmatrix} 0 & 1 \\ 0 & 0 \end{pmatrix}$. Then 0 is the only eigenvalue but A is not the zero matrix.

In summary, to diagonalize a matrix, one should perform the following steps:

(1) Compute the eigenvalues of A and the corresponding n linearly independent eigenvectors.

(2) Form the matrix P by taking its columns to be the eigenvectors found in step (1).

(3) The diagonalization is done and given by $D = P^{-1} AP$.

Example 3.33 Consider the matrix

$$A = \begin{pmatrix} 5 & 2 & 3 \\ 0 & 8 & 3 \\ 0 & 0 & 4 \end{pmatrix}.$$

Expanding the determinant along the first row we obtain the third degree equation

$$(5 - \lambda)(8 - \lambda)(4 - \lambda) = 0,$$

which has the three distinct eigenvalues

$$\lambda_1 = 5, \quad \lambda_2 = 8, \quad \text{and} \quad \lambda_3 = 4.$$

The corresponding eigenvectors are

$$\mathbf{p_1} = \begin{pmatrix} 1 \\ 0 \\ 0 \end{pmatrix}, \quad \mathbf{p_2} = \begin{pmatrix} 2 \\ 3 \\ 0 \end{pmatrix}, \quad \text{and} \quad \mathbf{p_3} = \begin{pmatrix} 6 \\ 3 \\ -4 \end{pmatrix}.$$

Thus,

$$P = \begin{pmatrix} 1 & 2 & 6 \\ 0 & 3 & 3 \\ 0 & 0 & -4 \end{pmatrix}.$$

One can easily check that

$$P^{-1}AP = \begin{pmatrix} 5 & 0 & 0 \\ 0 & 8 & 0 \\ 0 & 0 & 4 \end{pmatrix}.$$

$\square$

Note that the diagonalization of a matrix is not unique, since you may rename the eigenvalues or remix the columns of the matrix P.

As we have said before, one of the most important application to diagonalization is the computation of matrix powers. Suppose the matrix A is diagonalizable. Then there exists a matrix P such that $D = P^{-1}AP$, or $PDP^{-1} = A$, where

$$D = \begin{pmatrix} d_{11} & 0 & \cdots & 0 \\ 0 & d_{22} & \ddots & \vdots \\ \vdots & \ddots & \ddots & \vdots \\ 0 & \cdots & 0 & d_{nn} \end{pmatrix}.$$ Then for a positive integer k we have

$$A^k = \underbrace{PDP^{-1} \cdots PDP^{-1} \cdots PDP^{-1}}_{k-\text{times}} = PD^kP^{-1},$$

$$D^k = \begin{pmatrix} d_{11}^k & 0 & \cdots & 0 \\ 0 & d_{22}^k & \ddots & \vdots \\ \vdots & \ddots & \ddots & \vdots \\ 0 & \cdots & 0 & d_{nn}^k \end{pmatrix}.$$

Another advantage is that once a matrix is diagonalized, then it is easy to find its inverse if it has one. To see this, let $PDP^{-1} = A$. Then $A^{-1} = \left(PDP^{-1}\right)^{-1} = PD^{-1}P^{-1}$, where

$$D^{-1} = \begin{pmatrix} \frac{1}{d_{11}} & 0 & \cdots & 0 \\ 0 & \frac{1}{d_{22}} & \ddots & \vdots \\ \vdots & \ddots & \ddots & \vdots \\ 0 & \cdots & 0 & \frac{1}{d_{nn}} \end{pmatrix}.$$

Example 3.34 Consider the 2×2 matrix

$$A = \begin{pmatrix} 1 & 4 \\ 4 & 3 \end{pmatrix}.$$

Then the eigenpairs are

$$\lambda_1 = -1, \quad \mathbf{p_1} = \begin{pmatrix} 1 \\ -1 \end{pmatrix}, \quad \lambda_2 = 5; \quad \mathbf{p_2} = \begin{pmatrix} 1 \\ 2 \end{pmatrix}.$$

Hence,

$$P = \begin{pmatrix} 1 & 1 \\ -1 & 2 \end{pmatrix}, \quad \text{and} \quad P^{-1} = \frac{1}{3}\begin{pmatrix} 2 & -1 \\ 1 & 1 \end{pmatrix}.$$

Finally, for positive integer k, we have

$$\begin{aligned} A^k &= PD^k P^{-1} = \begin{pmatrix} 1 & 1 \\ -1 & 2 \end{pmatrix}\begin{pmatrix} -1 & 0 \\ 0 & 5 \end{pmatrix}^k \frac{1}{3}\begin{pmatrix} 2 & -1 \\ 1 & 1 \end{pmatrix} \\ &= \frac{1}{3}\begin{pmatrix} 2(-1)^k + 5^k & -1 + 5^k \\ -2 + 2\cdot 5^k & 1 + 2\cdot 5^k \end{pmatrix}. \end{aligned}$$

In particular, for $k = 100$, we have

$$A^{100} = \frac{1}{3}\begin{pmatrix} 2 + 5^{100} & (-1)^k + 5^{100} \\ (-2)^k + 2\cdot 5^{100} & 1 + 2\cdot 5^{100} \end{pmatrix}.$$

$\square$

Definition 3.25 *Let A be an $n \times n$ matrix. We say that A is orthogonally diagonalizable if there exists an orthogonal matrix P such that*

$$D = P^{-1}AP$$

where D is a diagonal matrix. The matrix P is said to orthogonally diagonalize A.

Theorem 3.20 *Let A be an $n \times n$ matrix. The following are equivalent.*

(a) The matrix A is orthogonally diagonalizable.

(b) The matrix A has set of n orthonormal eigenvectors.

(c) The matrix A is symmetric.

The proof follow along the lines of Theorem 3.20. The only change is to apply the Gram-Schmidt process to obtain orthonormal basis for each eigenspace. Then form P whose columns are the orthonormal basis. This matrix orthogonally diagonalizes A. Note that the proof of (c) implies (a) is a bit demanding and we refer to [8]. As for the proof of (a) implies (c), we have that $D = P^{-1}AP$, or $A = PDP^{-1}$. Since P is orthogonal, we have $A = PDP^T$. Therefore,

$$A^T = (PDP^T)^T = PD^T P^T = PDP^T = A.$$

Example 3.35 Consider the matrix

$$A = \begin{pmatrix} 8 & -2 & 2 \\ -2 & 5 & 4 \\ 2 & 4 & 5 \end{pmatrix}.$$

Expanding the determinant along the first row we obtain the cubic equation

$$\lambda(\lambda - 9)^2 = 0,$$

which has the eigenvalues $\lambda_1 = 0$ of multiplicity one and $\lambda_2 = 9$ of multiplicity two. The corresponding eigenpairs are

$$\lambda_1 = 0, \quad \mathbf{p_1} = \begin{pmatrix} 1 \\ 2 \\ -2 \end{pmatrix}; \quad \lambda_2 = 9, \quad \mathbf{p_2} = \begin{pmatrix} -2 \\ 1 \\ 0 \end{pmatrix}, \quad \text{and} \quad \mathbf{p_3} = \begin{pmatrix} 2 \\ 0 \\ 1 \end{pmatrix}.$$

By applying the Gram-Schmidt process we obtain the orthonormal basis

$$x_1 = \frac{1}{3}\begin{pmatrix} 1 \\ 2 \\ -2 \end{pmatrix}, \quad x_2 = \frac{1}{\sqrt{5}}\begin{pmatrix} -2 \\ 1 \\ 0 \end{pmatrix}, \quad \text{and} \quad x_3 = \frac{1}{3\sqrt{5}}\begin{pmatrix} 2 \\ 4 \\ 5 \end{pmatrix}.$$

Thus,

$$P = \begin{pmatrix} \frac{1}{3} & -\frac{2}{\sqrt{5}} & \frac{2}{3\sqrt{5}} \\ \frac{2}{3} & \frac{1}{\sqrt{5}} & \frac{4}{3\sqrt{5}} \\ -\frac{2}{3} & 0 & \frac{5}{3\sqrt{5}} \end{pmatrix}.$$

□

3.8.1 Exercises

Exercise 3.75 *Diagonalize each of the following matrices and find* A^{100}.
(a). $A = \begin{pmatrix} 1 & 4 \\ 4 & 3 \end{pmatrix}$, (b). $A = \begin{pmatrix} 5 & -3 \\ 6 & -4 \end{pmatrix}$ (c). $A = \begin{pmatrix} 5 & -3 \\ -6 & 2 \end{pmatrix}$.

Exercise 3.76 *Diagonalize each of the following matrices.*
(a). $A = \begin{pmatrix} 2 & 0 & 0 \\ 0 & 2 & 2 \\ 0 & 0 & 4 \end{pmatrix}$, (b). $A = \begin{pmatrix} 5 & 0 & 0 \\ 2 & 6 & 0 \\ 3 & 2 & 1 \end{pmatrix}$.

Exercise 3.77 *Explain why this matrix* $A = \begin{pmatrix} 2 & 4 & 6 \\ 0 & 2 & 2 \\ 0 & 0 & 4 \end{pmatrix}$ *is not diagonalizable.*

Exercise 3.78 *Show that if B is diagonalizable and invertible, then so is* B^{-1}.

Exercise 3.79 *Let*

$$P = \begin{pmatrix} p_{11} & p_{12} \\ p_{21} & p_{22} \end{pmatrix}.$$

Show that:

(a) P *is diagonalizable if* $(p_{11} - p_{22})^2 + 4p_{12}p_{21} > 0$.

(b) P *is not diagonalizable if* $(p_{11} - p_{22})^2 + 4p_{12}p_{21} < 0$.

Exercise 3.80 *Show that if A and B are orthogonal matrices, then AB is also orthogonal.*

Exercise 3.81 *Find the matrix P that orthogonally diagonalizes each of the following matrices.*

(a). $A = \begin{pmatrix} 2 & 1 & -1 \\ 0 & 1 & 1 \\ 1 & -1 & 1 \end{pmatrix}$, *(b).* $A = \begin{pmatrix} 3 & 2 & 6 \\ -6 & 3 & 2 \\ 2 & 6 & -3 \end{pmatrix}$.

Exercise 3.82 *Show that if P is orthogonal, then aP is orthogonal if and only if* $a = 1$ *or* $a = -1$.

3.9   Quadratic Forms

In this section, we examine the concepts of *quadratic forms* and their applications to diagonalizing matrices. We begin with the following definition.

Definition 3.26 *A general quadratic form with n variables is of the form*

$$Q(x_1, x_2, \ldots, x_n) = \sum_{i,j=1}^{n} a_{ij}x_i x_j, \tag{3.45}$$

where a_{ij} *are constants.*

It will be more convenient at times to write (3.45) in matrix forms as we do next. Let x be an $n \times 1$ column vector and A be a symmetric matrix, then (3.45) is equivalent to

$$Q = x^T A x, \tag{3.46}$$

where

$$x = \begin{pmatrix} x_1 \\ x_2 \\ \vdots \\ x_n \end{pmatrix} \quad \text{and} \quad A = \begin{pmatrix} a_{11} & a_{12} & \cdots & a_{1n} \\ a_{21} & a_{22} & \cdots & a_{2n} \\ \vdots & \vdots & \ddots & \vdots \\ a_{n1} & a_{n2} & \cdots & a_{nn} \end{pmatrix}.$$

Example 3.36 Consider the quadratic form

$$Q(x_1, x_2) = 3x_1^2 - 8x_1 x_2 + x_2^2.$$

Then Q can be represented by the following 2×2 matrices

$$\begin{pmatrix} 3 & -2 \\ -6 & 1 \end{pmatrix}, \quad \begin{pmatrix} 3 & -1 \\ -7 & 1 \end{pmatrix}, \quad \begin{pmatrix} 3 & -4 \\ -4 & 1 \end{pmatrix}.$$

However, only the third matrix among them is symmetric. □

Recall (3.8) says, generally, one can find symmetrization **R** of a square matrix A by

$$\mathbf{R} = \frac{1}{2}(A + A^T).$$

We state this fact as a theorem.

Theorem 3.21 *Any quadratic form can be represented by a symmetric matrix.*

Proof *Let $A = (a_{ij})$. If $a_{ij} \neq a_{ji}$ then replace those entries with $r_{ij} = \frac{a_{ij} + a_{ji}}{2}$. Then it is evident that $r_{ij} = r_{ji}$, and this will not change the corresponding quadratic form. This completes the proof.*

Example 3.37 Find the symmetric matrix that corresponds to the quadratic form

$$Q(x_1, x_2, x_3) = -x_1^2 + 3x_2^2 + x_3^2 - 2x_1 x_2 + 4x_2 x_3 + 7x_1 x_3.$$

The matrix is

$$A = \begin{pmatrix} -1 & -1 & 7/2 \\ -1 & 3 & 2 \\ 7/2 & 2 & 1 \end{pmatrix}.$$

It is clear that $A^T = A$ and $x^T A x = Q$. □

Definition 3.27 *Let $x \in \mathbb{R}^n$ and suppose A is an $n \times n$ constant symmetric matrix. Then the quadratic form*

$$Q(x) = x^T A x$$

is

(a) *positive definite if $Q(x) > 0$, for all $x \neq 0$,*

(b) *negative definite if $Q(x) < 0$, for all $x \neq 0$,*

(c) postive semidefinite if $Q(x) \geq 0$, for all $x \neq 0$,

(d) negative semidefinite if $Q(x) \leq 0$, for all $x \neq 0$,

(e) indefinite if $Q(x)$ changes sign.

(f) If Q is positive definite, then the symmetric matrix A is called positive definite matrix.

Example 3.38 Let

$$A = \begin{pmatrix} k_1 & 0 & \cdots & 0 \\ 0 & k_2 & \ddots & \vdots \\ \vdots & \ddots & \ddots & \vdots \\ 0 & \cdots & 0 & k_n \end{pmatrix} \quad \text{and} \quad x = \begin{pmatrix} x_1 \\ x_2 \\ \vdots \\ x_n \end{pmatrix}.$$

Then

$$x^T A x = (x_1, x_2, \ldots, x_n) \begin{pmatrix} k_1 & 0 & \cdots & 0 \\ 0 & k_2 & \ddots & \vdots \\ \vdots & \ddots & \ddots & \vdots \\ 0 & \cdots & 0 & k_n \end{pmatrix} \begin{pmatrix} x_1 \\ x_2 \\ \vdots \\ x_n \end{pmatrix}$$

$$= k_1 x_1^2 + k_2 x_2^2 + \ldots + k_n x_n^2.$$

Thus, for any $n \times n$ diagonal matrix A, $Q(x)$ is positive definite for $x \neq 0$ and provided that $k_i \geq 0$ and $k_i \neq 0$ for at least one $i = 1, 2, \ldots$.   □

Theorem 3.22 *Let $A = \begin{pmatrix} a & b \\ b & c \end{pmatrix}$ and consider the quadratic form*

$$Q(x, y) = (x, y) A \begin{pmatrix} x \\ y \end{pmatrix} = ax^2 + 2bxy + cy^2.$$

(a) Q is positive definite if and only if $a > 0$, and $ac - b^2 > 0$.

(b) Q is negative definite if and only if $a < 0$, and $ac - b^2 > 0$.

(c) Q is indefinite, if and only if $a > 0$, and $ac - b^2 < 0$.

(d) Q is indefinite and only if $a < 0$, and $ac - b^2 < 0$.

The proof follows immediately from the fact that

$$Q(x, y) = a\left(x + \frac{b}{a}y\right)^2 + \frac{(ac - b^2)}{a} y^2.$$

Next we turn our attention to the characterization of eigenvalues of matrices that are symmetric.

Theorem 3.23 *Let A be an $n \times n$ symmetric matrix with eigenvalues $\lambda_1 \geq \lambda_2 \geq \ldots \geq \lambda_n$. Let*

$$\mathscr{S}^{n-1} = \{x \in \mathbb{R}^n, \text{ such that its Euclidean norm } \sqrt{(x,x)} = ||x|| = 1\}.$$

Then,

(a) $\lambda_n \leq x^T A x \leq \lambda_1$ *for all* $x \in \mathscr{S}^{n-1}$.

(b) *Let $y_1, y_2 \in \mathscr{S}^{n-1}$. Then if $y_1 \in \mathbb{R}^n$ is the corresponding eigenvector for λ_1, then $y_1^T A y_1 = \lambda_1$. Similarly, If $y_2 \in \mathbb{R}^n$ is the corresponding eigenvector for λ_n, then $y_2^T A y_2 = \lambda_n$.*

The proof of the theorem is based on the fact that

$$x^T A x = (Ax, x) = (x, Ax).$$

Note that, $\mathscr{S}^{n-1}$ denotes the unit $(n-1)$-dimensional sphere in $\mathbb{R}^n$. Moreover, since the set $\mathscr{S}^{n-1}$ is closed and bounded, continuous functions on $\mathscr{S}^{n-1}$ attain their maximum and minimum values. Thus, if $x \in \mathscr{S}^{n-1}$, then the maximum and minimum of the quadratic form $Q = x^T A x$ can be easily computed using Theorem 3.23, as the next example shows.

Example 3.39 Consider the quadratic form

$$Q(x_1, x_2) = x_1 x_2 = \begin{pmatrix} x_1 & x_2 \end{pmatrix} \begin{pmatrix} 0 & 1/2 \\ 1/2 & 0 \end{pmatrix} \begin{pmatrix} x_1 \\ x_2 \end{pmatrix}.$$

The eigenvalues of the matrix $A = \begin{pmatrix} 0 & 1/2 \\ 1/2 & 0 \end{pmatrix}$ are $\lambda_1 = \frac{1}{2}$ and $\lambda_2 = -\frac{1}{2}$. Then the eigenpairs are

$$\lambda_1 = 1/2, \ v_1 = \begin{pmatrix} \frac{1}{\sqrt{2}} \\ \frac{1}{\sqrt{2}} \end{pmatrix}; \quad \lambda_2 = -1/2, \ v_2 = \begin{pmatrix} -\frac{1}{\sqrt{2}} \\ \frac{1}{\sqrt{2}} \end{pmatrix}.$$

Thus, the maximum $\lambda_1 = 1/2$ of Q occurs at $\pm v_1$, and its minimum $\lambda_2 = -1/2$ of Q occurs at $\pm v_2$. In fact, one may uses Lagrange multiplier to extremize the function $f(x_1, x_2) = x_1 x_2$ subject to the constraint function $g(x_1, x_2) = x_1^2 + x_2^2 - 1$. $\square$

Theorem 3.24 *If A is a real symmetric matrix, then there exists an orthogonal matrix T such that the transformation $x = T\bar{x}$ will reduce the quadratic form (3.46) to the canonical or diagonal form*

$$Q = \lambda_1 \bar{x}_1^2 + \lambda_2 \bar{x}_2^2 + \ldots + \lambda_n \bar{x}_n^2, \tag{3.47}$$

where $\bar{x} = \begin{pmatrix} \bar{x}_1 \\ \bar{x}_2 \\ \vdots \\ \bar{x}_n \end{pmatrix}$, and λ_i, $i = 1, 2, \ldots, n$ are the eigenvalues of A.

Proof *The proof is a direct consequence of Theorems 3.19 and 3.20. Let* **T** *be the orthogonal matrix P in Theorem 3.20 and assume* $\lambda_1, \lambda_2, \ldots, \lambda_n$ *are the eigenvalues of the symmetric matrix A. Let the columns of* **T** *be the obtained orthonormal vectors* $\frac{y_i}{||y_i||}$, $i = 1, 2, \ldots, n$. *Then we have*

$$\mathbf{T} = \left(\frac{y_1}{||y_1||} \ \frac{y_2}{||y_2||} \ \cdots \ \frac{y_n}{||y_n||} \right).$$

As a consequence,

$$A\mathbf{T} = \left(A\frac{y_1}{||y_1||} \ A\frac{y_2}{||y_2||} \ \cdots A\frac{y_n}{||y_n||} \right) = \left(\lambda_1 \frac{y_1}{||y_1||} \ \lambda_2 \frac{y_2}{||y_2||} \ \cdots \lambda_n \frac{y_n}{||y_n||} \right).$$

This yields

$$\begin{aligned} \mathbf{T}^T A\mathbf{T} &= \left(\frac{y_1}{||y_1||} \ \frac{y_2}{||y_2||} \ \cdots \ \frac{y_n}{||y_n||} \right)^T \left(\lambda_1 \frac{y_1}{||y_1||} \ \lambda_2 \frac{y_2}{||y_2||} \ \cdots \lambda_n \frac{y_n}{||y_n||} \right) \\ &= \begin{pmatrix} \lambda_1 & 0 & \cdots & 0 \\ 0 & \lambda_2 & \ddots & \vdots \\ \vdots & \ddots & \ddots & \vdots \\ 0 & \cdots & 0 & \lambda_n \end{pmatrix} = D. \end{aligned}$$

Clearly **T** *is orthogonal and hence* $\mathbf{T}^T = \mathbf{T}$. *Let*

$$x = \mathbf{T}\bar{x}.$$

Then,

$$\begin{aligned} Q &= x^T A x = (\mathbf{T}\bar{x})^T A \mathbf{T} x \\ &= \bar{x}^T \mathbf{T}^T A \mathbf{T} x = \bar{x}^T D x \\ &= \lambda_1 \bar{x}_1^2 + \lambda_2 \bar{x}_2^2 + \ldots + \lambda_n \bar{x}_n^2. \end{aligned}$$

This completes the proof.

Example 3.40 Consider the quadratic form that we wish to put in canonical form,

$$Q(x_1, x_2, x_3) = 3x_1^2 + 2x_2^2 + 3x_3^2 + 2x_1 x_3.$$

Now Q is equivalent to

$$Q = x^T A x,$$

where

$$A = \begin{pmatrix} 3 & 0 & 1 \\ 0 & 2 & 0 \\ 1 & 0 & 3 \end{pmatrix}, \quad x = \begin{pmatrix} x_1 \\ x_2 \\ x_3 \end{pmatrix}.$$

It is clear that A is symmetric. The eigenvalues of A satisfy

$$(2 - \lambda)^2 (\lambda - 4) = 0.$$

Thus the eigenvalues are

$$\lambda_1 = \lambda_2 = 2, \quad \text{and} \quad \lambda_3 = 4.$$

Let $K = \begin{pmatrix} k_1 \\ k_2 \\ k_3 \end{pmatrix}$. Then using (3.31) we have

$$\begin{aligned} (3-\lambda)k_1 + k_3 &= 0 \\ (2-\lambda)k_2 &= 0 \\ k_1 + (3-\lambda)k_3 &= 0. \end{aligned} \quad (3.48)$$

Substituting $\lambda_3 = 4$ for λ we get $k_2 = 0$, and $k_1 = k_3$. Letting $k_1 = 1$, we get the corresponding eigenvector

$$y_1 = \begin{pmatrix} 1 \\ 0 \\ 1 \end{pmatrix}.$$

Next we substitute $\lambda_1 = 2$ for λ and obtain $k_1 = -k_3$, and k_2 is free. Setting $k_3 = b$ and $k_2 = a$ we arrive at

$$y = \begin{pmatrix} -b \\ a \\ b \end{pmatrix} = a \begin{pmatrix} 0 \\ 1 \\ 0 \end{pmatrix} + b \begin{pmatrix} -1 \\ 0 \\ 1 \end{pmatrix}.$$

By choosing $a = 1$, $b = 0$ and then $a = 0$, $b = 1$, we arrive at the other two corresponding eigenvectors

$$y_2 = \begin{pmatrix} 0 \\ 1 \\ 0 \end{pmatrix}, \quad y_3 = \begin{pmatrix} -1 \\ 0 \\ 1 \end{pmatrix}.$$

After normalizing the eigenvectors we form the matrix

$$T = \begin{pmatrix} \frac{1}{\sqrt{2}} & 0 & -\frac{1}{\sqrt{2}} \\ 0 & 1 & 0 \\ \frac{1}{\sqrt{2}} & 0 & \frac{1}{\sqrt{2}} \end{pmatrix},$$

which is orthogonal. Let $\bar{x} = \begin{pmatrix} \bar{x}_1 \\ \bar{x}_2 \\ \bar{x}_3 \end{pmatrix}$. Then,

$$x = T\bar{x} = \begin{pmatrix} \frac{1}{\sqrt{2}} & 0 & -\frac{1}{\sqrt{2}} \\ 0 & 1 & 0 \\ \frac{1}{\sqrt{2}} & 0 & \frac{1}{\sqrt{2}} \end{pmatrix} \begin{pmatrix} \bar{x}_1 \\ \bar{x}_2 \\ \bar{x}_3 \end{pmatrix},$$

from which we arrive at

$$x_1 = \frac{1}{\sqrt{2}}\bar{x}_1 - \frac{1}{\sqrt{2}}\bar{x}_3$$

$$x_2 = \bar{x}_2$$

$$x_3 = \frac{1}{\sqrt{2}}\bar{x}_1 + \frac{1}{\sqrt{2}}\bar{x}_3.$$

Substituting x_1, x_2, and x_3 back into $Q(x_1, x_2, x_3)$ confirms that

$$Q(\bar{x}_1, \bar{x}_2, \bar{x}_3) = 4\bar{x}_1^2 + 2\bar{x}_2^2 + 2\bar{x}_3^2 = \lambda_1 \bar{x}_1^2 + \lambda_2 \bar{x}_2^2 + \lambda_3 \bar{x}_3^2.$$

$\square$

The above results extend to cover quadratic forms of the form $Q = c$, where c is constant. Here is an example.

Example 3.41 Consider the quadratic form

$$Q(x_1, x_2, x_3) = 4x_1^2 + 4x_2^2 + 4x_3^2 + 4x_1 x_2 + 4x_1 x_3 + 4x_2 x_3 - 3.$$

Now Q is equivalent to

$$Q = x^T A x - 3,$$

where

$$A = \begin{pmatrix} 4 & 2 & 2 \\ 2 & 4 & 2 \\ 2 & 2 & 4 \end{pmatrix}, \quad x = \begin{pmatrix} x_1 \\ x_2 \\ x_3 \end{pmatrix}.$$

The eigenvalues are

$$\lambda_1 = \lambda_2 = 2, \quad \text{and} \quad \lambda_3 = 8.$$

The corresponding normalized eigenvectors are

$$y_1 = \begin{pmatrix} -\frac{1}{\sqrt{2}} \\ \frac{1}{\sqrt{2}} \\ 0 \end{pmatrix}, \quad y_2 = \begin{pmatrix} -\frac{1}{\sqrt{6}} \\ -\frac{1}{\sqrt{6}} \\ \frac{2}{\sqrt{6}} \end{pmatrix}, \quad y_3 = \begin{pmatrix} \frac{1}{\sqrt{3}} \\ \frac{1}{\sqrt{3}} \\ \frac{1}{\sqrt{3}} \end{pmatrix}.$$

$$\mathbf{T} = \begin{pmatrix} -\frac{1}{\sqrt{2}} & -\frac{1}{\sqrt{6}} & \frac{1}{\sqrt{3}} \\ \frac{1}{\sqrt{2}} & -\frac{1}{\sqrt{6}} & \frac{1}{\sqrt{3}} \\ 0 & \frac{2}{\sqrt{6}} & \frac{1}{\sqrt{3}} \end{pmatrix},$$

which is orthogonal. Let $\bar{x} = \begin{pmatrix} \bar{x}_1 \\ \bar{x}_2 \\ \bar{x}_3 \end{pmatrix}$. Then,

$$x = \mathbf{T}\bar{x},$$

reduces Q to the form

$$Q(\bar{x}_1, \bar{x}_2, \bar{x}_3) = 2\bar{x}_1^2 + 2\bar{x}_2^2 + 8\bar{x}_3^2 - 3.$$

$\square$

Theorem 3.25 *A quadratic form $Q = x^T A x$ is positive definite, if and only if, all the eigenvalues of A are positive.*

Proof *Suppose Q is positive definite and let λ be an eigenvalue of the matrix A. If $\lambda = 0$, then there is an eigenvector x so that $Ax = 0$. But then $Q = x^T A x = 0$, which implies Q is not positive definite. Similarly, if $\lambda < 0$, then there is an eigenvector x so that $Ax = \lambda x$. But then $Q = x^T A x = \lambda ||x||^2 < 0$, since $\lambda < 0$. This implies Q is not positive definite. On the other hand, if all the eigenvalues $\lambda_i, i = 1, 2, \ldots, n$ are positive, then there is a transformation $x = \mathbf{T}\bar{x}$ that will reduce the quadratic form Q to the canonical form*

$$Q = \lambda_1 \bar{x}_1^2 + \lambda_2 \bar{x}_2^2 + \ldots + \lambda_n \bar{x}_n^2 > 0,$$

unless $\bar{x} = 0$. But then $x = \mathbf{T}0 = 0$. This completes the proof.

We have the following example.

Example 3.42 The quadratic form

$$Q(x_1, x_2, x_3) = 4x_1^2 + 4x_2^2 + 4x_3^2 + 4x_1 x_2 + 4x_1 x_3 + 4x_2 x_3$$

is positive definite since the eigenvalues of

$$A = \begin{pmatrix} 4 & 2 & 2 \\ 2 & 4 & 2 \\ 2 & 2 & 4 \end{pmatrix}$$

are

$$\lambda_1 = \lambda_2 = 2, \quad \text{and} \quad \lambda_3 = 8.$$

□

Another characterization of the positive definiteness of a quadratic form is with respect to the determinant of all *principal submatrices*. We begin with the following definition.

Definition 3.28 *Let $A = (a_{ij})$ be an $n \times n$ matrix. For $1 \leq k \leq n$, the kth principal submatrix of A is*

$$\begin{pmatrix} a_{11} & a_{12} & \cdots & a_{1k} \\ a_{21} & a_{22} & \cdots & a_{2k} \\ \vdots & \vdots & \ddots & \vdots \\ a_{k1} & a_{k2} & \cdots & a_{kk} \end{pmatrix}.$$

Now we have the following theorem.

Theorem 3.26 *A quadratic form $Q = x^T A x$ is positive definite if and only if the determinant of every principal submatrix is positive.*

We furnish the following simple example.

Example 3.43 Consider the matrix A in Example 3.42. Then, the principal subma-trices of A are

$$B = (4), \quad \text{and} \quad C = \begin{pmatrix} 4 & 2 \\ 2 & 4 \end{pmatrix}.$$

Since

$$\det(B) = 4, \quad \text{and} \quad \det(C) = 12,$$

are all positive, the quadratic form $Q = x^T A x$ is positive definite. □

The next theorem is about reducing two quadratic forms simultaneously to canonical forms when one of them is positive definite.

Theorem 3.27 *If at least one of the quadratic forms*

$$Q_1 = x^T A x, \quad Q_2 = x^T B x \tag{3.49}$$

is positive definite, it is always possible to reduce the two forms simultaneously to linear combinations of only squares of new variables, that is, to canonical forms, by a nonsingular real transformation.

Proof *Suppose Q_2 is positive definite. Then by Theorem 3.24, there exists $\mathbf{T}$ such that*

$$x = \mathbf{T}y \tag{3.50}$$

that reduces Q_2 to the form

$$Q_2 = \mu_1 y_1^2 + \mu_2 y_2^2 + \ldots + \mu_n y_n^2, \tag{3.51}$$

where $\mu_i, i = 1, 2, \ldots, n$ are the eigenvalues of the symmetric matrix B. Since B is positive definite, all of its eigenvalues are positive. Hence we may set

$$\eta_i = \sqrt{\mu_i}\, y_i, \quad i = 1, 2, \ldots, n, \tag{3.52}$$

which reduces Q_2 to the form

$$Q_2 = \eta_1^2 + \eta_2^2 + \ldots + \eta_n^2 = \eta^T \eta, \tag{3.53}$$

where $\eta = \begin{pmatrix} \eta_1 \\ \vdots \\ \eta_n \end{pmatrix}$. At the same time (3.50) reduces Q_1 to the form

$$\begin{aligned} Q_1 &= x^T A x = (\mathbf{T}y)^T A \mathbf{T}y = y^T \mathbf{T}^T A \mathbf{T}y \\ &= y^T (\mathbf{T}^T A \mathbf{T})y. \end{aligned} \tag{3.54}$$

Now (3.52) reduces (3.54) to

$$Q_1 = \eta^T (\mathbf{T}'^T A \mathbf{T}')\eta \tag{3.55}$$

where $\mathbf{T}'$ *is the matrix obtained from* $\mathbf{T}$ *by dividing each element of the* i^{th} *column by* $\sqrt{\mu_i}$. *Hence we may write* Q_1 *as*

$$Q_1 = \eta^T G \eta, \quad where \quad G = \mathbf{T}'^T A \mathbf{T}'. \tag{3.56}$$

Then the matrix G is symmetric since

$$G^T = (\mathbf{T}'^T A \mathbf{T}')^T = (A\mathbf{T}')^T (\mathbf{T}')^T = \mathbf{T}'^T A \mathbf{T}' = G.$$

Thus (3.56) maybe reduced to canonical form by setting

$$\eta = S\alpha, \quad where \quad \alpha = \begin{pmatrix} \alpha_1 \\ \vdots \\ \alpha_n \end{pmatrix},$$

and S is made up of the normalized eigenvectors of G. Thus

$$Q_1 = \lambda_1 \alpha_1^2 + \lambda_2 \alpha_2^2 + \ldots + \lambda_n \alpha_n^2, \tag{3.57}$$

where λ_i, $i = 1, 2, \ldots, n$ *are the eigenvalues of G. At the same time using* $\eta = S\alpha$ *in (3.53) gives*

$$\begin{aligned} Q_2 &= \eta^T \eta = (S\alpha)^T (S\alpha) = \alpha^T S^T S\alpha = \alpha^T \alpha \\ &= \alpha_1^2 + \alpha_2^2 + \ldots + \alpha_n^2 \quad (since \ S \ is \ orthogonal). \end{aligned} \tag{3.58}$$

Thus, the change of variables

$$x = \mathbf{T}y = \mathbf{T}'\eta = \mathbf{T}'S\alpha, \tag{3.59}$$

will simultaneously reduce Q_1 *and* Q_2 *to diagonal forms, or to the canonical forms (3.57) and (3.58), respectively. This completes the proof.*

We provide the following example.

Example 3.44 Find the real transformation that will simultaneously reduce the quadratic forms

$$Q_1 = 3x_1^2 + 3x_2^2 - 2x_1 x_2 \quad and \quad Q_2 = 2x_1^2 + 2x_2^2,$$

to canonical forms. It is clear that Q_2 is positive and definite. Moreover, in matrix notation, we have that

$$Q_2 = x^T Bx,$$

where $B = \begin{pmatrix} 2 & 0 \\ 0 & 2 \end{pmatrix}$ with eigenvalues $\mu_1 = \mu_2 = 2$. Let $K = \begin{pmatrix} k_1 \\ k_2 \end{pmatrix}$. Then

$$(2 - \mu_1)k_1 + 0k_3 = 0$$
$$0k_1 + (2 - \mu_1)k_2 = 0.$$

Substituting $\mu_1 = 2$, we get $0k_1 + 0k_2 = 0$. Letting $k_1 = a, k_2 = b$ we arrive at

$$k = \begin{pmatrix} a \\ b \end{pmatrix} = a \begin{pmatrix} 1 \\ 0 \end{pmatrix} + b \begin{pmatrix} 0 \\ 1 \end{pmatrix}.$$

Let $a = 1$, $b = 0$ and get the eigenvector $K_1 = \begin{pmatrix} 1 \\ 0 \end{pmatrix}$. Similarly, if we set $a = 0$, $b = 1$ we arrive at the second eigenvector $K_2 = \begin{pmatrix} 0 \\ 1 \end{pmatrix}$. So $\mathbf{T} = \begin{pmatrix} 1 & 0 \\ 0 & 1 \end{pmatrix}$, and hence the transformation $x = \mathbf{T}y$, $y = \begin{pmatrix} y_1 \\ y_2 \end{pmatrix}$ reduces Q_2 to

$$Q_2 = \mu_1 y_1^2 + \mu_2 y_2^2 = 2y_1^2 + 2y_2^2.$$

Or,

$$Q_2 = \eta_1^2 + \eta_2^2, \quad \text{where} \quad \eta_i = \sqrt{\mu_i} y_i = \sqrt{2} y_i, \quad i = 1, 2.$$

Thus,

$$\mathbf{T}' = \begin{pmatrix} \frac{1}{\sqrt{2}} & 0 \\ 0 & \frac{1}{\sqrt{2}} \end{pmatrix},$$

and $Q_1 = \eta^T (\mathbf{T}'^T A \mathbf{T}') \eta$, where $A = \begin{pmatrix} 3 & -1 \\ -1 & 3 \end{pmatrix}$. In particular,

$$\begin{aligned} Q_1 &= \eta^T \begin{pmatrix} \frac{1}{\sqrt{2}} & 0 \\ 0 & \frac{1}{\sqrt{2}} \end{pmatrix} \begin{pmatrix} 3 & -1 \\ -1 & 3 \end{pmatrix} \begin{pmatrix} \frac{1}{\sqrt{2}} & 0 \\ 0 & \frac{1}{\sqrt{2}} \end{pmatrix} \eta \\ &= \eta^T \begin{pmatrix} 3/2 & -1/2 \\ -1/2 & 3/2 \end{pmatrix} \eta := \eta^T G \eta. \end{aligned}$$

The matrix G has the normalized eigenpairs

$$\lambda_1 = 1, \ v_1 = \begin{pmatrix} \frac{1}{\sqrt{2}} \\ \frac{1}{\sqrt{2}} \end{pmatrix}; \quad \lambda_2 = 2, \ v_2 = \begin{pmatrix} \frac{1}{\sqrt{2}} \\ -\frac{1}{\sqrt{2}} \end{pmatrix}.$$

Thus,

$$S = \begin{pmatrix} \frac{1}{\sqrt{2}} & \frac{1}{\sqrt{2}} \\ \frac{1}{\sqrt{2}} & -\frac{1}{\sqrt{2}} \end{pmatrix}.$$

Setting

$$\eta = S\alpha, \quad \text{where} \quad \alpha = \begin{pmatrix} \alpha_1 \\ \alpha_2 \end{pmatrix},$$

gives

$$\eta_1 = \frac{1}{\sqrt{2}} (\alpha_1 + \alpha_2), \quad \eta_2 = \frac{1}{\sqrt{2}} (\alpha_1 - \alpha_2).$$

This implies that

$$
\begin{aligned}
Q_1 &= \alpha^T S^T G S \alpha = \alpha^T \begin{pmatrix} \frac{1}{\sqrt{2}} & \frac{1}{\sqrt{2}} \\ \frac{1}{\sqrt{2}} & -\frac{1}{\sqrt{2}} \end{pmatrix} \begin{pmatrix} 3/2 & -1/2 \\ -1/2 & 3/2 \end{pmatrix} \begin{pmatrix} \frac{1}{\sqrt{2}} & \frac{1}{\sqrt{2}} \\ \frac{1}{\sqrt{2}} & -\frac{1}{\sqrt{2}} \end{pmatrix} \alpha \\
&= \alpha_1^2 + 2\alpha_2^2,
\end{aligned}
$$

as expected. Thus the transformation that will simultaneously transform Q_1 and Q_2 into canonical forms is

$$
\begin{aligned}
x &= T'S\alpha = \begin{pmatrix} \frac{1}{\sqrt{2}} & 0 \\ 0 & \frac{1}{\sqrt{2}} \end{pmatrix} \begin{pmatrix} \frac{1}{\sqrt{2}} & \frac{1}{\sqrt{2}} \\ \frac{1}{\sqrt{2}} & -\frac{1}{\sqrt{2}} \end{pmatrix} \begin{pmatrix} \alpha_1 \\ \alpha_2 \end{pmatrix} \\
&= \begin{pmatrix} \frac{1}{2}(\alpha_1 + \alpha_2) \\ \frac{1}{2}(\alpha_1 - \alpha_2) \end{pmatrix}.
\end{aligned}
$$

Componentwise, the transformation is

$$
x_1 = \frac{1}{2}(\alpha_1 + \alpha_2), \quad x_2 = \frac{1}{2}(\alpha_1 - \alpha_2).
$$

□

3.9.1 Exercises

Exercise 3.83 *Write the quadratic forms in matrix forms with symmetric matrices.*

(a) $Q(x_1,x_2) = 3x_1^2 + 3x_2^2 - x_1x_2$.

(b) $Q(x_1,x_2,x_3) = x_1^2 + x_2^2 + x_3^2 - 8x_1x_2 + 4x_2x_3 + 10x_1x_3$.

Exercise 3.84 *For each of the given matrices, write down the corresponding quadratic form and then find a symmetric matrix which determines the same quadratic form. (a)* $A = \begin{pmatrix} 2 & 1 \\ 3 & 4 \end{pmatrix}$, *(b)* $B = \begin{pmatrix} 5 & -1 & 2 \\ 3 & 4 & 1 \\ 1 & 6 & 2 \end{pmatrix}$, *(c)* $C = \begin{pmatrix} 1 & 2 & 0 \\ 3 & 4 & 5 \\ 0 & 7 & 6 \end{pmatrix}$.

Exercise 3.85 *Let A be an $n \times n$ matrix. We say $A = (a_{ij})$ is positive definite if $x^T A x > 0$ for nonzero $n \times 1$ vector x. Show that if A is positive definite, then $a_{ii} > 0$, $i = 1, 2, \ldots, n$.*

Exercise 3.86 *Give an example of a quadratic form in 2 variables $Q(x_1,x_2)$, which is*

(a) positive definite,

(b) negative definite,

(c) postive semidefinite,

(d) negative semidefinite,

(e) indefinite.

Exercise 3.87 *Find an orthogonal transformation which will reduce each of the quadratic forms given below to canonical form.*

(a) $Q(x_1,x_2,x_3) = 3x_1^2 + 3x_2^2 + 3x_3^2 - x_1x_2 - x_2x_3.$

(b) $Q(x_1,x_2,x_3) = 3x_1^2 + 4x_2^2 + 3x_3^2 + 4x_1x_2 - 4x_2x_3.$

(c) $Q(x_1,x_2) = 5x_1^2 + 8x_2^2 - 4x_1x_2 - 36$

(d) $Q(x_1,x_2,x_3,x_4) = 5x_1x_4 + 5x_2x_3.$

(e) $Q(x_1,x_2,x_3) = x_1^2 + x_2^2 + x_3^2 - 2x_1x_2.$

(f) $Q(x_1,x_2,x_3) = 2x_1^2 + x_2^2 + x_3^2 + 2x_1x_2 - 2x_1x_3 - 4x_2x_3.$

Exercise 3.88 *Let $\mathscr{S}^{n-1}$ be defined as in Theorem 3.23. Find the maximum and minimum of each of the multivariable functions on $\mathscr{S}^{n-1}$.*

(a)
$$f(x_1,x_2,x_3) = 3x_1^2 + 2x_2^2 + 3x_3^2 + 2x_1x_3.$$

(b)
$$f(x_1,x_2,x_3) = 4x_1^2 + 4x_2^2 + 4x_3^2 + 4x_1x_2 + 4x_1x_3 + 4x_2x_3.$$

Exercise 3.89 *Use Theorem 3.26 to show the quadratic forms in Example 3.89 are positive definite.*

Exercise 3.90 *Show the matrix*

$$A = \begin{pmatrix} 1 & -1 & 2 & 0 \\ -1 & 4 & -1 & 1 \\ 2 & -1 & 6 & -2 \\ 0 & 1 & -2 & 4 \end{pmatrix}$$

is positive definite.

Exercise 3.91 *Find all values of x so that the matrix*

$$A = \begin{pmatrix} 2 & -1 & x \\ -1 & 2 & -1 \\ x & -1 & 2 \end{pmatrix}$$

is

(a) positive semidefinite,

(b) positive definite.

Exercise 3.92 *Let A and B be symmetric matrices and consider the two quadratic forms*

$$Q_1 = x^T A x \quad and \quad Q_2 = x^T B x.$$

Show that if there is a matrix P that simultaneously diagonalizes Q_1 and Q_2 then $A^{-1}B$ is diagonalizable.

Exercise 3.93 *Use Exercise 3.92 to show the two quadratic forms*

$$Q_1 = x_1^2 + x_1 x_2 - x_2^2 \quad and \quad Q_2 = x_1^2 - 2x_1 x_2$$

can not be simultaneously diagonalized.

Exercise 3.94 *Find the real transformation that will simultaneously reduce the quadratic forms*

$$Q_1 = x_1 x_2 \quad and \quad Q_2 = 3x_1^2 - 2x_1 x_2 + 2x_2^2.$$

to canonical forms.

Exercise 3.95 *Find the real transformation that will simultaneously reduce the quadratic forms*

$$Q_1 = 4x_1^2 + 4x_2^2 + 4x_3^2 + 4x_1 x_2 + 4x_1 x_3 + 4x_2 x_3,$$

and

$$Q_2 = 3x_1^2 + 3x_3^2 + 4x_1 x_2 + 8x_1 x_3 + 4x_2 x_3,$$

to canonical forms.

3.10   Functions of Symmetric Matrices

In this section, we restrict our study to symmetric matrices. We have already seen that if A and B are square matrices and symmetric, then AB is symmetric only if $AB = BA$. However, $A + B$ is symmetric. Also, if A is any square matrix then,

$$A^2 = AA, \quad A^3 = AA^2, \quad \ldots, \quad A^{n+1} = AA^n,$$

and consequently, for any positive integers r and s we have

$$A^r A^s = A^s A^r = A^{r+s}. \tag{3.60}$$

On the other hand,

$$A^{-n} = (A^{-1})^n$$

provided that A is non-singular and A^{-1} is unique. If we adopt the notation

$$A^0 = I,$$

then equation (3.60) holds for any integers r and s. In addition, if A is a square symmetric matrix, then A^r is also symmetric for positive integer r.

Theorem 3.28 *Suppose A is an $n \times n$ real symmetric matrix. If $\lambda_i \neq 0$ is an eigenvalue of A with corresponding eigenvector u_i, then for an integer r, the value λ_i^r is an eigenvalue of A^r with the same eigenvector u_i.*

Proof *Since $Au_i = \lambda_i u_i$, the outcome is the result of repeating*

$$A^2 u_i = A(Au_i) = A\lambda_i u_i = \lambda_i Au_i = \lambda_i \lambda_i u_i = \lambda_i^2 u_i.$$

By repeating the above process, we deduce the relation

$$A^r u_i = \lambda_i^r u_i,$$

for any positive integer r. By multiplying $Au_i = \lambda_i u_i$ from the left with A^{-1} we get $u_i = A^{-1}\lambda_i u_i = \lambda_i A^{-1} u_i$. After multiplication by λ_i^{-1} yields

$$A^{-1} u_i = \lambda_i^{-1} u_i.$$

A similar argument leads to $A^r u_i = \lambda_i^r u_i$, for negative integer r. This completes the proof.

We note that if A is symmetric, then A^r can not possess additional eigenvalues to those obtained from A, nor can it possess eigenvectors which are linearly independent of those of A. However, A^r may have eigenvectors not possessed by A, as the next example shows.

Example 3.45 Suppose A is a 2×2 symmetric matrix with eigenvalues 2 and -1 and corresponding eigenvectors $u_1 = \begin{pmatrix} 1 \\ 2 \end{pmatrix}$ and $u_2 = \begin{pmatrix} 3 \\ 1 \end{pmatrix}$. Let $w = \begin{pmatrix} 4 \\ 2 \end{pmatrix}$. Compute $A^3 w$. First we write w as a combination of u_1 and u_2. That is, we need to find constants c_1 and c_2 such that $w = c_1 u_1 + c_2 u_2$. Or,

$$4 = c_1 + 3c_2, \ 2 = 2c_1 + c_2.$$

Solving the system, we arrive at $c_1 = \frac{2}{5}$ and $c_2 = \frac{6}{5}$. Now by Theorem 3.28, we see that

$$A^3 u_1 = 2^3 u_1 \quad \text{and} \quad A^3 u_2 = (-1)^3 u_2.$$

Hence,

$$
\begin{aligned}
A^3 w &= A^3 \left(\frac{2}{5} u_1 + \frac{6}{5} u_2 \right) \\
&= \frac{2}{5} A^3 u_1 + \frac{6}{5} A^3 u_2 \\
&= \frac{2}{5} (2^3 u_1) + \frac{6}{5} ((-1)^3 u_2) \\
&= \frac{16}{5} u_1 - \frac{6}{5} u_2 \\
&= \begin{pmatrix} -\frac{2}{5} \\ \frac{26}{5} \end{pmatrix}.
\end{aligned}
$$

$\square$

The next theorem plays an important role in the proof of the Cayley-Hamilton Theorem.

Theorem 3.29 *Let A be an $n \times n$ symmetric matrix. For constants α_i, $i = 1, 2, \ldots, n$ let*

$$P(A) = \alpha_n A^n + \alpha_{n-1} A^{n-1} + \ldots + \alpha_1 A + \alpha_0 I$$

be the characteristic polynomial of A. Then all eigenvectors of A are eigenvectors of $P(A)$ and if the eigenvalues of A are $\lambda_1, \ldots, \lambda_n$, then those of $P(A)$ are

$$P(\lambda_1), \quad P(\lambda_2), \quad \ldots, \quad P(\lambda_n).$$

Proof *Let λ_i be an eigenvalue of A with corresponding eigenvector $u_i, i = 1, 2, \ldots, n$. Then from Theorem 3.28, we have,*

$$
\begin{aligned}
P(A)u_i &= \alpha_n A^n u_i + \alpha_{n-1} A^{n-1} u_i + \ldots + \alpha_1 A u_i + \alpha_0 u_i \\
&= \alpha_n \lambda^n u_i + \alpha_{n-1} \lambda^{n-1} u_i + \ldots + \alpha_1 \lambda u_i + \alpha_0 u_i \\
&= P(\lambda)u_i.
\end{aligned}
$$

Thus,

$$\Big(P(A) - P(\lambda)\Big)u_i = 0,$$

which implies $P(\lambda)$ is an eigenvalue of $P(A)$ with corresponding eigenvector u_i. This completes the proof.

The next theorem, known as the Cayley-Hamilton Theorem, sheds light on an interesting relationship between a matrix and its characteristic polynomial.

Theorem 3.30 *(Cayley-Hamilton Theorem) Let A be an $n \times n$ symmetric matrix. If*

$$P(\lambda) = |A - \lambda I| = 0,$$

then A satisfies $P(A) = 0$ (zero matrix).

Proof *We know*

$$P(\lambda) = |A - \lambda I| = (-1)^n [\lambda^n + \beta_{n-1} \lambda^{n-1} + \ldots + \beta_1 \lambda + (-1)^n \beta_0].$$

By definition, we have that $P(\lambda_i) = 0$, $i = 1, 2, \ldots, n$. By Theorem 3.29, we have

$$P(A) = P(\lambda_i)u_i, \quad i = 1, 2, \ldots, n,$$

where λ_i is an eigenvalue of $P(A)$ and u_i is its corresponding eigenvector. Let $B = P(A)$. Then $Bx = 0$ possesses the n linearly independent solutions $x = u_1, u_2, \ldots, u_n$. But since B is a square matrix of order n, B must be of rank $n - n = 0$. Hence $B = P(A)$ must be a zero matrix and it follows that $B = P(A) = 0$. This completes the proof.

We note that if A is an $n \times n$ symmetric matrix, then Theorem 3.30 enables us to express any polynomial in A as a linear combination of $I, A, A^2, \ldots, A^n$ as the next example shows.

Example 3.46 Let

$$A = \begin{pmatrix} 1 & -2 & 4 \\ 0 & -1 & 2 \\ 2 & 0 & 3 \end{pmatrix}.$$

Then its characteristic polynomial is given by

$$P(\lambda) = \lambda^3 - 3\lambda^2 - 9\lambda + 3.$$

Now, by Cayley-Hamilton Theorem we have

$$A^3 - 3A^2 - 9A + 3I = 0. \tag{3.61}$$

Thus,

$$A^3 = 3A^2 + 9A - 3I.$$

Multiplying (3.61) by A we arrive at

$$A^4 = 3A^3 + 9A^2 - 3A = 3(3A^2 + 9A - 3I) + 9A^2 - 3A = 18A^2 + 24A - 9I.$$

On the other hand, if we multiply (3.61) by A^{-1} we obtain

$$A^{-1} = \frac{1}{3}\left(-A^2 + 3A + 9I\right).$$

Similarly, if we multiply the preceding equation by A^{-1} again it yields,

$$A^{-2} = \frac{1}{3}\left(-A + 3I + 9A^{-1}\right) = \frac{1}{3}[-A + 3I + 3(-A^2 + 3A + 9I)],$$

or

$$A^{-2} = \frac{1}{3}[-3A^2 + 8A + 30I].$$

$\square$

Another application to the Cayley-Hamilton Theorem is *Sylvester's formula* which we discuss next. Assume all the eigenvalues of the $n \times n$ symmetric matrix A are distinct. We attempt to write any polynomial in A of degree $n - 1$ as

$$P(A) = \alpha_1 A^{n-1} + \alpha_2 A^{n-2} + \ldots + \alpha_{n-1} A + \alpha_n I.$$

Equivalently, we seek $C_1, C_2, \ldots, C_n$ such that

$$\begin{aligned} P(A) &= C_1\left[(A - \lambda_2 I)(A - \lambda_3 I) \ldots (A - \lambda_n I)\right] \\ &+ C_2\left[(A - \lambda_1 I)(A - \lambda_3 I) \ldots (A - \lambda_n I)\right] \\ &\vdots \\ &+ C_n\left[(A - \lambda_1 I)(A - \lambda_2 I) \ldots (A - \lambda_{n-1} I)\right]. \end{aligned} \tag{3.62}$$

Note that the right-hand side of (3.62) is of degree $n - 1$ in A. To determine C_i, $i = 1, 2, \cdots n$, we multiply (3.62) by u_k and use $Au_k = \lambda_k u_k$ to observe that the coefficients of all C_i except C_k contains the term $\lambda_k - \lambda_k$, and hence vanish. Thus, it follows after some calculations that

$$P(A)u_k = C_k\left[(\lambda_k - \lambda_1)\cdots(\lambda_k - \lambda_{k-1})(\lambda_k - \lambda_{k+1})\cdots(\lambda_k - \lambda_n)\right]u_k, \qquad (3.63)$$

where $k = 1, 2, \ldots, n$. By Theorem 3.29, we have $P(A)u_i = P(\lambda_i)u_i$, and as a consequence, (3.63) becomes

$$P(\lambda_k)u_k = C_k\left[(\lambda_k - \lambda_1)\cdots(\lambda_k - \lambda_{k-1})(\lambda_k - \lambda_{k+1})\cdots(\lambda_k - \lambda_n)\right]u_k.$$

This yields the relation

$$C_k = \frac{P(\lambda_k)}{\prod_{r \neq k}(\lambda_k - \lambda_r)}, \quad k = 1, 2, \ldots, n \qquad (3.64)$$

where the notation $\prod$ denotes the product of those factors for which r takes on nonzero values, through n, excluding k. Substituting (3.63) and (3.64) into (3.62) we obtain

$$P(A) = \sum_{k=1}^{n} P(\lambda_k)Z_k(A), \qquad (3.65)$$

where

$$Z_k(A) = \frac{\prod_{r \neq k}(A - \lambda_r I)}{\prod_{r \neq k}(\lambda_k - \lambda_r)}. \qquad (3.66)$$

We furnish the following example.

Example 3.47 Compute A^m for positive integers m where

$$A = \begin{pmatrix} 2 & 1 \\ 1 & 2 \end{pmatrix}.$$

The eigenvalues of A are $\lambda_1 = 3$, $\lambda_2 = 1$. We are interested in calculating $P(A) = A^m$, $m > 1$. From (3.66) we have

$$Z_1(A) = \frac{(A - \lambda_2 I)}{(\lambda_1 - \lambda_2 I)} = \frac{1}{2}(A - I).$$

Similarly,

$$Z_2(A) = \frac{(A - \lambda_1 I)}{(\lambda_2 - \lambda_1 I)} = -\frac{1}{2}(A - 3I). \text{ Thus}$$

$$\begin{aligned} P(A) &= A^m = \sum_{k=1}^{2} P(\lambda_k)Z_k(A) \\ &= P(\lambda_1)Z_1(A) + P(\lambda_2)Z_2(A) \\ &= P(3)Z_1(A) + P(1)Z_2(A) \end{aligned}$$

$$= 3^m \left[\frac{1}{2}(A - I) \right] + (1)^m \left[-\frac{1}{2}(A - 3I) \right]$$

$$= \frac{3^m}{2}(A - I) - \frac{1}{2}(A - 3I).$$

Hence, $A^{100} = \frac{3^{100}}{2}(A - I) - \frac{1}{2}(A - 3I)$.   □

Next, we extend the application of Sylvester's formula to linear systems of ordinary differential equations. Recall that $e^x = \sum_{n=0}^{\infty} \frac{x^n}{n!}$ converges for all x. If A is a matrix of order n, the sum $e^A = \sum_{n=0}^{\infty} \frac{A^n}{n!}$ is a polynomial of order $n - 1$ in A. So if A has distinct eigenvalues, then we can use Sylvester's formula to calculate e^A. For simplicity, suppose A is of order two with distinct eigenvalues λ_1 and λ_2. Then,

$$Z_1(A) = \frac{(A - \lambda_2 I)}{(\lambda_1 - \lambda_2 I)}, \quad Z_2(A) = \frac{(A - \lambda_1 I)}{(\lambda_2 - \lambda_1 I)}.$$

Setting, $P(A) = e^A$, we obtain

$$
\begin{aligned}
e^A = P(A) &= \sum_{k=1}^{2} P(\lambda_k) Z_k(A) \\
&= P(\lambda_1) Z_1(A) + P(\lambda_2) Z_2(A) \\
&= e^{\lambda_1} \frac{A - \lambda_2 I}{\lambda_1 - \lambda_2} + e^{\lambda_2} \frac{A - \lambda_1 I}{\lambda_2 - \lambda_1} \\
&= \frac{1}{\lambda_1 - \lambda_2} \left[(e^{\lambda_1} - e^{\lambda_2}) A - (\lambda_2 e^{\lambda_1} - \lambda_1 e^{\lambda_2}) I \right].
\end{aligned}
$$

Note that if we replace A with At then we have

$$e^{At} = \frac{1}{\lambda_1 - \lambda_2} \left[(e^{\lambda_1 t} - e^{\lambda_2 t}) A - (\lambda_2 e^{\lambda_1 t} - \lambda_1 e^{\lambda_2 t}) I \right]. \tag{3.67}$$

We have the following definition.

Definition 3.29 *Let A be an $n \times n$ constant matrix. Then we define the exponential matrix function by $e^{A(t-t_0)}$ and is the solution of $x' = Ax$, $x(t_0) = I$ (identity matrix).*

More precisely,

$$x(t) = e^{A(t-t_0)} x_0 \tag{3.68}$$

is the unique solution of

$$x'(t) = Ax(t), \quad x(t_0) = x_0,$$

for all $t \in \mathbb{R}$.

Example 3.48 Solve for $t \geq 0$,

$$x' = \begin{pmatrix} 2 & 3 \\ 3 & 2 \end{pmatrix} \begin{pmatrix} x_1 \\ x_2 \end{pmatrix}, \quad x(0) = \begin{pmatrix} 2 \\ -3 \end{pmatrix}.$$

The matrix $A = \begin{pmatrix} 2 & 3 \\ 3 & 2 \end{pmatrix}$ has the eigenvalues $\lambda_1 = 5$, $\lambda_2 = -1$. The solution of the system is $x(t) = e^{At}x_0$. By (3.67) we have

$$
\begin{aligned}
e^{At} &= \frac{1}{6}\left[(e^{5t} - e^{-t})A - (-e^{5t} - 5e^{-t})I\right] \\
&= \frac{1}{6}\begin{pmatrix} 3e^{5t} + 3e^{-t} & 3e^{5t} - 3e^{-t} \\ 3e^{5t} - 3e^{-t} & 3e^{5t} + 3e^{-t} \end{pmatrix}.
\end{aligned}
$$

Finally the solution is given by

$$
x(t) = e^{At}x_0 = \frac{1}{6}\begin{pmatrix} 3e^{5t} + 3e^{-t} & 3e^{5t} - 3e^{-t} \\ 3e^{5t} - 3e^{-t} & 3e^{5t} + 3e^{-t} \end{pmatrix}\begin{pmatrix} 2 \\ -3 \end{pmatrix}.
$$

□

3.10.1 Exercises

Exercise 3.96 *Find the eigenvalues of A and A^5 where,*

$$
A = \begin{pmatrix} 3 & -12 & 4 \\ -1 & 0 & -2 \\ -1 & 5 & -1 \end{pmatrix}.
$$

Exercise 3.97 *Suppose A is a 3×3 symmetric matrix with eigenvalues 2 and -1 and corresponding eigenvectors $u_1 = \begin{pmatrix} 1 \\ 0 \\ -1 \end{pmatrix}$ and $u_2 = \begin{pmatrix} 2 \\ 1 \\ 0 \end{pmatrix}$. Compute A^5w, where $w = \begin{pmatrix} 7 \\ 2 \\ -3 \end{pmatrix}$.*

Exercise 3.98 *Compute A^5 and A^{-4} in Example 3.46.*

Exercise 3.99 *Verify the statement of the Cayley-Hamilton Theorem for the matrix $A = \begin{pmatrix} 1 & 2 \\ 4 & 3 \end{pmatrix}$, and then compute A^4 and A^{-3}.*

Exercise 3.100 *Verify the statement of the Cayley-Hamilton Theorem for the matrix $A = \begin{pmatrix} 1 & -2 & 4 \\ 0 & -1 & 2 \\ 2 & 0 & 3 \end{pmatrix}$, and then compute A^3 and A^{-3}.*

Exercise 3.101 *Use Sylvester's formula to compute A^{100} where,*

$$
A = \begin{pmatrix} 1 & 4 & 16 \\ 18 & 20 & 4 \\ -12 & -14 & -7 \end{pmatrix}.
$$

Exercise 3.102 *For $t \geq 0$, solve*

(a) $x' = \begin{pmatrix} 1 & 2 \\ 2 & 1 \end{pmatrix} \begin{pmatrix} x_1 \\ x_2 \end{pmatrix}, \quad x(0) = \begin{pmatrix} -2 \\ -3 \end{pmatrix}.$

(b) $x' = \begin{pmatrix} 0 & 2 & -2 \\ 0 & 1 & 0 \\ 1 & -1 & 3 \end{pmatrix} \begin{pmatrix} x_1 \\ x_2 \\ x_3 \end{pmatrix}, \quad x(0) = \begin{pmatrix} -2 \\ 2 \\ -3 \end{pmatrix}.$

Exercise 3.103 *Write the second-order differential equation*

$$y'' + 4y' + 3y = 0, \ y(0) = -1, y'(0) = 3,$$

as a system and find its solution.

4

<hr>

Calculus of Variations

This chapter is devoted to the study of the calculus of variations. The subject of calculus of variations is a wide field in mathematics that is devoted to minimizing or maximizing functionals. The calculus of variations has a rampant application in physics, engineering, and applied mathematics. In addition, the calculus of variations naturally makes its presence felt in the field of partial differential equations. In this chapter, we will consider many applications, such as distance between two points, Brachistochrone problem, surfaces of revolution, navigation, Catenary and others. The chapter covers a wide range of classical topics on the subject of the calculus of variations. Our aim is to cover the topics in a way that strikes a balance between the development of theory and applications. The chapter is suitable for advanced undergraduate and graduate students. In most sections, we limit ourselves to smooth solutions of the Euler-Lagrange equations and finding explicit solutions to classical problems. We will generalize the concept to systems and functionals that contain higher derivatives of the unknown functions. The chapter contains a long but interesting section on the sufficient conditions for the existence of an extremal.

<hr>

4.1 Introduction

Let $f : \mathbb{R} \to \mathbb{R}$ be a real valued function that is continuous. Then we know from calculus that if f has a local minimum or maximum value at an interior point c, and if $f'(c)$ exists, then

$$f'(c) = 0. \tag{4.1}$$

Condition (4.1) is a necessary condition for maximizing or minimizing the function f. Let $f(x) = x^3$. Then, $f'(0) = 0$. However, the function has neither a maximum nor minimum at $c = 0$, as the graph in Fig. 4.1 shows. This shows that condition (4.1) is not sufficient.

Before we commence on formal definitions, we must be precise when talking about maximum or minimum in the sense of distances. This brings us to the notion of a *norm*.

Definition 4.1 *(Normed spaces) Let V denote a linear space over the field $\mathbb{R}$. A functional $\|x\|$, which is defined on V is called the norm of $x \in V$, if it has the following*

DOI: 10.1201/9781003449881-4

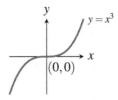

FIGURE 4.1
$f'(0) = 0$, but f has neither a maximum nor a minimum.

properties:

1. $\|x\| > 0$ *for all* $x \neq 0$, $x \in V$.

2. $\|x\| = 0$ *if* $x = 0$.

3. $\|\alpha x\| = |\alpha| \|x\|$ *for all* $x \in V$, $\alpha \in \mathbb{R}$.

4. $\|x + y\| \leq \|x\| + \|y\|$ *(triangle inequality)*

Example 4.1 The space $(\mathbb{R}^n, +, \cdot)$ over the field $\mathbb{R}$ is a vector space (with the usual vector addition, $+$ and scalar multiplication, $\cdot$) and there are many suitable norms for it. For example, if $x = (x_1, x_2, \ldots, x_n)$ then

1. $\|x\| = \max_{1 \leq i \leq n} |x_i|$,

2. $\|x\| = \sqrt{\sum_{i=1}^{n} x_i^2}$, or

3. $\|x\| = \sum_{i=1}^{n} |x_i|$,

4. $\|x\|_p = \left(\sum_{i=1}^{n} |x_i|^p \right)^{1/p}$, $\quad p \geq 1$

are all suitable norms. Norm 2. is the Euclidean norm: the norm of a vector is its Euclidean distance to the zero vector and the metric defined from this norm is the usual Euclidean metric. Norm 3. generates the "taxi-cab" metric on $\mathbb{R}^2$ and Norm 4. is the l^p norm. $\qquad \square$

Let $D \subset \mathbb{R}^n$ and define a function $f : D \to \mathbb{R}^n$. Let c be a point in the interior of D. We define a neighborhood of c by

$$N(\delta, c) = \{x : \|x - c\| < \delta\} \quad \text{for some } \delta > 0.$$

Thus, a point $c \in D$ is said to be a *relative, or local minimum point* of the function f over D if for all $x \in N(\delta, c)$ we have $f(c) \leq f(x)$. On the other hand, if $f(c) < f(x)$ for all $x \in N(\delta, c)$ with $x \neq c$, then c is said to be a *strict relative minimum point of f* over D.

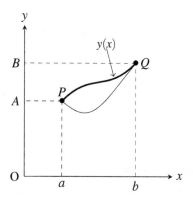

FIGURE 4.2
Shortest path between two points.

Assume the function is scalar. That is $f : \mathbb{R} \to \mathbb{R}$. Then the Taylor series expansion of f at c is

$$f(x) = f(c) + (x-c)f'(c) + \frac{1}{2}(x-c)^2 f''(c) + O((x-c)^3).$$

By making the change of variables $x = c + \varepsilon$, the above expression takes the form

$$f(c+\varepsilon) = f(c) + \varepsilon f'(c) + \frac{1}{2}\varepsilon^2 f''(c) + O(\varepsilon^3). \tag{4.2}$$

The proofs of the next two theorems are based on (4.2) and we urge the interested readers to consult any calculus textbook.

Theorem 4.1 *A necessary condition for a function f to have a relative minimum at a point c in its domain is (i) $f'(c) = 0$ and (ii) $f''(c) \geq 0$.*

Theorem 4.2 *A sufficient condition for a function f to have a strict relative minimum at a point c in its domain is (i) $f'(c) = 0$ and (ii) $f''(c) > 0$.*

Our main purpose is to extend the above discussion to the calucus of variations. Suppose we have two points $P(a,A)$ and $Q(b,B)$ in the xy-plane and we are interested in finding the shortest path between them, see Fig. 4.2. Let $f(x)$ be a candidate for being the shortest path between the two points. We know from calculus that if f' is continuous on $[a,b]$, then the length of the curve $y = f(x)$, $a \leq x \leq b$, is given by

$$L = \int_a^b \sqrt{1 + (f'(x))^2}\, dx = \int_a^b \sqrt{1 + (y')^2}\, dx. \tag{4.3}$$

Note that the integral in (4.3) is a *functional* since the integrand depends on the unknown function y. Since the right hand side of (4.3) depends on the unknown function y we write

$$L(y) = \int_a^b \sqrt{1 + (f'(x))^2}\, dx = \int_a^b \sqrt{1 + (y')^2}\, dx, \tag{4.4}$$

to emphasize that it is a functional relation. Our work now is to develop a necessary condition parallel to condition (4.1) that will enable us to compute the minimum function y so that $L(y)$ is minimized, which in turns will yield the shortest path, or distance between the two pints P and Q. Of course, the lucky function will have to satisfy the boundary conditions $y(a) = A$ and $y(b) = B$.

4.2   Euler-Lagrange Equation

In this section we develop the Euler-Lagrange equation, which is a necessary for minimizing or maximizing functionals. We begin by considering the functional or variational

$$L(y) = \int_a^b F(x, y, y')dx, \tag{4.5}$$

where $y'(x) = \frac{dy}{dx}$. We are interested in finding a particular function $y(x)$ that maximizes or minimizes (4.5) subject to the boundary conditions $y(a) = A$ and $y(b) = B$. Such a function will be called extremal of $L(y)$.

Definition 4.2 *Let* $\mathbb{S}$ *be a vector space (space that has algebraic structures under multiplication and addition). Our main problem in calculus of variations is to find* $y = y_0(x) \in \mathbb{S}[a, b]$ *for which the functional* $L(y)$ *takes an extremal value (maximum or minimum) with respect to all* $y(x) \in \mathbb{S}[a, b]$.

The set $C^k[a, b]$ denotes the set of functions that are continuous on $[a, b]$ with their k-th derivatives also being continuous on $[a, b]$. The vector space $\mathbb{S}[a, b]$ can be thought of as the space of *competing functions*. To be precise, let Σ be the set of all competing functions for the variational problem (4.5), then

$$\Sigma = \{y : y \in C^2([a, b]),\ y(a) = A,\ y(b) = B\}.$$

Note that this space is not linear because if $y, w \in \Sigma$, then $y(a) + w(a) = 2A \neq A$ unless $A = 0$. The same is true for the boundary condition at b. Next we define relative minimum and relative maximum for a functional.

Definition 4.3 *A competing function* $y_0 \in \Sigma$ *is said to yield relative minimum (maximum) for* $L(y)$ *in* Σ *if*

$$L(y) - L(y_0) \geq 0 \ (\leq 0)$$

for all

$$y \in N(y_0, \varepsilon) := \{y \in \Sigma : ||y - y_0|| < \varepsilon\}, \quad \text{for some} \quad \varepsilon > 0,$$

where $N(y_0, \varepsilon)$ *is neighborhood of* y_0.

Below, we build upon the notion of competing functions to define the so-called *space of admissible functions*.

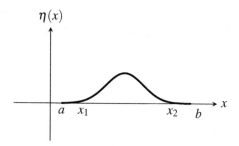

FIGURE 4.3
The function $\eta(x)$ with $a, b > 0$.

Definition 4.4 *The space of admissible functions $\mathscr{C}$ is defined as*

$$\mathscr{C} = \{\zeta : \zeta \in C^2([a,b]), \ \zeta(a) = \zeta(b) = 0\}.$$

This way, if $y_0 \in \Sigma$, then $y_0 + \eta \in \Sigma$, for $\eta \in \mathscr{C}$. Before we can obtain the Euler-Lagrange equation, we state and prove one of the most important results, which is called the *fundamental lemma of calculus of variations*.

Lemma 10 *[Fundamental lemma of calculus of variations] Assume $f(x)$ is continuous in $[a,b]$ such that*

$$\int_a^b f(x)\eta(x)dx = 0 \tag{4.6}$$

for every continuous function $\eta \in \mathscr{C}$. Then $f(x) = 0$ for all $x \in [a,b]$.

Proof *Suppose the contrary. That is, $f(x)$ is not zero over its entire domain $[a,b]$. Then, without loss of generality (w.l.o.g), let us assume it is positive for some interval $[x_1, x_2]$ that is contained in $[a,b]$. Define*

$$\eta(x) = \begin{cases} (x-x_1)^3(x_2-x)^3, & x_1 < x < x_2 \\ 0, & otherwise. \end{cases}$$

see Fig. 4.3. Then, the term $(x-x_1)^3(x_2-x)^3 > 0$, for $x \in (x_1,x_2)$. We must make sure that $\eta \in C^2([a,b])$.

$$\lim_{x \to x_1^+} \frac{\eta(x) - \eta(x_1)}{x - x_1} = \lim_{x \to x_1^+} \frac{(x-x_1)^3(x_2-x)^3 - 0}{x - x_1}$$

$$= \lim_{x \to x_1^+} (x-x_1)^2(x_2-x)^3 = 0.$$

Moreover,

$$\lim_{x \to x_1^-} \frac{\eta(x) - \eta(x_1)}{x - x_1} = \lim_{x \to x_1^-} \frac{0 - 0}{x - x_1} = 0.$$

It follows that $\eta'(x_1) = 0$, and hence η is continuously differentiable at x_1. The prove of η is continuously differentiable at x_2 follows along the same lines. Next we show the second derivative η exists at x_1.

$$\lim_{x \to x_1^+} \frac{\eta'(x) - \eta'(x_1)}{x - x_1} = \lim_{x \to x_1^+} \frac{3(x-x_1)^2(x_2-x)^2(x_2+x_1-2x) - 0}{x - x_1}$$

$$= \lim_{x \to x_1^+} 3(x-x_1)(x_2-x)^2(x_2+x_1-2x) = 0.$$

In addition,

$$\lim_{x \to x_1^-} \frac{\eta'(x) - \eta'(x_1)}{x - x_1} = \lim_{x \to x_1^-} \frac{0-0}{x - x_1} = 0.$$

Hence, $\eta''(x_1) = 0$. It follows along the lines of the previous work that $\eta''(x_2) = 0$. Thus, the second derivative of η exists and is given by

$$\eta''(x) = \begin{cases} (x-x_1)(x_2-x)\{(x-x_1)^2 + (x_2-x)^2\} \\ -3(x-x_1)(x_2-x), & x_1 < x < x_2 \\ 0, & \text{otherwise.} \end{cases}$$

It is evident that

$$\lim_{x \to x_1} \eta''(x) = \eta''(x_1) = 0,$$

and

$$\lim_{x \to x_2} \eta''(x) = \eta''(x_2) = 0.$$

This shows that $\eta \in C^2([a,b])$. To get a contradiction, we integrate $f(x)\eta(x)$ from $x = a$, to $x = b$.

$$\int_a^b f(x)\eta(x)dx = \int_a^{x_1} f(x)\eta(x)dx + \int_{x_1}^{x_2} f(x)\eta(x)dx + \int_{x_2}^b f(x)\eta(x)dx$$

$$= 0 + \int_{x_1}^{x_2} f(x)\eta(x)dx + 0$$

$$= \int_{x_1}^{x_2} f(x)(x-x_1)^3(x_2-x)^3 dx > 0,$$

which contradicts (4.6). Thus, $f(x)$ can not be non-zero anywhere in its domain $[a,b]$. We conclude that $f(x)$ is zero on its entire domain $[a,b]$. The proof of taking $f < 0$ is similar, so we omit it. This completes the proof.

Our aim is to find the path $y(x)$ that minimizes or maximizes the functional. We will consider all possible functions by adding a function $\eta(x) \in \mathscr{C}$.

Theorem 4.3 *[Euler-Lagrange equation] Assume F in (4.5) is twice differentiable with respect to its arguments. Let $y \in C^2[a,b]$ such that $y(a) = A$, and $y(b) = B$. That*

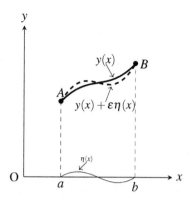

FIGURE 4.4
Possible extremal.

*is $y \in \Sigma$. Then $y = y(x)$ is an extremal function for the functional (4.5) if it satisfies
the Euler-Lagrange second-order differential equation*

$$\frac{d}{dx}\left(\frac{\partial F}{\partial y'}\right) - \frac{\partial F}{\partial y} = 0. \tag{4.7}$$

Proof *Let $\eta(x)$ be defined as in Lemma 10 and $y \in \Sigma$. For $\varepsilon > 0$ set*

$$y(x) + \varepsilon \eta(x) \in \Sigma,$$

*where y is an extremal function for the functional $L(y)$ given by (4.5). See Fig. 4.4.
In the functional $L(y)$ we replace y by $y + \varepsilon\eta$ and obtain*

$$L(\varepsilon) = \int_a^b F(x, y + \varepsilon\eta, y' + \varepsilon\eta')dx. \tag{4.8}$$

*Once y and η are assigned, then $L(\varepsilon)$ has extremum when $\varepsilon = 0$. But this possible
only when*

$$\frac{dL(\varepsilon)}{d\varepsilon} = 0 \quad when \quad \varepsilon = 0.$$

Suppress the arguments in F and compute $\frac{dL(\varepsilon)}{d\varepsilon}$.

$$\begin{aligned}
\frac{dL(\varepsilon)}{d\varepsilon} &= \frac{d}{d\varepsilon} \int_a^b F(x, y + \varepsilon\eta, y' + \varepsilon\eta')dx \\
&= \int_a^b \left[\frac{\partial F}{\partial x}\frac{dx}{d\varepsilon} + \frac{\partial F}{\partial y}\frac{d(y + \varepsilon\eta)}{d\varepsilon} + \frac{\partial F}{\partial y'}\frac{d(y' + \varepsilon\eta')}{d\varepsilon}\right]dx \\
&= \int_a^b \left[\frac{\partial F}{\partial y}\eta + \frac{\partial F}{\partial y'}\eta'\right]dx,
\end{aligned}$$

since $\dfrac{dx}{d\varepsilon} = 0$. *Setting*

$$\dfrac{dL(\varepsilon)}{d\varepsilon}\Big|_{\varepsilon=0} = 0$$

we arrive at

$$\int_a^b \left[\dfrac{\partial F}{\partial y}(x, y + \varepsilon\eta, y' + \varepsilon\eta')\eta + \dfrac{\partial F}{\partial y'}(x, y + \varepsilon\eta, y' + \varepsilon\eta')\eta'\right]dx\Big|_{\varepsilon=0} = 0.$$

Thereupon, we obtain the necessary condition

$$\int_a^b \left[\dfrac{\partial F}{\partial y}(x, y, y')\eta(x) + \dfrac{\partial F}{\partial y'}(x, y, y')\eta'(x)\right]dx = 0. \qquad (4.9)$$

We perform an integration by parts on the second term in the integrand of (4.9). Let $dv = \eta'(x)dx$ *and* $u = \frac{\partial F}{\partial y'}$. *Then*

$$v = \eta(x), \quad \text{and} \quad du = \dfrac{d}{dx}\left(\dfrac{\partial F}{\partial y'}\right)dx.$$

It follows that

$$
\begin{aligned}
\int_a^b \dfrac{\partial F}{\partial y'}\eta' dx &= \dfrac{\partial F}{\partial y'}\eta(x)\Big|_a^b - \int_a^b \dfrac{d}{dx}\left(\dfrac{\partial F}{\partial y'}\right)\eta(x)dx \\
&= \dfrac{\partial F}{\partial y'}\eta(b) - \dfrac{\partial F}{\partial y'}\eta(a) - \int_a^b \dfrac{d}{dx}\left(\dfrac{\partial F}{\partial y'}\right)\eta(x)dx \\
&= -\int_a^b \dfrac{d}{dx}\left(\dfrac{\partial F}{\partial y'}\right)\eta(x)dx,
\end{aligned}
$$

since $\eta(a) = \eta(b) = 0$. *Substituting back into (4.9) we arrive at*

$$\int_a^b \left[\dfrac{\partial F}{\partial y} - \dfrac{d}{dx}\left(\dfrac{\partial F}{\partial y'}\right)\right]\eta(x)dx = 0.$$

It follows from Lemma 10 that

$$\dfrac{d}{dx}\left(\dfrac{\partial F}{\partial y'}\right) - \dfrac{\partial F}{\partial y} = 0, \qquad (4.10)$$

for all functions $\eta(x)$. *Equation (4.10) is referred to as Euler-Lagrange equation. This completes the proof.*

Remark 14   *1. Equation (4.10) is a second-order ordinary differential equation.*

  2. The function y satisfying the Euler-Lagrange equation is a necessary, but not sufficient, condition for L(y) to be an extremum. In other words, a function y(x) may satisfy the Euler-Lagrange equation even when L(y) is not an extremum.

For simpler notations, we may write F_y, F'_y to denote $\frac{\partial F}{\partial y}$ and $\frac{\partial F}{\partial y'}$, respectively. We have the following simple examples.

Example 4.2 Find the extremal function for

$$L(y) = \int_0^1 \left((y')^2 + xy + y^2 \right) dx, \quad y(0) = 1, \ y(1) = 2.$$

Here,

$$F(x,y,y') = (y')^2 + xy + y^2, \text{ with}$$

$$F_{y'} = 2y', \ F_y = x + 2y, \quad \text{and} \quad \frac{d}{dx}F_{y'} = 2y''.$$

It follows that

$$\frac{d}{dx}F_{y'} - F_y = 2y'' - x - 2y = 0,$$

which is the second-order ODE

$$2y'' - 2y = x,$$

and can be solved using the method of Section 1.9. Thus the solution is

$$y(x) = c_1 e^x + c_2 e^{-x} - \frac{1}{2}x.$$

Using the given boundary conditions we end up with the system

$$c_1 + c_2 = 1, \ c_1 e + c_2 e^{-1} = \frac{5}{2}, \text{ with}$$

$$c_1 = \frac{2e^{-1} - 5}{2(e^{-1} - e)}, \quad c_2 = \frac{5 - 2e}{2(e^{-1} - e)}.$$

Finally, the extremal function is given by

$$y(x) = \frac{2e^{-1} - 5}{2(e^{-1} - e)}e^x + \frac{5 - 2e}{2(e^{-1} - e)}e^{-x} - \frac{1}{2}x.$$

$\square$

Example 4.3 Find the extremal function for

$$L(y) = \int_0^{\pi/4} \left((y')^2/2 - 2y^2 \right) dx, \quad y(0) = 1, \ y(\pi/4) = 2.$$

Here,

$$F(x,y,y') = (y')^2/2 - 2y^2, \text{ with}$$

$$F_{y'} = y', \ F_y = -4y, \quad \text{and} \quad \frac{d}{dx}F_{y'} = y''.$$

It follows that

$$\frac{d}{dx}F_{y'} - F_y = y'' + 4y = 0,$$

which can be solved using the method of Section 1.8 . It follows that the solution is

$$y(x) = c_1 \cos(2x) + c_2 \sin(2x).$$

Using the given boundary conditions, we arrive at $c_1 = 1$ and $c_2 = 2$. Hence, the extremal function is given by

$$y(x) = \cos(2x) + 2\sin(2x).$$

For fun, we evaluate the functional L at the extremal function. After some calculations we arrive at

$$(y')^2/2 - 2y^2 = 6\cos(4x) - 8\sin(4x).$$

As a result, we see that

$$L(\cos(2x) + 2\sin(2x)) = \int_0^{\pi/4} [6\cos(4x) - 8\sin(4x)]dx = -4.$$

□

Example 4.4 Find the extremal function for

$$L(y) = \int_1^2 \left(\frac{x^2(y')^2}{2} + y^2\right)dx, \quad y(1) = 1,\ y(2) = -1.$$

It follows that

$$\frac{d}{dx}F_{y'} - F_y = x^2 y'' + 2xy' - 2y = 0,$$

which is a Cauchy-Euler equation. Using Section 1.11, we arrive at the solution

$$y(x) = c_1 x + c_2 \frac{1}{x^2}.$$

Applying the given boundary conditions, we obtain $c_1 = -\frac{5}{7}$ and $c_2 = \frac{12}{7}$. Finally, the extremal function is given by

$$y(x) = -\frac{5}{7}x + \frac{12}{7}\frac{1}{x^2}.$$

□

Notice that the Euler-Lagrange equation depends on the nature of the function $F(x,y,y')$. Different forms, or alternate forms of (4.7) could be obtained based on the dependence of F on the variables x,y, or y'. We have the following corollaries.

Corollary 6 *If $F = F(x, y')$, that is F does not explicitly depend on the variable y, then the Euler-Lagrange equation (4.7) becomes*

$$F_{y'} = C,$$ (4.11)

where C is a constant

Proof *From (4.7), the term F_y is zero since F is independent of y. Hence we are left with*

$$\frac{d}{dx} F_{y'} = 0.$$

An integration with respect to x gives the result.

Corollary 7 *If $F = F(y, y')$, that is F does not explicitly depend on the variable x, then the Euler-Lagrange equation (4.7) is reduced to*

$$F - y' F_{y'} = C,$$ (4.12)

where C is a constant

Proof *It suffices to show that*

$$\frac{d}{dx}\left(F - y' F_{y'}\right) = 0.$$

Thus,

$$
\begin{aligned}
\frac{d}{dx}\left(F - y' F_{y'}\right) &= F_x + F_y\, y' + F_{y'}\, y'' - y'' F_{y'} - y' \frac{d}{dx} F_{y'} \\
&= F_x + F_y\, y' - y' \frac{d}{dx} F_{y'} \\
&= F_x + y'\left(F_y - \frac{d}{dx} F_{y'}\right) \\
&= F_x \ \left(\text{since } y \text{ satisfies Euler-Lagrange equation}\right) \\
&= 0 \ \left(\text{since } F \text{ does not include } x.\right)
\end{aligned}
$$

An integration with respect to x gives the result. This completes the proof.

Now we are in a good place to find the shortest distance or path between two points in the plane.

Example 4.5 Consider the functional

$$L(y) = \int_a^b \sqrt{1 + (y')^2}\, dx,$$

that was given by (4.4). Then,

$$F = \sqrt{1 + (y')^2},$$

which is independent of x, and y. Using

$$F_{y'} = C$$

that is given in Corollary 4.12, it follows that

$$\frac{y'}{\sqrt{1+(y')^2}} = C.$$

Solving for y' we end up with

$$y' = constant = K,$$

or by noticing that the left-hand of $\frac{y'}{\sqrt{1+(y')^2}} = C$ can be constant only if $y' = K$, where K is some function of C (another constant). Hence,

$$y(x) = Kx + D.$$

Using the boundary conditions $A = y(a)$ and $B = y(b)$ we obtain

$$K = \frac{B-A}{b-a}, \quad D = \frac{Ab-Ba}{b-a}.$$

Of course the shortest path is a straight line, as we have expected.   □

We make the following definition regarding smoothness of a function.

Definition 4.5 *Let $\Omega \subset \mathbb{R}^n$. Then the function $f : \Omega \to \mathbb{R}^n$ is said to be smooth on Ω if $f(x) \in C^n(\Omega)$, in the sense that $f(x)$ has n derivatives in the entire domain Ω and the nth derivative of $f(x)$ is continuous.*

For example, the function

$$f(x) = \begin{cases} 0, & x \leq 0 \\ x^2, & x > 0 \end{cases}$$

is in $C^1(\mathbb{R})$ but not in $C^2(\mathbb{R})$. Recall, Theorem 4.3 asks for $y \in C^2([a,b])$, which may not be the case in some situations. To better illustrate the requirement, we look at the next example.

Example 4.6 Consider the variational

$$L(y) = \int_1^3 (y-2)^2(x-y')^2 dx, \quad y(1) = 2, \, y(3) = \frac{9}{2}.$$

The integrand is positive and hence the variational is minimized when its value is zero at the extremal. This is achieved for

$$y(x) = \begin{cases} 2, & 1 \leq x \leq 2 \\ \frac{x^2}{2}, & 2 < x \leq 3 \end{cases}$$

Note that $y''(2)$ does not exist and hence $y \notin C^2([1,3])$, or not smooth. Nevertheless, $y = y(x)$ satisfies the corresponding Euler-Lagrange equation

$$\frac{d}{dx}[-2(y-2)^2(x-y')] - 2(y-2)(x-y')^2 = 0.$$

Thus, we have found an example in which y satisfies the corresponding Euler-Lagrange equation and yet its second derivative does not exist at every point in $[1,3]$. □

The next theorem guarantees when an extremal of a variation is indeed in $C^2([a,b])$.

Theorem 4.4 *Assume F in (4.5) is twice differentiable with respect to its arguments. Let $y \in C^1[a,b]$ and satisfies the Euler-Lagrange second-order differential equation*

$$\frac{d}{dx}\left(\frac{\partial F}{\partial y'}\right) - \frac{\partial F}{\partial y} = 0.$$

Then $y(x)$ has a continuous second derivatives at all points (x,y), where

$$F_{y'y'}(x,y(x),y'(x)) \neq 0.$$

We will further discuss $F_{y'y'}$ in the next two sections. Now we try to connect the concept of extremum of functionals with functions that we discussed in Section 4.1. Consider the variational problem (4.8). An expansion of Maclaurin series of the first term on the right hand side about ε gives

$$
\begin{aligned}
L(\varepsilon) &= L(y) + \varepsilon \int_a^b [F_y \eta + F_{y'} \eta']dx \\
&+ \int_a^b \left(F_{yy}\eta^2 + 2F_{yy'}\eta\eta' + F_{y'y'}(\eta')^2\right)dx \frac{\varepsilon^2}{2!} + O(\varepsilon^3) \\
&:= L(y) + \varepsilon\delta L(y) + \frac{\varepsilon^2}{2!}\delta^2 L(y) + O(\varepsilon^3).
\end{aligned}
$$

Let

$$L(0) = L(y), \quad L'(0) = \delta L(y), \quad \text{and} \quad L''(0) = \delta^2 L(y).$$

Then, we may write $L(\varepsilon)$ in the form

$$L(\varepsilon) = L(0) + \varepsilon L'(0) + \frac{\varepsilon^2}{2!}L''(0) + O(\varepsilon^3).$$

The terms $\delta L(y)$ and $\delta^2 L(y)$ are called the first variation and second variation, respectively, and they will be discussed in detail in Section 4.4.

Example 4.7 Consider the variational in Example 4.5 with $y(0) = 0$, and $y(1) = 3$. Then,

$$y(x) = 3x.$$

Moreover,

$$F_y = 0, \quad F_{y'} = \frac{y'}{\sqrt{1+(y')^2}}, \quad F_{yy'} = F_{y'y} = 0 \quad \text{and} \quad F_{y'y'} = (1+(y')^2)^{-3/2}.$$

Thus,

$$
\begin{aligned}
\delta L(y) &= \int_0^1 [F_y \eta + F_{y'} \eta'] dx = \int_0^1 \frac{3}{\sqrt{10}} \eta'(x) dx \\
&= \frac{3}{\sqrt{10}} (\eta(1) - \eta(0)) = 0, \text{ and}
\end{aligned}
$$

$$
\begin{aligned}
\delta^2 L(y) &= \int_0^1 \left(F_{yy} \eta^2 + 2 F_{yy'} \eta \eta' + F_{y'y'} (\eta')^2 \right) dx \\
&= \int_0^1 (10)^{-3/2} (\eta'(x))^2 dx \geq 0.
\end{aligned}
$$

$\square$

4.2.1   Exercises

Exercise 4.1 *Assume $f(x)$ is continuously differentiable in $[a,b]$ such that*

$$\int_a^b f(x)\eta'(x)dx = 0$$

for every continuous function $\eta \in C^2([a,b])$ such that $\eta(a) = \eta(b) = 0$. Show then $f(x) = $ constant for all $x \in [a,b]$.

Exercise 4.2 *Assume $f(x)$ is $C^2([a,b])$ such that*

$$\int_a^b f(x)\eta''(x)dx = 0$$

for every continuous function $\eta \in C^3([a,b])$ such that $\eta(a) = \eta(b) = 0$. Show then $f(x) = c_1 + c_2 x$ for all $x \in [a,b]$, where c_1 and c_2 are constants.

Exercise 4.3 *Assume $f(x)$ and $g(x)$ are continuous in $[a,b]$ and*

$$\int_a^b [f(x)\eta(x) + g(x)\eta'(x)]dx = 0$$

for every function $\eta \in C^1([a,b])$ such that $\eta(a) = \eta(b) = 0$. Show then $g(x)$ is differentiable and $g'(x) - f(x) = 0$ for all $x \in [a,b]$.

Exercise 4.4 *Let*

$$\eta(x) = \begin{cases} (x-\alpha)(\beta-x), & \alpha < x < \beta \\ 0, & otherwise. \end{cases}$$

Show that $\eta(x) \in C(\mathbb{R})$.

Exercise 4.5 *In Example 4.6, find all points such that*

$$F_{y'y'}(x,y(x),y'(x)) \neq 0$$

fails to hold.

Exercise 4.6 *Find the extremal function $y_0 = y_0(x)$ for*

$$L(y) = \int_1^2 x^2(y')^2 dx, \quad y(1) = 1, \ y(2) = 3.$$

Exercise 4.7 *Find the extremal function for*

$$L(y) = \int_0^1 ((y')^2 - 2yy' + y^2)dx, \quad y(0) = 0, \ y(1) = 2.$$

Exercise 4.8 *Find the extremal function for*

$$L(y) = \int_0^1 ((y')^2/2 + 4y)dx, \quad y(0) = 0, \ y(1) = 2.$$

Exercise 4.9 *Find the extremal function for*

$$L(y) = \int_0^{\pi/4} ((y')^2/2 - 3y'y - 2y^2)dx, \quad y(0) = 1, \ y(\pi/4) = -1.$$

Exercise 4.10 *Find the extremal function for*

$$L(y) = \int_0^{\pi/4} ((y')^2/2 - 3y'y - 2y^2 - xy)dx, \quad y(0) = 1, \ y(\pi/4) = -1.$$

Exercise 4.11 *Find the extremal function for*

$$L(y) = \int_1^2 (2x^2(y')^2 - y^2/2)dx, \quad y(1) = 1, \ y(2) = -1.$$

Exercise 4.12 *Find the extremal function for*

$$L(y) = \int_0^{\pi} ((y')^2 - y^2 + 4y\cos(x))dx, \quad y(0) = 0, \ y(\pi) = 0.$$

Exercise 4.13 *Find the extremal function for*

$$L(y) = \int_1^2 (x^2(y')^2 + 4xy + \frac{3}{4}y^2)dx, \quad y(1) = 1, \ y(2) = 2.$$

Exercise 4.14 *(a) Show that*

$$g_y - y'g_x - \frac{y''g}{1+(y')^2} = 0$$

is the Euler-Lagrange equation for the functional

$$L(y) = \int_a^b g(x,y)\sqrt{1+(y')^2}\, dx.$$

(b) *Find the extremal of*

$$L(y) = \int_a^b x\sqrt{1+(y')^2}\,dx, \quad y(a) = A,\ y(b) = B.$$

You don't need to find the constants of integration.

Exercise 4.15 *Show that the extremal for*

$$L(y) = \int_a^b \frac{1}{y}\sqrt{1+(y')^2}\,dx, \quad y > 0$$

is

$$(x-B)^2 + y^2 = R^2,$$

for appropriate constants B and R.

Exercise 4.16 *Find the extremal for*

$$L(y) = \int_1^2 \frac{1}{x}\sqrt{1+(y')^2}\,dx, \quad y(1) = 0,\ y(2) = 1.$$

Exercise 4.17 *(a) Show that the functional*

$$L(y) = \int_1^2 (x(y')^2 - xy + y)dx, \quad y(1) = 0,\ y(2) = 1$$

has the Euler-Lagrange equation

$$2xy'' + 2y' + x - 1 = 0.$$

(b) Use the substitution $y' = u$ to find the extremal $y(x)$.

Exercise 4.18 *(a) Find the extremal function $y_0(x)$ for*

$$L(y) = \int_0^1 ((y')^2 + 1)dx, \quad y(0) = 0,\ y(1) = 1.$$

(b) Let

$$\varphi(\varepsilon) = \int_0^1 \left[F(x, y_0 + \varepsilon\eta, y_0' + \varepsilon\eta').\right.$$

Use the y_0 from part (a) and $\eta(x) = x(1-x)$ to show

$$\varphi'(0) = \frac{d\varphi(\varepsilon)}{d\varepsilon}\bigg|_{\varepsilon=0} = 0.$$

Exercise 4.19 *Let p and q be known constants. Find the extremal $y = y(x)$ that minimizes or maximizes the functional*

$$L(y) = \int_a^b (y^2 + pyy' + q(y')^2)dx, \quad y(a) = A,\ y(b) = B,$$

by considering the three cases:

(1). q = 0,

(2). q > 0,

(3). q < 0.

In each case, explain the effect of p on the solution.

Exercise 4.20 *Find the general form of the extremal y = y(x) and show it is a relative minimum for*

$$L(y) = \int_a^b (x(y')^2 - yy' + y)dx, \quad a > b > 0.$$

Exercise 4.21 *Consider the variational*

$$L(y) = \int_0^1 xyy'dx.$$

Does it have an extremal if:

(a) y(0) = 0, y(1) = 1;

(b) y(0) = 0, y(1) = 0.

Exercise 4.22 *Consider the variational*

$$L(y) = \int_0^1 yy'dx.$$

Does it have an extremal if:

(a) y(0) = 0, y(1) = 1;

(b) y(0) = 0, y(1) = 0;

Exercise 4.23 *Find the extremal for the functional*

$$L(y) = \int_0^1 y'^2 f(x)dx, \quad y(0) = 0, \, y(1) = 1,$$

where

$$f(x) = \begin{cases} -1, & 0 \le x < \frac{1}{2} \\ 1, & \frac{1}{2} < x \le 1 \end{cases}$$

Exercise 4.24 *Display a function y(x) that minimizes the functional*

$$L(y) = \int_{-1}^1 y^2(2x - y')^2 dx, \quad y(-1) = 0, \, y(1) = 1$$

and yet y fails to have a second derivative at x = 0.

Exercise 4.25 *Consider the variational*

$$L(y) = \int_0^1 ((1 - y'^2)^2 + y^2)dx, \quad y(0) = 0, \, y(1) = 0.$$

(a) *Find the corresponding Euler-Lagrange equation.(Hard to solve and so don't bother).*

(b) *Set $y_0(x) = 0$. Clearly it satisfies both boundary conditions. Find another function $y_1(x)$ that satisfies both boundary conditions and $y_1(x) \neq 0$ for all $x \in (0,1)$.*

(c) *Compare $L(y_0)$ and $L(y_1)$, and decide which one is likely to be a "better" candidate to minimize L, and explain why.*

Exercise 4.26 *Compute*

$$\delta L(y_0) \text{ and } \delta^2 L(y_0),$$

of the variational of Example 4.17.

4.3   Impact of y' on Euler-Lagrange Equation

Consider the functional

$$L(y) = \int_a^b F(x,y,y')dx, \quad y(a) = A, \ y(b) = B.$$

So far we have encountered functionals $F = F(x,y,y')$ in which y' has entered non-linearly. This resulted in a second-order differential equations with two linearly independent solutions where the two constants are found using the provided boundary conditions. To be precise, if y' enters nonlinearly in F, then $F_{y'}$ is a function, possibly in x, y and y'. In this case we denote it by $\phi = \phi(x,y,y')$. Then

$$\frac{d}{dx}\phi = \phi_x + \phi_y\frac{dy}{dx} + \phi_{y'}\frac{dy'}{dx},$$

and the corresponding Euler-Lagrange equation becomes

$$\frac{d}{dx}F_{y'} - F_y = \phi_x + \phi_y\frac{dy}{dx} + \phi_{y'}\frac{dy'}{dx} - F_y = 0,$$

which is a second-order differential equation. On the other hand, if y' enters linearly in F, then $F_{y'}$ is a function in x and y only. Then $\phi = \phi(x,y)$ and $\frac{d}{dx}\phi = \phi_x + \phi_y\frac{dy}{dx}$, which implies that

$$\frac{d}{dx}F_{y'} - F_y = \phi_x + \phi_y\frac{dy}{dx} - F_y = 0.$$

The last expression is a first-order differential equation with its solution having only one constant to be computed, based on two boundary conditions. In most cases, such solution will not exist. To enforce this notion, we consider

$$L(y) = \int_1^2 x^2yy'dx, \quad y(1) = 1, \ y(2) = -1. \tag{4.13}$$

Then the corresponding Euler-Lagrange Equation gives $2xy = 0$. Since $x \neq 0$ we must have $y(x) = 0$, which can not satisfy any of the given boundary conditions. Boundary value problems, in general, are extremely sensitive to boundary conditions. This leads us to the next point, where $y′$ enters linearly and the accompanying Euler-Lagrange equation does not result in a well-posed differential equation. There are no constants to compute in this scenario utilizing the specified boundary conditions in order for the solution to satisfy them.

Let's examine variationals of the form

$$L(y) = \int_a^b \left(N(x,y)y′ + M(x,y) \right) dx, \quad y(a) = A, \ y(b) = B. \tag{4.14}$$

Then,

$$
\begin{aligned}
\frac{d}{dx}F_{y′} - F_y &= \frac{d}{dx}N(x,y) - \frac{\partial}{\partial y}\left(N(x,y)y′ + M(x,y) \right) \\
&= N_x + y′N_y - \left(y′N_y + M_y \right) \\
&= N_x - M_y. \text{ Thus,}
\end{aligned}
$$

$$\frac{d}{dx}F_{y′} - F_y = 0, \text{ implies}$$

$$N_x - M_y = 0. \tag{4.15}$$

Relation (4.15) is not even a differential equation, but rather a relation that, in most cases, can not satisfy both boundary conditions. For example, if $F(x,y,y) = 2xy′ + y^2$, then the corresponding Euler-Lagrange equation is $\frac{d}{dx}F_{y′} - F_y = 2 - 2y = 0$, only when $y(x) = 1$, which may not satisfy any of the given two boundary conditions. However, there is a useful result in the case $N_x = M_y$ for all x and y, that we state and prove in the following theorem.

Theorem 4.5 *[Path independent] Let $y(x) \in C^1([a,b])$ be an extremal function for the functional (4.14). If (4.15) holds for all $N_x = M_y$ then the value of L is path independent. That is, there is a function $f(x,y)$ such that*

$$L(y) = f(b,y(b)) - f(a,y(a)).$$

Proof *Suppose (4.15) holds for all x and y. Then there is a function $f(x,y)$ such that $f_y = N$ and $f_x = M$. As a consequence, we have*

$$F = N(x,y)y′ + M(x,y) = f_y\frac{dy}{dx} + f_x.$$

This is saying that $F = \dfrac{df}{dx}$. Thus

$$F\,dx = df.$$

More precisely, we have

$$\big(N(x,y)y' + M(x,y)\big)dx = df.$$

Integrating both from $x = a$ to $x = b$ we get

$$\begin{aligned}
L(y) &= \int_a^b \Big(N(x,y)y' + M(x,y)\Big)dx \\
&= \int_a^b df = f(x,y(x))\Big|_{x=a}^{x=b} \\
&= f(b,y(b)) - f(a,y(a)).
\end{aligned}$$

This shows the value of L is independent of the extremal $y = y(x)$, and so L is path independent. This completes the proof.

Example 4.8 Consider

$$L(y) = \int_1^2 \big[(x^3 + y^2)y' + 3x^2y\big]dx, \quad y(1) = 1, \ y(2) = -1.$$

The corresponding Euler-Lagrange equation is $-2yy' = 0$. Either $y(x) = 0$ or $y(x) = $ constant. Neither one satisfies both boundary conditions, as was expected. Here $N(x,y) = x^3 + y^2$ and $M(x,y) = 3x^2y$. Moreover,

$$N_x = 3x^2 = M_y$$

and so condition (4.15) is satisfied for all x and y. So there exists a function $f(x,y)$ such that $N = x^3 + y^2 = f_y$ and $M = 3x^2y = f_x$. This gives $f(x,y) = \int 3x^2y\,dx = x^3y + g(y)$, for some function g. In addition, $f_y = x^3 + g'(y) = N = x^3 + y^2$, which implies that $g'(y) = y^2$. An integration yields $g(y) = \dfrac{y^3}{3} + c$. Hence, $f(x,y) = yx^3 + \dfrac{y^3}{3} + c$. Finally, according to Theorem 4.5 we have

$$\begin{aligned}
L(y) &= \big(f(2,y(2)) + c\big) - \big(f(1,y(1)) + c\big) \\
&= f(2,-1) - f(1,1) = -\frac{29}{3}.
\end{aligned}$$

□

4.3.1 Exercises

Exercise 4.27 *Show each of the functionals is path independent and evaluate L.*

(a) $L(y) = \displaystyle\int_0^2 \big[(3y + 7)y' + (2x - 1)\big]dx, \quad y(0) = 1, \ y(2) = 0,$

(b) $L(y) = \displaystyle\int_1^2 \big[(2yx^2 + 7)y' + (2y^2x - 3)\big]dx, \quad y(1) = 1, \ y(2) = -1,$

(c) $L(y) = \int_1^2 [3xy^2 y' + (x^3 + y^3)] dx, \quad y(1) = 1, \ y(2) = 0,$

(d) $L(y) = \int_0^{\pi/2} [(x\cos(xy) + e^y)y' + (y\cos(xy) + 1)] dx, \quad y(0) = 0, \ y(\pi/2) = 1.$

Exercise 4.28 *Determine $M(x,y)$ so that the functional*

$$L(y) = \int_a^b \left[(xe^{xy} + 2xy + \frac{1}{x})y' + M(x,y)\right] dx, \quad a > b > 0$$

with fixed end points, is path independent.

Exercise 4.29 *Develop a parallel theory for the variational with fixed end points*

$$L(y) = \int_a^b [N(x,y) + y'M(x,y)] dx.$$

4.4 Necessary and Sufficient Conditions

In this section we are interested in obtaining sufficient conditions under which the variational (4.16) has a relative minimum or a relative extremum. For $\eta(x) \in \mathscr{C}$ and $y(x) \in \Sigma$, consider the functional or variational

$$L(y) = \int_a^b F(x,y,y') dx, \quad y(a) = A, \ y(b) = B. \tag{4.16}$$

Let $F = F(x,y,y')$ and replace y by $y + \varepsilon\eta(x)$ and y' by $y' + \varepsilon\eta'(x)$. Define

$$\triangle F = F(x, y + \varepsilon\eta, y' + \varepsilon\eta') - F(x,y,y'),$$

which is the change in F. An expansion of Maclaurin series of the first term on the right hand side about ε gives

$$
\begin{aligned}
F(x, y + \varepsilon\eta, y' + \varepsilon\eta') &= F(x,y,y') + (F_y\eta + F_{y'}\eta')\varepsilon \\
&+ \left(F_{yy}\eta^2 + 2F_{yy'}\eta\eta' + F_{y'y'}(\eta')^2\right)\frac{\varepsilon^2}{2!} + O(\varepsilon^3).
\end{aligned}
$$

Or,

$$\triangle F = \varepsilon(F_y\eta + F_{y'}\eta') + \frac{\varepsilon^2}{2!}\left(F_{yy}\eta^2 + 2F_{yy'}\eta\eta' + F_{y'y'}\eta'^2\right) + O(\varepsilon^3). \tag{4.17}$$

We make the following definition.

Definition 4.6 *Let $y \in \Sigma$ and $\eta \in \mathscr{C}$. If $F = F(x,y,y')$ then*
a) the first variation of F is

$$\delta F = F_y\eta + F_{y'}\eta', \tag{4.18}$$

b) the second variation of F is

$$\delta^2 F = F_{yy}\eta^2 + 2F_{yy'}\eta\,\eta' + F_{y'y'}\eta'^2. \tag{4.19}$$

In a similar way, we may obtain the first and second variations of the functional L. Let

$$\triangle L = L(y + \varepsilon\eta) - L(y),$$

which is the change in L. An expansion of Maclaurin series of the first term on the right hand side about ε gives

$$
\begin{aligned}
L(y + \varepsilon\eta) \;=\;& L(y) + \varepsilon\int_a^b [F_y\eta + F_{y'}\eta']dx \\
&+ \int_a^b \left(F_{yy}\eta^2 + 2F_{yy'}\eta\eta' + F_{y'y'}(\eta')^2\right)dx\frac{\varepsilon^2}{2!} + O(\varepsilon^3).
\end{aligned}
$$

So,

$$\triangle L(y) \;=\; \varepsilon\delta L(y) + \frac{\varepsilon^2}{2!}\delta^2 L(y) + O(\varepsilon^3), \tag{4.20}$$

where $O(\varepsilon^3)$ can be written as

$$\int_a^b (\varepsilon_1\eta^2 + \varepsilon_2\eta\eta' + \varepsilon_3\eta'^2)dx. \tag{4.21}$$

Due to the continuity of F_{yy}, $F_{yy'}$, and $F_{y'y'}$, it follows that $\varepsilon_1, \varepsilon_2, \varepsilon_3 \to 0$ as $||\eta||_1 \to 0$, where

$$||\eta||_1 = \max_{a\le x\le b} |\eta(x)| + \max_{a\le x\le b} |\eta'(x)|.$$

So we have another definition.

Definition 4.7 *Let $y \in \Sigma$, and $\eta \in \mathscr{C}$. If $F = F(x,y,y')$ then*
a)   the first variation of L at $y = y(x)$ along the direction of $\eta(x)$ is

$$\delta L(y) = \int_a^b [F_y\eta + F_{y'}\eta']dx, \tag{4.22}$$

b) the second variation of L at $y = y(x)$ along the direction of $\eta(x)$ is

$$\delta^2 L(y) = \int_a^b \left(F_{yy}\eta^2 + 2F_{yy'}\eta\eta' + F_{y'y'}(\eta')^2\right)dx. \tag{4.23}$$

Recall from calculus, if a function f has a local minimum or maximum value at an interior point x_0 of its domain and if f' is defined at x_0, then $f'(x_0) = 0$. In the next lemma we give an analogous result to functionals.

Lemma 11 *If $y \in C^1([a,b])$ is an extremal for the functional (4.16), then*

$$\delta L(y) = 0.$$

Proof *We have from (4.22) that*

$$\delta L(y) = \int_a^b [F_y \eta + F_{y'} \eta'] dx.$$

Perform an integration by parts on the second term in the integrand. Let $dv = \eta'(x)dx$ and $u = \dfrac{\partial F}{\partial y'}$. Then

$$v = \eta(x) \quad and \quad du = \frac{d}{dx}\left(\frac{\partial F}{\partial y'}\right)dx.$$

Since $\eta(b) = \eta(a) = 0$, we arrive at

$$\int_a^b [F_{y'}\eta'(x)]dx = -\int_a^b \frac{d}{dx}\left(\frac{\partial F}{\partial y'}\right)\eta(x)dx.$$

It follows that

$$\delta L(y) = \int_a^b \left[F_y - \frac{d}{dx}\left(\frac{\partial F}{\partial y'}\right)\right]\eta(x)dx = 0,$$

by Euler-Lagrange equation. This completes the proof.

Note that the variation of L is

$$L(y+\varepsilon\eta) - L(y) = \varepsilon\delta L(y) + \frac{\varepsilon^2}{2!}\delta^2 L(y) + O(\varepsilon^3),$$

and since $y = y(x)$ is an extremal of L, we have $\delta L(y) = 0$ by Lemma 11. Hence,

$$L(y+\varepsilon\eta) - L(y) = \delta^2 L(y) + O(\varepsilon^3).$$

As a direct consequence we state the following.

Theorem 4.6 *[Legendre necessary condition] Let $\delta L(y)$ and $\delta^2 L(y)$ be given by (4.22) and (4.23), respectively, for the functional defined in (4.16).*

1. *If the extremal $y = y(x)$ of L is a local minimum, then $\delta^2 L(y) \geq 0$,*

2. *Similarly, a necessary condition for the extremal $y = y(x)$ of L to be a local maximum is that $\delta^2 L(y) \leq 0$.*

3. *If $\delta^2 L(y)$ changes signs, then L can not have minima or maxima.*

For the next example we need the following inequality.

Lemma 12 *[Poincare inequality] If $f(x)$ is continuous on $[a,b]$ with $f(a) = f(b) = 0$, then*

$$\int_a^b |f(x)|^2 dx \leq \frac{(b-a)^2}{\pi^2} \int_a^b |f'(x)|^2 dx.$$

It takes some ingenuity to apply Theorem 4.6 as the next example shows.

Example 4.9 For a fixed $b > 0$ consider the functional

$$L(y) = \int_0^b \left((y')^2 - y^2 \right) dx, \quad y(0) = y(b) = 0.$$

Then, $F_{yy} = -2$, $F_{y'y} = 0$, and $F_{y'y'} = 2$. It follows that

$$\frac{1}{2}\delta^2 L(y) = \int_0^b \left((\eta'(x))^2 - \eta^2(x) \right) dx.$$

Since $\eta(0) = \eta(b) = 0$, we have from Lemma 12 that

$$\int_0^b \eta^2(x) dx \leq \frac{b^2}{\pi^2} \int_0^b (\eta'(x))^2 dx.$$

As a consequence we obtain

$$\frac{1}{2}\delta^2 L(y) \geq \left(1 - \frac{b^2}{\pi^2} \right) \int_0^b (\eta'(x))^2 dx.$$

It is evident from the above inequality that $\delta^2 L(y) \geq 0$, for all such functions η and extremal y if $b \leq \pi$. This implies that y is a candidate for minimizing L.

As for the case $b > \pi$, we carefully choose η by

$$\eta_k(x) = \sin(\frac{k\pi x}{b}), \quad k = 1, 2, \ldots.$$

It is evident that $\eta_k(0) = \eta_k(b) = 0$. A direct substitution of η and η' into $\delta^2 L(y)$ yields

$$\frac{1}{2}\delta^2 L(y) = \int_0^b \left(\frac{k^2\pi^2}{b^2}\cos^2(\frac{k\pi x}{b}) - \sin^2(\frac{k\pi x}{b}) \right) dx.$$

Using trigonometric substitutions, one can compute the definite integral and find

$$\frac{1}{2}\delta^2 L(y) = \frac{k^2\pi^2 - b^2}{2b^2}.$$

Since $b > \pi$, we have $\delta^2 L(y) < 0$ for $k = 1$, and $\delta^2 L(y) > 0$ for $k^2 > \frac{b^2}{\pi^2}$. This shows that $\delta^2 L(y)$ changes signs and therefore in this case $(b > \pi)$ the considered functional L can not have either relative minimum or relative maximum. $\qquad\square$

In practice, if the functional of interest contain all of x, y, and y' then it is difficult to study the sign of $\delta^2 L(y)$, as it was evident from Example 4.9. So, it is in our interest to explore the relation between the second variation and $F_{y'y'}$. Integrate by parts the second term in the integrand of $\delta^2 L(y)$ that is defined by (4.23). Then

$$\int_a^b 2F_{yy'}\eta(x)\eta'(x)dx = -\int_a^b \eta^2 \frac{d}{dx}F_{yy'}dx.$$

Substituting into (4.23) gives the alternate form

$$\delta^2 L(y) = \int_a^b \left[\eta^2\left(F_{yy} - \frac{d}{dx}F_{yy'}\right) + (\eta')^2 F_{y'y'}\right]dx. \tag{4.24}$$

Now one can choose η so that the sign of $(\eta')^2 F_{y'y'}$ dominates the sign of the integrand of $\delta^2 L(y)$ that is given by (4.24). In particular, in order to have $\delta^2 L(y) \geq 0$ for all η, it is necessary that $F_{y'y'} \geq 0$. As a consequence we have the following theorem.

Theorem 4.7 *[Legendre necessary condition]*

1. *If $y = y(x)$ is a local minimum of L in Σ, then*

$$F_{y'y'} \geq 0 \quad \text{for all} \quad x \in [a,b].$$

2. *If $y = y(x)$ is a local maximum of L in Σ, then*

$$F_{y'y'} \leq 0 \quad \text{for all} \quad x \in [a,b].$$

3. *If $F_{y'y'}$ changes signs, then L cannot have minima or maxima.*

Proof *We will only prove **1.** since the proof of **2.** follows along the lines. In addition, our argument here is inspired by the one given in [12] or [21]. The idea of the proof is to display a function η with $\eta(a) = \eta(b) = 0$, so that $|\eta|$ is uniformly bounded and at the same time $|\eta'|$ can be made as large as we want it. One of the logical choice of such η is in term of sine functions. We accomplish our proof by contradiction. That is, assume there is a point $x_1 \in (a,b)$ such that*

$$F_{y'y'}(x_1) := F_{y'y'}(x_1, y(x_1), y'(x_1)) < 0.$$

By the continuity of $F_{y'y'}$, we can find a number $\zeta > 0$ such that $[x_1 - \zeta, x_1 + \zeta] \subset [a,b]$ with $F_{y'y'} < \frac{F_{y'y'}(x_1)}{2}$ for all $x \in (x_1 - \zeta, x_1 + \zeta)$.

The idea is to chose η so that the term $\eta'^2 F_{y'y'}$ dominates the other terms in the integrand of $\delta^2 L(y)$. In other words, it is imperative that $F_{y'y'} \geq 0$ in order for $\delta^2 L(y) \geq 0$. Let $k > 2$ be an integer and set

$$\eta(x) = \begin{cases} \sin^{2k}\left(\frac{\pi(x-x_1)}{\zeta}\right), & x \in [x_1 - \zeta, x_1 + \zeta] \\ 0, & x \notin [x_1 - \zeta, x_1 + \zeta]. \end{cases}$$

Then

$$\eta'(x) = \begin{cases} \frac{2k\pi}{\zeta}\sin^{2k-1}(\frac{\pi(x-x_1)}{\zeta})\cos(\frac{\pi(x-x_1)}{\zeta}), & x \in [x_1 - \zeta, x_1 + \zeta] \\ 0, & x \notin [x_1 - \zeta, x_1 + \zeta]. \end{cases}$$

Observe that $\eta(x_1 - \zeta) = \eta(x_1 + \zeta) = 0$, *and* $\eta'(x_1 - \zeta) = \eta'(x_1 + \zeta) = 0$, *and* $\eta''(x_1 - \zeta) = \eta''(x_1 + \zeta) = 0$. *Since* η *is zero outside* $[x_1 - \zeta, x_1 + \zeta]$ *and* $F_{y'y'} < \frac{F_{y'y'}(x_1)}{2}$ *for all* $x \in (x_1 - \zeta, x_1 + \zeta)$, *we have that*

$$\int_a^b \eta'^2 F_{y'y'}\,dx = \int_{x_1-\zeta}^{x_1+\zeta} \eta'^2 F_{y'y'}\,dx$$

$$\leq \frac{F_{y'y'}(x_1)}{2}\frac{4k^2\pi^2}{\zeta^2}\int_{x_1-\zeta}^{x_1+\zeta}\sin^{4k-2}(\frac{\pi(x-x_1)}{\zeta})\cos^2(\frac{\pi(x-x_1)}{\zeta})dx$$

$$= \frac{F_{y'y'}(x_1)}{2}\frac{4k^2\pi^2}{\zeta^2}\frac{\zeta}{\pi}\int_{-\pi}^{\pi}\sin^{4k-2}(z)\cos^2(z)dz$$

$$\left(by\ letting\ z = \frac{\pi(x-x_1)}{\zeta}\right)$$

$$= 2k^2 p_0\pi\frac{F_{y'y'}(x_1)}{\zeta},$$

for a fixed $p_0 = \int_{-\pi}^{\pi}\sin^{4k-2}(z)\cos^2(z)dz > 0$. *Since the term*

$$\left|\eta^2\left(F_{yy} - \frac{d}{dx}F_{yy'}\right)\right|$$

is bounded independent of η, *we take* ζ *small enough so that*

$$\int_a^b \eta'^2 F_{y'y'}\,dx < 2k^2 p_0\pi\frac{F_{y'y'}(x_1)}{\zeta}$$

can be made as negative as we want, which in turns will make

$$\delta^2 L(y) = \int_a^b \left[\eta^2\left(F_{yy} - \frac{d}{dx}F_{yy'}\right) + (\eta')^2 F_{y'y'}\right]dx < 0.$$

This is a contradiction to the fact that $\delta^2 L(y) \geq 0$. *This completes the proof.*

Warning:

Be aware that we only know when $\delta^2 L(y) \geq 0$ for all functions η we see that $F_{y'y'} \geq 0$ (the reverse is not true). Also, so far we only know that if $y = y(x)$ is a relative minimum, then $\delta^2 L(y) \geq 0$ (Necessary condition).

Our ultimate goal is to have results that assure our solution is indeed the relative minimum. That is, $\delta^2 L(y) \geq 0$ for all functions η implies that $y = y(x)$ is a relative minimum of L. This will be established after the next examples.

Example 4.10 Consider the functional

$$L(y) = \int_{-1}^{1} x\sqrt{(y')^2/2 + 1}\, dx, \quad y(-1) = y(1) = 1.$$

It follows that

$$F_{y'y'} = \frac{x}{2((y')^2/2 + 1)^{3/2}},$$

which changes signs for $x \in [-1,1]$. So by Legendre's Theorem, this functional has neither a local minimum nor a local maximum. One can easily verify that $y(x) = 1$ is the only extremal. ☐

Example 3 Consider the functional

$$L(y) = \int_{0}^{1} ((y')^2/2 + y)\, dx, \quad y(0) = 0, \, y(1) = 1.$$

Here,

$$F(x,y,y') = (y')^2/2 + y.$$

Since $F_{yy} = F_{y'y} = 0$, and $F_{y'y'} = 1$, it follows that

$$\delta^2 L(y) = \int_{0}^{1} (\eta')^2(x) > 0.$$

Thus, the necessary condition for a relative minimum is met. In particular, L can not have a local maximum which requires $\delta^2 L(y) \le 0$. The equation $y = x^2/2 + x/2$ can be computed to identify a potential minimizer, and we'll show later that it does, in fact, minimize the functional. ☐

Sufficient Conditions

Our next task is to obtain conditions that are sufficient for a function y to be a relative minimum or a relative maximum for the functional L. Let $y(x)$ be an extremal of the functional (4.16). We have established that if $\delta^2 L(y) \ge 0$ for all functions η then $F_{y'y'} \ge 0$. We will be in a great shape if we can show that

$$\delta^2 L(y) \ge 0$$

for all functions η *if and only if*

$$F_{y'y'} \ge 0 \quad \text{for} \quad x \in [a,b].$$

The next lemma plays a crucial role in proving our results regarding sufficient conditions.

Lemma 13 *If $\alpha(x) > 0$, and the ordinary differential equation*

$$z' + \beta(x) - \frac{z^2}{\alpha(x)} = 0 \text{ for } x \in [a,b], \tag{4.25}$$

has a solution $z = z(x)$, then $\delta^2 L(y) > 0$, where

$$\alpha(x) = F_{y'y'}(x, y(x), y'(x)),\tag{4.26}$$

and

$$\beta(x) = F_{yy}(x, y(x), y'(x)) - \frac{d}{dx}F_{y'y}(x, y(x), y'(x)).\tag{4.27}$$

Proof *Let $y \in \Sigma$ and $\eta \in \mathscr{C}$. Remember our functional is given by (4.16). Using the terms α and β, $\delta^2 L(y)$ can be put in the simplified form*

$$\delta^2 L(y) = \int_a^b \left(\alpha(x)\eta'^2 + \beta(x)\eta^2 \right) dx.\tag{4.28}$$

Jacobi brilliantly recognized that for any continuous function $z = z(x)$, one has

$$\int_a^b (z\eta^2)' dx = 0, \quad \text{for all functions} \quad \eta \in \mathscr{C}.$$

He also observed that

$$(z\eta^2)' = 2z\eta\eta' + z'\eta^2.$$

With these two observations in mind, $\delta^2 L(y)$ given in (4.28) takes the form

$$\delta^2 L(y) = \int_a^b \left(\alpha\eta'^2 + 2z\eta\eta' + (z' + \beta)\eta^2 \right) dx.\tag{4.29}$$

We already know that if $\alpha(x) \geq 0$, $x \in [a, b]$ then $y = y(x)$ is a relative minimum. So it is safe to assume $\alpha(x) > 0$ for $x \in [a, b]$. Consider the integrand in (4.29). After some manipulations we arrive at

$$\begin{aligned}
\alpha\eta'^2 + 2z\eta\eta' + (z' + \beta)\eta^2 &= \alpha\left(\eta'^2 + 2\frac{z}{\alpha}\eta\eta' + \frac{z^2}{\alpha^2}\eta^2\right) \\
&\quad + \left(z' + \beta - \frac{z^2}{\alpha}\right)\eta^2 \\
&= \alpha\left(\eta' + \frac{z}{\alpha}\eta\right)^2 + \left(z' + \beta - \frac{z^2}{\alpha}\right)\eta^2.
\end{aligned}$$

Thus, if

$$z' + \beta(x) - \frac{z^2}{\alpha(x)} = 0,$$

has a solution z, then (4.29) reduces to

$$\delta^2 L(y) = \int_a^b \alpha(x)\left(\eta' + \frac{z}{\alpha}\eta\right)^2 dx \geq 0,$$

for any $\eta \in \mathscr{C}$. Furthermore,

$$\delta^2 L(y) = 0$$

if and only if the initial value problem

$$\eta' + \frac{z}{\alpha}\eta = 0, \quad \eta(a) = 0,$$

if and only if

$$\eta(x) = 0,$$

due to the uniqueness of the solution. This would violates the fact that $\eta(x)$ can not be zero on the whole interval $[a,b]$. This tells us that $(\eta' + \frac{z}{\alpha}\eta)^2 > 0$, and hence

$$\delta^2 L(y) = \int_a^b \alpha(x)(\eta' + \frac{z}{\alpha}\eta)^2 dx > 0.$$

This completes the proof.

The million-dollar question is, when does the differential equation given by (4.25) have a solution? We adopt the following terminology:

Definition 4.8 *The second variation $\delta^2 L(y)$ of the functional $L(y)$ is said to be positive definite if*

$$\delta^2 L(y) > 0 \quad \text{for all} \quad \eta \in \mathscr{C} \quad \text{and} \quad \eta \neq 0.$$

The results of Lemma 13 depend on the existence of a solution for the Ricatti nonlinear first-order differential equation given by (4.25). We introduce a new function $h = h(x)$ and use the transformation

$$z(x) = -\frac{\alpha(x)h'(x)}{h(x)}. \tag{4.30}$$

Then (4.25) is transformed to the Jacobi differential equation

$$\left(\alpha(x)h'\right)' - \beta(x)h = 0 \quad \text{for} \quad x \in [a,b]. \tag{4.31}$$

We already know from Chapter 1 that (4.31) has a solution defined on the whole interval $[a,b]$ as long as $\alpha(x) > 0$ and $\beta(x)$ is continuous. However, our next headache stems from the fact of inverting the transformation to go back from $z(x)$ to $h(x)$. In other words, we can not have the solution $h(x)$ of (4.31) to vanish or have zeros in $[a,b]$. The next definition regarding conjugacy plays an important role in deciding whether or not the Jacobi equation (4.31) vanishes in $[a,b]$ or not.

Definition 4.9 *Two points $x = \xi_1$ and $x = \xi_2$, $\xi_1 \neq \xi_2$, are said to be conjugate points for the Jacobi differential equation (4.31) if it has solution h such that $h \neq 0$ between ξ_1 and ξ_2, and $h(\xi_1) = h(\xi_2) = 0$.*

Notice that (4.31) has the general solution of the form

$$h(x) = c_1 h_1(x) + c_2 h_2(x)$$

where h_1 and h_2 are two linearly independent solutions on $[a,b]$.

Remark 15 *The following statements are equivalent:*

i. *There is no conjugate points to a in $(a,b]$.*

ii. *The solution $h = h(x)$ of the initial value problem*

$$\frac{d}{dx}[\alpha(x)h'(x)] - \beta(x)h(x) = 0, \quad h(a) = 0 \quad and \quad h'(a) = 1 \qquad (4.32)$$

has no zero in $(a,b]$.

Thus, if the interval $[a,b]$ contains no conjugate points, then the Jacobi equation (4.31) admits a solution h that does not vanish at any points in $[a,b]$. We have the following theorem.

Theorem 4.8 *The Jacobi equation (4.31) has a nonzero solution for all $x \in [a,b]$ if $\alpha(x) > 0$ and there are no conjugate points to a in $(a,b]$.*

The implication of Lemma 13 and Theorem 4.8 is that (4.31) will have a nonzero solution, which is a necessary condition for $\delta^2 L(y)$ to be positive definite. Thus we have the next theorem.

Theorem 4.9 *Let $y \in C^1([a,b])$ be an extremal for the functional (4.16). Suppose that $\alpha(x) > 0$ for all $x \in [a,b]$. If there are no conjugate points to a in $(a,b]$, then the second variation $\delta^2 L(y)$ is positive definite.*

Example 4.11 Consider the functional

$$L(y) = \int_0^1 ((y')^2 + y^2 - yy')dx, \quad y(0) = 0, \ y(1) = 1.$$

We compute the Jacobi equation given by (4.31). It turns out that

$$\alpha = F_{y'y'} = 2 \quad and \quad \beta = F_{yy} - \frac{d}{dx}F_{yy'} = 2.$$

Then, the Jacobi differential equation is $\left(2h'(x)\right)' - 2h = 0$, which reduces to $h''(x) - h(x) = 0$, and has the general solution $h(x) = c_1 e^x + c_2 e^{-x}$. Applying $h(0) = 0$, and $h'(0) = 1$ we arrive at

$$h(x) = \frac{1}{2}(e^x - e^{-x}),$$

which has no zeros in $(0,1]$. Thus, by Theorem 4.8 it follows that the second variation $\delta^2 L(y)$ is positive definite on $[0,1]$. $\qquad\square$

Finally we have the following result that would yield Jacobi Necessary condition along with Theorem 4.7.

Theorem 4.10 *Let $y \in C^1([a,b])$ be an extremal for the functional (4.16). Suppose that $\alpha(x) > 0$ for all $x \in [a,b]$.*

(1). *If $\delta^2 L(y)$ is positive definite, then there is no conjugate points to a in $(a,b]$.*

(2). *If $\delta^2 L(y) \geq 0$ for all $\eta \in \mathcal{C}$, then there is no conjugate points to a in (a,b).*

Note that the statement $\delta^2 L(y) \geq 0$ for all $\eta \in \mathcal{C}$, permits the possibility that $\delta^2 L(y) = 0$ for some $\eta \neq 0 \in \mathcal{C}$.

Proof *We begin by proving* **(1).** *by first showing $x = b$ can not be a conjugate point to a. We do this by contradiction. Assume b is a conjugate point to a. Then there is a function h_* depending on x such that $h_*(a) = h_*(b) = 0$ and satisfying the Jacobbi equation (4.31). That is*

$$\left(\alpha(x)h'_*\right)' - \beta(x)h_* = 0. \tag{4.33}$$

Just a reminder that

$$\alpha(x) = F_{y'y'}(x, y(x), y'(x)),$$

and

$$\beta(x) = F_{yy}(x, y(x), y'(x)) - \frac{d}{dx}F_{y'y}(x, y(x), y'(x)).$$

Multiplying (4.33) with $h_*(x)$ followed by an integration by parts on the first term in the integrand yields

$$\int_a^b h_*(x)\left(\alpha(x)h'_*(x)\right)' dx = h_*(x)\alpha(x)h'_*(x)\Big|_{x=a}^b - \int_a^b \alpha(x)h'^2_*(x)dx.$$

Then the full integral becomes

$$
\begin{aligned}
\int_a^b \left\{(\alpha(x)h'_*(x))' - \beta(x)h_*\right\}h_* dx &= h_*(x)\alpha(x)h'_*(x)\Big|_{x=a}^b \\
&\quad - \int_a^b \alpha(x)h'^2_*(x)dx - \int_a^b \beta(x)h^2_*(x)dx \\
&= -\int_a^b \left[\alpha(x)h'^2_*(x) + \beta(x)h^2_*(x)\right]dx = 0,
\end{aligned}
$$

as a consequence of $\left(\alpha(x)h'_*\right)' - \beta(x)h_* = 0$. Thus,

$$\int_a^b \left[\alpha(x)h'^2_*(x) + \beta(x)h^2_*(x)\right]dx = 0.$$

Our aim is to show the above integral is $\delta^2 L(y)$. Notice that

$$\int_a^b [\alpha(x)h'_*(x)]'dx = \int_a^b [F_{y'y'}h'_*(x)]'dx.$$

Performing an integration by parts yields

$$\int_a^b [F_{y'y'}h_*'(x)]'h_*(x)dx = h_*(x)h_*'(x)F_{y'y'}\Big|_{x=a}^b - \int_a^b F_{y'y'}h_*'^2(x)dx$$

$$= -\int_a^b F_{y'y'}h_*'^2(x)dx. \tag{4.34}$$

Since h_* is a solution of (4.33) we have

$$\int_a^b \left[\left(\alpha(x)h_*'\right)' - \beta(x)h_*\right]h_*(x)dx = 0. \tag{4.35}$$

Substituting (4.34) into (4.35) gives

$$\delta^2 L(y) = \int_a^b \left(F_{y'y'}h_*'^2(x) + \beta(x)h_*^2(x)\right)dx = 0.$$

This implies there is a nontrivial $\eta \in \mathscr{C}$ such that the second variation vanishes, contradicting the fact that $\delta^2 L(y)$ is positive definite. Hence b can not be conjugate to a. Left to show that there is no conjugate points to a in (a,b). We follow the proof given by Gelfand and Fomin ([12], p. 109). The plan is to build a family of positive definite functionals $K(\mu)$, which depend on the parameter $\mu \in [0,1]$, such that $K(1)$ is the second variation and $K(0)$ is unconstrained by conjugate points to a. This means that any solution to the Jacobi equation for K will be a continuous function of μ. This continuity is then used by to demonstrate that the absence of a conjugate point for $K(0)$ implies that for $K(\mu)$, and in particular $K(1)$. Let K represent the functional as defined by

$$K(\mu) = \mu \delta^2 L(y) + (1-\mu)\int_a^b \eta'^2(x)dx.$$

It can be easily shown that $\int_a^b \eta'^2(x)dx$ has no conjugate points in $(a,b]$. Moreover, $K(\mu)$ is positive definite for all $\mu \in [0,1]$. The Jacobi Equation associated to $K(\mu)$ is

$$(J)_\mu := \left[\left(\mu\alpha(x) + (1-\mu)\right)u'\right]' - \mu\beta(x)u = 0. \tag{4.36}$$

Every solution $u(x;\mu)$ to (4.36), however, is continuous with regard to $\mu \in [0,1]$. As a result, we may state that $u(x,\mu)$, has a continuous derivative with respect to μ for all μ in an open interval including $[0,1]$ because $\mu\alpha(x) + (1-\mu) > 0$ for all $\mu \in [0,1]$. Therefore, the solution $u(x;\mu)$ with $u(a;\mu) = 0$ and $u'(a;\mu) = 1$ depends on μ continuously for $x \in (a,b]$. Let's begin by the value $\mu = 0$. Then $(J)_0$ of (4.36) gives $u'' = 0$, with the solution

$$u(x;0) = x - a$$

which has no conjugate points in (a,b). Next we deal with $\mu = 1$, and assume the contrary. That is there is a conjugate point $c^* \in (a,b]$, that is, $u(c^*;1) = 0$. Then there is $\mu_0 \in (0,1)$ *so that corresponding solution* $u(x;\mu_0)$ *satisfies* $u(b;\mu_0) = 0$. *This implies* $(J)_{\mu_0}$ *has a nonzero solution* $u(x;\mu_0)$ *with*

$$u(a;\mu_0) = u(b;\mu_0) = 0.$$

We will try to get a contradiction by concluding $\delta^2 L(y) = 0$. *Multiply* $(J)_\mu$ *with* $u = u(x;\mu_0)$ *and then integrate the resulting equation from* $x = a$ *to* $x = b$. *After some calculations we arrive at*

$$\int_a^b \left[\left(\mu_0 \alpha(x) + (1 - \mu_0) \right) u'^2 + \mu_0 \beta(x) u^2 \right] dx = 0,$$

which is equivalent to

$$\mu_0 \delta^2 L(y) + (1 - \mu_0) \int_a^b \eta'^2(x) dx = 0$$

with $\eta(x) = u(x;\mu_0) \neq 0$ *and* $\eta \in \mathscr{C}$. *This is a contradiction to the fact that* $\delta^2 L(y) > 0$ *and* $\int_a^b \eta'^2(x) dx > 0$ *for all* $\eta \neq 0 \in \mathscr{C}$.

The proof of (2). *follows along the same lines beginning with the statement "Left to show that there is no conjugate points to a in* (a,b)*." This completes the proof.*

The next result is known as *Jacobi Necessary Condition*, which is consequential of Theorems 4.7 and 4.10.

Theorem 4.11 *[Jacobi necessary condition] Let* $y \in C^1([a,b])$ *be an extremal for the functional (4.16) with* $\alpha(x) > 0$ *for* $x \in [a,b]$. *If* $y = y(x)$ *is a local minimum, then there is no conjugate point to a in* (a,b).

Remark 16 *As we shall see next that, if* $\alpha(x) > 0$ *for* $x \in [a,b]$ *and under the strong condition that there is no conjugate point to a in* $(a,b]$, *then* $y = y(x)$ *is a local minimum. In other words, the existence of no conjugate point to a in* $(a,b]$ *is equivalent to* $\delta^2 L(y) > 0$.

In the next theorem we provide sufficient conditions for relative minimum and relative maximum.

Theorem 4.12 *[Legendre sufficient condition] Let* $y_0(x) \in C^1([a,b])$ *(smooth) be an extremal function for the functional*

$$L(y) = \int_a^b F(x,y,y') dx, \quad y(a) = A, \ y(b) = B.$$

Assume there is no conjugate point to a in $(a,b]$.

1. *If* $F_{y'y'}(x, y_0(x), y_0'(x)) > 0$, *then* $y_0(x)$ *is relative minimum for* $L(y)$.

2. *If* $F_{y'y'}(x, y_0(x), y_0'(x)) < 0$, *then* $y_0(x)$ *is relative maximum for* $L(y)$.

Proof *We follow the proof of Sagan, [21]. Let α be given by (4.26). Assume $\alpha(x) > 0$ and that the interval $[a,b]$ does not contain any conjugate points to a Then, due to the continuity of the Jacobi's equation (4.31), a bigger interval $[a, b + \varepsilon]$ exists that still has no conjugate points to a and is such that $\alpha(x) > 0$ in $[a, b + \varepsilon]$. For nonzero constant ζ, consider the variational*

$$\int_a^b \left(\alpha(x)\eta'^2 + \beta(x)\eta^2 \right) dx - \zeta^2 \int_a^b \eta'^2 dx. \tag{4.37}$$

Then the corresponding Euler-Lagrange equation of (4.37) is

$$\beta(x)\eta - \frac{d}{dx}[(\alpha - \zeta^2)]\eta'] = 0. \tag{4.38}$$

Given that $\alpha(x)$ is positive in $[a, b + \varepsilon]$ and so has a positive greatest lower bound, and that the solution to the equation (4.38) satisfying $\eta(a) = 0, \eta'(a) = 1$ depends continuously on ζ for all sufficiently small ζ, we have:

1. *$\alpha(x) - \zeta^2 > 0$, $a \leq x \leq b$;*

2. *The solution to the equation (4.38) satisfies $\eta(a) = 0$ and $\eta'(a) = 1$ does not vanish for $a \leq x \leq b$.*

Thus, by Theorem 4.8, these two conditions imply that the quadratic functional (4.37) is positive definite for all sufficiently small ζ. That is, there exists a positive constant d such that

$$\int_a^b \left(\alpha(x)\eta'^2 + \beta(x)\eta^2 \right) dx > d \int_a^b \eta'^2 dx. \tag{4.39}$$

As a consequence of (4.39), the functional or variational $L(y)$ has a minimum. In other words, if $y = y(x)$ is the extremal and $y = y(x) + \eta(x)$ is a sufficiently close neighboring curve, then from the notation of Definition 4.3, and equations (4.20) and (4.21) we have that

$$L(y + \eta) - L(y) = \int_a^b \left(\alpha(x)\eta'^2 + \beta(x)\eta^2 \right) dx + \int_a^b (\varepsilon_1 \eta^2 + \varepsilon_2 \eta'^2) dx, \tag{4.40}$$

where $\varepsilon_1(x), \varepsilon_2(x) \to 0$ as $||\eta||_1 \to 0$, uniformly for $a \leq x \leq b$. On the other hand, from Lemma 12 we see that

$$\int_a^b \eta^2(x) dx \leq \frac{(b-a)^2}{2} \int_a^b (\eta'(x))^2 dx.$$

This yields

$$\int_a^b (\varepsilon_1 \eta^2 + \varepsilon_2 \eta'^2) dx \leq \varepsilon \left(1 + \frac{(b-a)^2}{2} \right) \int_a^b (\eta'(x))^2 dx, \tag{4.41}$$

when $|\varepsilon_1(x)| \leq \varepsilon, |\varepsilon_2(x)| \leq \varepsilon$. Since we can chose $\varepsilon > 0$ arbitrarily small, it follows from (4.39) and (4.41) that

$$L(y+\eta) - L(y) = \int_a^b \left(\alpha(x)\eta'^2 + \beta(x)\eta^2 \right) dx + \int_a^b (\varepsilon_1 \eta^2 + \varepsilon_2 \eta'^2) dx > 0,$$

*for sufficiently small $\|\eta\|_1$. Therefore, we conclude that the extremal $y = y(x)$ is a relative minimum of the functional (4.16). This completes the proof of **1.** The proof of **2.** is not trivial, and it follows along the lines of the proof of **1.***

Example 4.12 Consider the functional

$$L(y) = \int_0^{\pi/2} ((y')^2 - y^2) dx, \quad y(0) = 1, \ y(\pi/2) = 0. \tag{4.42}$$

Then $y_0(x) = \cos(x)$ is an extremal for (4.42). Now,

$$F_{yy'} = 0, \quad F_{y'y'} = 2, \quad \text{and} \quad F_{yy} = -2.$$

Then, the Jacobi differential equation is

$$h''(x) + h(x) = 0,$$

has the nontrivial solution $h(x) = \sin(x)$. Clearly, $h(0) = 0$, and there are no other points $a^* \in (0, \pi/2]$ such that $h(a^*) = 0$. Therefore, the interval $[0, \pi/2]$ admits no conjugate points. More over, the Legendre condition

$$F_{y'y'}(x, y_0(x), y_0'(x)) = 2 > 0,$$

and by **1.** of Theorem 4.12, $y_0(x) = \cos(x)$ minimizes or a relative minimum of the functional (4.42). Note that if we make use of the equivalent condition for conjugacy **ii.** of Remark 15, we see that $h(x) = \sin(x)$ solves the Jacobi differential equation and it satisfies $h(0) = 0, h'(0) = 1$ and $h(x)$ does not vanish at any other points in the interval $(0, \pi/2]$. Therefore there are no conjugate points. □

Example 4.13 Show the extremal $y(x) = 6(1 - \frac{1}{x})$ is a relative minimum for the functional

$$L(y) = \int_1^2 x^2(y')^2 dx, \quad y(1) = 0, \ y(2) = 3.$$

We have $F_{y'} = 2x^2 y'$, $\frac{d}{dx}F_{y'} = 4xy' + 2x^2 y''$, and $F_y = 0$. This yields the Euler-Lagrange equation

$$x^2 y'' + 2xy' = 0,$$

which has the solution

$$y(x) = 6(1 - \frac{1}{x}),$$

by the method of Section 1.11. Next we compute the Jacobi equation. It turns out that

$$\alpha = F_{y'y'} = 2x^2, \text{ and } \beta = F_{yy} - \frac{d}{dx}F_{yy'} = 0.$$

Then, by (4.32) the Jacobi differential equation is

$$\left(2x^2h'(x)\right)' = 0.$$

A direct integration leads to $h(x) = -\dfrac{c_1}{2x} + c_2$. Applying $h(1) = 0$ and $h'(1) = 1$, we arrive at

$$h(x) = -\frac{1}{x} + 1,$$

which has no zeros in $(1,2]$. Finally

$$F_{y'y'} = 2x^2 > 0 \quad \text{for all} \quad x \in [1,2].$$

We conclude by Theorem 4.12, the extremal $y(x) = 6(1 - \frac{1}{x})$ is a relative minimum for the functional.   □

Remark 17 *In Theorem 4.12, if $y_0(x)$ is a solution of the Euler-Lagrange equation and*

$$F_{y'y'}(x, y_0(x), y_0'(x)) = 0,$$

then $y_0(x)$ is neither a maximum nor a minimum and it is said to be a saddle path.

It is unclear if extremals with conjugate points may be categorized in a way that is similar to how saddle points are classified in finite dimensions. Although many physical applications of such a classification may be of little interest, it turns out that it is unquestionably a lucrative area of research in topology and differential geometry. The Calculus of Variations in the Large, a large area of study invented by M. Morse [16], is founded on the classification of extremals with conjugate points. Morse theory is outside the scope of this book, and we urge the interested reader to consult the reference [16].

4.4.1 Exercises

Exercise 4.30 *Find the extremal function for*

$$L(y) = \int_0^1 \left((y')^2 + y^2 + 2ye^x\right)dx, \quad y(0) = 0, \ y(1) = 1$$

and show it minimizes the functional L.

Exercise 4.31 *Find the extremal function for*

$$L(y) = \int_0^{\pi/4} \left((y')^2/2 - 4y\right)dx, \quad y(0) = 0, \ y(\pi/4) = 1$$

and show it minimizes the functional L.

Exercise 4.32 *Find the extremal function for*

$$L(y) = \int_1^2 \left(x^2(y')^2 + y'\right)dx, \quad y(1) = 1, \ y(2) = 3$$

and show it minimizes the functional L.

Exercise 4.33 *Find the extremal function for*

$$L(y) = \int_0^1 \sqrt{1 + (y')^2}\, dx, \quad y(0) = 0,\ y(1) = 1$$

and show it minimizes the functional L.

Exercise 4.34 *Find the extremal $y = y(x)$ and show it minimizes*

$$L(y) = \int_1^2 x^3 (y')^2 dx, \quad y(1) = 0,\ y(2) = 1.$$

Exercise 4.35 *Let $g(x)$ be continuous and positive on the interval $[a, b]$ with $a > b > 0$. Show that if $y = y(x)$ is an extremal for the functional*

$$L(y) = \int_a^b g(x)(y')^2 dx,$$

with fixed end points, then it minimizes the functional.

Exercise 4.36 *Show that if $y = y(x)$ is an extremal for the functional*

$$L(y) = \int_0^\pi \left(y\sin(x) - (y')^2 + 2yy' + 1 \right) dx,$$

with fixed end points, then it is a relative maximum.

4.5   Applications

This section is devoted to the application of calculus of variations. We will look into familiar problems in physics such as *minimal surface, geodesics* on sphere, and the *histochrone problem*.

Minimal surface area

Suppose we have a curve y given by $y = f(x)$ that is continuous on $[a, b]$. For simplicity, we assume $f(x) > 0$ on $[a, b]$. The goal is to find the curve passing thorough the points $P(a, A)$ and $Q(b, B)$ which when rotated about the x-axis gives a minimum surface area. This is depicted in Fig. 4.5.

Let ds be the arc length of PQ. Then at any point on the curve, ds rotates through a distance $2\pi y$ around the x-axis. Hence the sectional area is $2\pi y ds = 2\pi y \dfrac{ds}{dx} dx$. Therefore, the total surface area is $\int_a^b 2\pi y \dfrac{ds}{dx} dx = \int_a^b 2\pi y \sqrt{1 + y'^2} dx$. We must minimize the functional

$$L(y) = \int_a^b 2\pi y \sqrt{1 + y'^2}\, dx, \quad y(a) = A,\ y(b) = B.$$

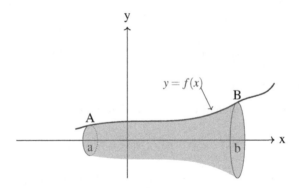

FIGURE 4.5
Surface of revolution; minimal surface area.

Since $F = y\sqrt{1+y'^2}$ is independent of x, we use

$$F - y'F_{y'} = C.$$

After simple calculations, it follows that

$$y\sqrt{1+y'^2} - y'\frac{yy'}{\sqrt{1+y'^2}} = C,$$

which simplifies to

$$\frac{y}{\sqrt{1+y'^2}} = C.$$

Solving for y' we arrive at $y' = \frac{1}{C}\sqrt{y^2 - C^2}$, and as a consequence, we are to solve

$$C\frac{dy}{\sqrt{y^2 - C^2}} = dx.$$

Let $y = C\cosh(t)$. Using the identity $\cosh^2(u) - \sinh^2(u) = 1$, and $\frac{dy}{dt} = C\sinh(t)$ we have

$$C\int \frac{dy}{\sqrt{y^2 - C^2}} = C\int \frac{C\sinh(t)}{C\sinh(t)} dt = Ct.$$

But $\frac{y}{C} = \cosh(t)$, which implies that $t = \cosh^{-1}(\frac{y}{C})$. Thus after integrating both sides we end up with

$$C\cosh^{-1}(\frac{y}{C}) = x + K,$$

or,

$$\cosh^{-1}(\frac{y}{C}) = \frac{x}{C} + K.$$

Taking cosine hyperbolic inverse on both sides leads to

$$y = C\cosh(\frac{x}{C} + K),$$

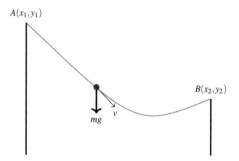

FIGURE 4.6
Brachistochrone curve.

where the constants C and K can be found using the boundary conditions. The graph of the solution represents catenary.

In engineering, catenaries are frequently used in designing bridges, roofs and arches.

Brachistochrone curve

A brachistochrone curve, also known as a curve of fastest descent in physics and mathematics, is the curve on a plane between a point A and a lower point B, where B is not directly below A, on which a bead slides frictionlessly under the influence of a uniform gravitational field to a given end point in the shortest amount of time. Johann Bernoulli posed the issue in 1696, asking: "Given two points A and B in a vertical plane, what is the curve sketched out by a point acting only under the influence of gravity, which starts at A and reaches B in the shortest time?" For the mathematical set up, we assume a mass m with initial velocity zero slides with no friction under the force of gravity g from a point $A(x_1, y_1)$ to a point $B(x_2, y_2)$ along a wire defined by a curve $y = f(x)$ in the xy-plane $(x_1 < x_2, y_1 > y_2)$. Which curve leads to the *fastest time* of descent? See Fig. 4.6.

A variational problem can be formulated by computing the time of descent t for a fixed curve connecting the points A and B. Let s denotes the distance traveled and $v = v(t)$ represents the velocity. Then $v = \dfrac{ds}{dt}$, which implies that $dt = \dfrac{ds}{v}$. The arc length ds of AB is $ds = \sqrt{1 + y'^2}$. To obtain an expression for v we use the fact that energy is conserved through the motion; that is

(kinetic energy at $t > 0$) + (potential energy at $t > 0$) = (kinetic energy at $t = 0$) + (potential energy at $t = 0$). This translate into

$$\frac{1}{2}mv^2 + mgy = 0 + mgy_1. \tag{4.43}$$

Solving for v we get

$$v = \sqrt{2g(y_1 - y(x))}.$$

Using the obtained values of ds and v in $dt = \dfrac{ds}{v}$ gives

$$dt = \frac{\sqrt{1+y'^2}}{\sqrt{2g(y_1 - y(x))}} dx.$$

Integrating both sides from $x = x_1$ to $x = x_2$ we obtain the total time of descent

$$t = \int_{x_1}^{x_2} \frac{\sqrt{1+y'^2}}{\sqrt{2g(y_1 - y(x))}} dx.$$

Thus our problem is to minimize the functional or variational

$$L(y) = \int_{x_1}^{x_2} \frac{\sqrt{1+y'^2}}{\sqrt{2g(y_1 - y(x))}} dx, \quad y(x_1) = y_1, \; y(x_2) = y_2. \qquad (4.44)$$

Notice that F is independent of x, and so we make use of the necessary Euler-equation

$$F - y' F_{y'} = C.$$

After some calculations we arrive at

$$\frac{\sqrt{1+y'^2}}{\sqrt{y_1 - y}} - y'^2 \frac{(1+y'^2)^{-1/2}}{\sqrt{y_1 - y}} = C,$$

which reduces to

$$\left(\frac{dy}{dx}\right)^2 = \frac{1 - C^2(y_1 - y)}{C^2(y_1 - y)}.$$

Solving for $\dfrac{dy}{dx}$ and separating the variables, it follows that

$$dx = -\frac{\sqrt{y_1 - y)}}{\sqrt{C_1 - (y_1 - y)}} dy, \quad C_1 = C^{-2}.$$

The negative sign is due to the fact that $\dfrac{dy}{dx} < 0$. Integrating both sides and using the transformation $y_1 - y = C_1 \sin^2(\varphi/2)$, we obtain $x = C_1/2\left(\varphi - \sin(\varphi)\right) + C_2$. The solution is then

$$y_1 - y = C_1 \sin^2(\varphi/2), \quad x = C_1/2\left(\varphi - \sin(\varphi)\right) + C_2,$$

which is the parametrization of a *cycloid*.

Great circle: Geodesic

In this problem, we are interested in finding the shortest path between two points on a sphere with radius $a > 0$. It turned out that the shortest surface path between them is an arc of a great circle between the two points. This problem is analogous to finding the shortest distance between two points in a plane. Next, we formulate the problem into a variational equation and find its solution. Let $a > 0$ and consider the sphere centered at the origin with radius a,

$$x^2 + y^2 + z^2 = a^2.$$

We will use *spherical coordinates*

$$x = r\sin(\theta)\cos(\phi), \quad y = r\sin(\theta)\sin(\phi), \quad z = r\cos(\theta),$$

where θ is the angle from the positive z-axis, ϕ is the angle from the positive x-axis and $r = a$ is constant. By the chain rules we have

$$dx = \frac{\partial x}{\partial \theta}d\theta + \frac{\partial x}{\partial \phi}d\phi,$$

$$dy = \frac{\partial y}{\partial \theta}d\theta + \frac{\partial y}{\partial \phi}d\phi,$$

and

$$dz = \frac{\partial z}{\partial \theta}d\theta + \frac{\partial z}{\partial \phi}d\phi.$$

As a consequence, we arrive at

$$dx = r\cos(\theta)\cos(\phi)d\theta - r\sin(\theta)\sin(\phi)d\phi,$$

$$dy = r\cos(\theta)\sin(\phi)d\theta + r\sin(\theta)\cos(\phi)d\phi,$$

and

$$dz = -r\sin(\theta)d\theta.$$

Using the identity $\cos^2(u) + \sin^2(u) = 1$, and after some calculations, we arrive at

$$
\begin{aligned}
(dx)^2 + (dy)^2 + (dz)^2 &= a^2\left((d\theta)^2 + \sin^2(\theta)(d\phi)^2\right)\\
&= a^2\left(1 + \sin^2(\theta)(\frac{d\phi}{d\theta})^2\right)(d\theta)^2.
\end{aligned}
$$

Let $P(a, \theta_1, \phi_1)$ and $Q(a, \theta_2, \phi_2)$ be any two points on the sphere. Then the arc length ds between the two points is given by

$$ds = \sqrt{(dx)^2 + (dy)^2 + (dz)^2} = a\sqrt{\left(1 + \sin^2(\theta)(\frac{d\phi}{d\theta})^2\right)}\, d\theta.$$

Knowing that $s = \int_P^Q ds$, we arrive at

$$s = a\int_{\theta_1}^{\theta_2}\sqrt{1 + \sin^2(\theta)(\frac{d\phi}{d\theta})^2}\, d\theta.$$

Setting $x = \theta$ and $y = \phi$. Then, $dx = d\theta$ and $y' = \dfrac{dy}{dx} = \dfrac{d\phi}{d\theta}$. Thus, the problem reduces to minimizing the functional

$$L(y) = \int_{x_1}^{x_2} \sqrt{1 + \sin^2(x)(y')^2} \; dx, \quad y(x_1) = y_1, y(x_2) = y_2.$$

Since $F = \sqrt{1 + \sin^2(x)(y')^2}$ is independent of y we use alternate form of Euler-Lagrange equation $(F_{y'})_x = 0$, which implies $F_{y'} = c$, for constant c. It can be obtained that

$$F_{y'} = \frac{y' \sin^2(x)}{\sqrt{1 + \sin^2(x)(y')^2}} = c.$$

Solving for y' we see that

$$y' = \frac{c \csc^2(x)}{\sqrt{1 - c^2(1 + \cot^2(x))}}.$$

Separating the variables and then integrating both sides yiels

$$y = \int \frac{c \csc^2(x)}{\sqrt{1 - c^2(1 + \cot^2(x))}} + \text{constant.}$$

Let $u = c \cot(x)$. Then $du = -c \csc^2(x)dx$ and the above integral reduces to

$$y = -\int \frac{du}{\sqrt{1 - c^2 - u^2}} = \cos^{-1}\left(\frac{u}{\sqrt{1 - c^2}}\right) + d,$$

for some constant d. This implies that

$$\frac{u}{\sqrt{1 - c^2}} = \cos(y - d),$$

or

$$c \cot(x) = \sqrt{1 - c^2} \cos(y - d).$$

Finally, replacing x by θ and y by ϕ, leads to the solution

$$c \cot(\theta) = \sqrt{1 - c^2} \cos(\phi - d).$$

Next we try to make some sense out of this solution. Multiply both sides by $a \sin(\theta)$, where a is the radius of the sphere and at the same time use $\cos(u - v) = \cos(u)\cos(v) + \sin(u)\sin(v)$ to get

$$ca \cos(\theta) = \sqrt{1 - c^2}\left(a \cos(d) \sin(\theta) \cos(\phi) + a \sin(d) \sin(\theta) \sin(\phi)\right).$$

Recall that, we are in spherical coordinates, and so the above equation takes the form in rectangular coordinates

$$cz = \sqrt{1 - c^2}\left(\cos(d)x + \sin(d)y\right), \quad c^2 \in (0, 1)$$

which represents an equation of the plane that intersects the sphere. Since the plane passes through the centre of the sphere, which is the origin, the section of the sphere by the plane is the great circle, or geodesic. All sections of other planes are small circles. This great circle has two arcs between P and Q; the major arc, and the minor has the minimum length. This is the geodesic on the surface of a sphere. Recall, a geodesic on a given surface is a curve lying on that surface along which distance between two points is as small as possible.

4.5.1 Exercises

Exercise 4.37 *Show that the shortest path between two points on a circular cylinder is along the circular helix joining them. Assume the two points are not on a generator. Hint: use cylindrical coordinates to parametrize the circular cylinder $x^2 + y^2 = a^2$. Let $P(a, \theta_1, z_1)$ and $Q(a, \theta_2, z_2)$. Compute ds and then integrate to obtain the variational that needs to be minimized.*

Hint: Let $x = a\cos(\theta), y = a\sin(\theta), z = z(\theta)$. Show $ds = \sqrt{a^2 + [z'(\theta)]^2} d\theta$.

Exercise 4.38 *Find the geodesics on a right circular cone. Use spherical coordinates*

$$x = u\sin(\alpha)\cos(v), \ y = u\sin(\alpha)\sin(v), \ z = u\cos(v),$$

to show $ds = \sqrt{1 + u^2 \sin^2(\alpha)(v')^2} \, du$ and minimize $\int \sqrt{1 + u^2 \sin^2(\alpha)(v')^2} \, du$, where α is the apex angle. If you replace u with x and v with y, then you are to minimize

$$L(y) = \int \sqrt{1 + x^2 \sin^2(\alpha)(y')^2} \, dx.$$

Exercise 4.39 *[Hanging chain] Let $y = y(x)$ be the curve configuration of a uniform inextensible heavy chain hanging from two fixed points $P(a, A)$ and $Q(b, B)$ at rest in a constant gravitational field. For mathematical convenience assume the rope density and gravity are both one. Show that the shape of the curve y is a catenary.*

Answer: $y(x) = c\cosh\left(\frac{x-d}{c}\right)$, *for constants c and d.*

Exercise 4.40 *[Minimal surface] Consider the solution of the Minimal surface problem*

$$y = C\cosh(\frac{x}{C} + K).$$

Show that under the boundary conditions

$$y(-\frac{L}{2}) = y(\frac{L}{2}) = 1,$$

the constant $K = 0$.

Hint: Make use of the identities $\cosh(x+y)$ and $\cosh(x-y)$.

4.6   Generalization of Euler-Lagrange Equation

In this section, we extend the development of Euler-Lagrange equations to variational with *higher-order derivatives* and variational involving *several variables*.

Generalizations to variational with higher-order derivatives

Let $y = y(x) \in C^4[a,b]$ and consider the variational with second-order derivative and given boundary conditions

$$L(y) = \int_a^b F(x,y,y',y'')dx \qquad (4.45)$$

$$y(a) = A_1, \quad y'(a) = A_2, \quad y(b) = B_1, \quad y'(b) = B_2.$$

Let $\eta = \eta(x) \in C^4[a,b]$ satisfying

$$\eta(a) = \eta'(a) = \eta(b) = \eta'(b) = 0.$$

We follow the same development as in Section 4.2. For $\varepsilon > 0$, set

$$y(x) + \varepsilon\eta(x),$$

where y is an extremal function for the functional $L(y)$ given by (4.45). In the functional $L(y)$ replace y by $y + \varepsilon\eta$ to arrive at

$$L(\varepsilon) = \int_a^b F(x,y+\varepsilon\eta,y'+\varepsilon\eta',y''+\varepsilon\eta'')dx.$$

Once y and η are assigned, then $L(\varepsilon)$ has extremum when $\varepsilon = 0$. But this possible only when

$$\frac{dL(\varepsilon)}{d\varepsilon} = 0 \quad \text{when} \quad \varepsilon = 0.$$

Suppress the arguments in F and compute $\frac{dL(\varepsilon)}{d\varepsilon}$ and notice that since $\frac{dx}{d\varepsilon} = 0$

$$
\begin{aligned}
\frac{dL(\varepsilon)}{d\varepsilon} &= \frac{\partial}{\partial\varepsilon}\int_a^b F(x,y+\varepsilon\eta,y'+\varepsilon\eta',y''+\varepsilon\eta'')dx\Big|_{\varepsilon=0}\\
&= \int_a^b \left[F_y\eta + F_{y'}\eta' + F_{y''}\eta''\right]dx.
\end{aligned}
$$

We perform an integration by parts on the second and third terms in the integrand. Let $dv = \eta'(x)dx$, and $u = \frac{\partial F}{\partial y'}$. Then

$$v = \eta(x) \quad \text{and} \quad du = \frac{d}{dx}\left(\frac{\partial F}{\partial y'}\right)dx.$$

It follows that

$$\int_a^b F_{y'} \eta' dx = -\int_a^b \frac{d}{dx} F_{y'} \eta(x) dx,$$

since $\eta(a) = \eta(b) = 0$. Performing integration by parts twice on the third term gives

$$\int_a^b F_{y''} \eta'' dx = \int_a^b \frac{d^2}{dx^2} F_{y''} \eta dx.$$

Consequently, we have

$$\int_a^b \left[F_y - \frac{d}{dx} F_{y'} + \frac{d^2}{dx^2} F_{y''} \right] \eta(x) dx = 0.$$

It follows from Lemma 10 that

$$F_y - \frac{d}{dx} F_{y'} + \frac{d^2}{dx^2} F_{y''} = 0. \tag{4.46}$$

for all functions $\eta(x)$. Equation (4.46) is referred to as Euler-Lagrange equation.

Remark 18 *1. Equation (4.46) is a fourth order ordinary differential equation.*

2. The function y satisfying the Euler-Lagrange equation is a necessary, but not sufficient, condition for L(y) to be an extremum. In other words, a function y(x) may satisfy the Euler-Lagrange equation even when L(y) is not an extremum.

We have the following theorem.

Theorem 4.13 *[Euler-Lagrange equation] If a function $y = y(x) \in C^4([a,b])$ is an extremal to the variational problem in (4.45), then y(x) must satisfy the Euler-Lagrange equation*

$$F_y - \frac{d}{dx} F_{y'} + \frac{d^2}{dx^2} F_{y''} = 0.$$

The prove of the results in Remark 19 are left as an exercise.

Remark 19 *Let $y = y(x) \in C^4[a,b]$ be an extremal of (4.45).*

(a) If F does not contain y, then the respective necessary Euler-Lagrange equation reduces to

$$\frac{d}{dx} F_{y''} - F_{y'} = constant. \tag{4.47}$$

(b) If F does not explicitly contain x, then the corresponding required Euler-Lagrange equation becomes

$$y'' F_{y''} - y' \left(\frac{d}{dx} F_{y''} - F_{y'} \right) - F = constant. \tag{4.48}$$

The aforementioned findings are easily generalized to functionals with *nth* order derivatives. Let $y = y(x) \in C^n[a,b]$ and consider the variational with *nth* order derivatives

$$L(y) = \int_a^b F(x,y,y',y'',y''',\ldots,y^{(n-1)},y^{(n)})dx$$

and boundary conditions

$$y(a) = A_1, \quad y'(a) = A_2, \quad \ldots \quad ,y^{(n-1)}(a) = A_n,$$

$$y(b) = B_1, \quad y'(b) = B_2, \quad \ldots \quad ,y^{(n-1)}(b) = B_n.$$

Then it can be easily shown that $y(x)$ satisfies the necessary Euler-Lagrange equation

$$F_y - \frac{d}{dx}F_{y'} + \frac{d^2}{dx^2}F_{y''} + \ldots + (-1)^n\frac{d^n}{dx^n}F_{y^{(n)}} = 0.$$

Example 4.14 Find the extremal $y = y(x)$ for the functional

$$L(y) = \int_0^1 (x + y'^2 + (y'')^2)dx$$

subject to

$$y(0) = 0, \quad y'(0) = 1, \quad y(1) = -1, \quad y'(1) = 2.$$

The corresponding necessary Euler-Legandre condition is

$$y^{(4)} - y'' = 0.$$

Using the method of Section 1.8, we obtain the general solution

$$y(x) = c_1 + c_2 x + c_3 e^x + c_4 e^{-x}.$$

Applying the given boundary conditions yields

$$c_1 = \frac{1 + e(2 - e) - 2e^{-1}}{e - e^{-1}}, \quad c_2 = \frac{e(e - 1) + e^{-1} - 1}{e - e^{-1}},$$

$$c_3 = \frac{3 - 2e^{-1}}{e - e^{-1}}, \quad c_4 = \frac{e + 2(e^{-1} - 1)}{1 - e^{-2}}.$$

□

Generalizations to variational involving several variables.

Let $y, z \in C^2[a,b]$, and consider the variational with two variables y and z

$$L(y,z) = \int_a^b F(x,y,y',z,z')dx, \tag{4.49}$$

with boundary conditions

$$y(a) = A_1, \quad y(b) = B_1, \quad z(a) = A_2, \quad z(b) = B_2.$$

Let $\eta_1 = \eta_1(x) \in C([a,b])$ and $\eta_2 = \eta_2(x) \in C([a,b])$, such that

$$\eta_1(a) = \eta_1(b) = \eta_2(a) = \eta_2(b) = 0.$$

By imitating the derivation of previous work we arrive at

$$
\begin{aligned}
\frac{dL(\varepsilon)}{d\varepsilon} &= \frac{\partial}{\partial \varepsilon} \int_a^b F(x, y + \varepsilon\eta_1, y' + \varepsilon\eta', z + \varepsilon\eta_2, z' + \varepsilon\eta_2') dx \Big|_{\varepsilon=0} \\
&= \int_a^b \left[F_y \eta_1 + F_{y'} \eta_1' + F_z \eta_2 + F_{z'} \eta_2' \right] dx.
\end{aligned}
$$

since $\frac{dx}{d\varepsilon} = 0$. Setting $\frac{dL(\varepsilon)}{d\varepsilon}\big|_{\varepsilon=0}$ and integrating by parts the terms that involves η_1' and η_2' we arrive at

$$\int_a^b \left[\left(F_y - \frac{d}{dx} F_{y'} \right) \eta_1(x) dx + \left(F_z - \frac{d}{dx} F_{z'} \right) \eta_2(x) dx \right] = 0,$$

that must hold for all $\eta_1(x), \eta_2(x)$. So without loss of generality, we assume it holds for $\eta_2(x) = 0$. Then, we have

$$\int_a^b \left(F_y - \frac{d}{dx} F_{y'} \right) \eta_1(x) dx = 0$$

and by Lemma 10, we arrive at

$$F_y - \frac{d}{dx} F_{y'} = 0.$$

Substituting this back into the above integral gives

$$\int_a^b \left(F_z - \frac{d}{dx} F_{z'} \right) \eta_2(x) dx = 0,$$

and by Lemma 10, we see that $F_z - \frac{d}{dx} F_{z'} = 0$. As a consequence, we state the following theorem.

Theorem 4.14 *[Euler-Lagrange equation] If the functions $y = y(x), z = z(x)$ are extremal of the variational problem in (4.49), then $y(x), z(x)$ must satisfy the the pair of Euler-Lagrange equations*

$$F_y - \frac{d}{dx} F_{y'} = 0, \quad F_z - \frac{d}{dx} F_{z'} = 0.$$

Again, the above discussion can be generalized to a variational with n variable functions. To see this, we assume each of $y_i = y_i(x) \in C([a,b])$, $i = 1,2,\ldots n$ is an extremal for the variational

$$L(y_1, y_2, \ldots, y_n) = \int_a^b F(x, y_1, y_2, \ldots, y_n, y_1', y_2', \ldots, y_n') dx,$$

with

$$y_i(a) = A_i, \quad y_i(b) = B_i, \quad i = 1, 2, \ldots, n.$$

Then each of $y_i = y_i(x)$, $i = 1, 2, \ldots n$ must satisfy the necessary Euler-Lagrange equation

$$F_{y_i} - \frac{d}{dx} F_{y_i'} = 0, \quad i = 1, 2, \ldots n.$$

Example 4.15 Consider the functional

$$L(y, z) = \int_0^{\pi/2} (x + y^2 - z^2 + y'^2 + z'^2) dx$$

with boundary conditions

$$y(0) = 1, \quad y(\pi/2) = 2, \quad z(0) = -1, \quad z(\pi/2) = 4.$$

The corresponding pairs of Euler-Lagrange equations given in Theorem 4.14 are

$$y'' - y = 0, \quad z'' + z = 0,$$

with the general solutions

$$y(x) = c_1 e^x + c_2 e^{-x}, \quad z(x) = c_3 \sin(x) + c_4 \cos(x),$$

where

$$c_1 = \frac{2 - e^{-\pi/2}}{e^{\pi/2} - e^{-\pi/2}}, \quad c_2 = 1 - \frac{2 - e^{-\pi/2}}{e^{\pi/2} - e^{-\pi/2}}, \quad c_3 = 4, \quad c_4 = -1.$$

$\square$

4.6.1 Exercises

Exercise 4.41 *Find the extremal $y(x)$ for the variational*

$$L(y) = \int_0^1 (1 + y''^2) dx, \quad y(0) = 0, \ y'(0) = 1, \ y(1) = 1, \ y'(1) = 1.$$

Exercise 4.42 *Find the extremals $y = y(x), z = z(x)$ for the variational*

$$L(y, z) = \int_0^{\pi/4} (4y^2 + z^2 - y'^2 - z'^2) dx$$

subject to

$$y(0) = 1, \ y(\pi/4) = 0, \ z(0) = 0, \ z(\pi/4) = 1.$$

Exercise 4.43 *Find the extremals $y = y(x), z = z(x)$ for the variational with boundary conditions*

$$L(y,z) = \int_0^{\pi/4} (4y^2 + z^2 + y'z')dx,$$

$$y(0) = 1, \quad y(\pi/4) = 0, \quad z(0) = 0, \quad z(\pi/4) = 1.$$

Hint: Solving for the constants will be messy.

Exercise 4.44 *Prove parts (a) and (b) of Remark 19.*

Exercise 4.45 *Use Exercise 4.44 to show that for constants c_1 and c_2 the Euler-Lagrange equation of the variational*

$$L(y) = \int_a^b \frac{(1+y'^2)^2}{y''}dx$$

is

$$y'' \frac{c_1 y' + c_2}{(1+y'^2)^2} = 1$$

and solve the differential equation.

Hint: Use the transformation $y' = \tan(u)$ to solve the differential equation.

Exercise 4.46 *Find the extremals $y = y(x)$, $z = z(x)$ (no need to solve for the constants) for the variational*

$$L(y,z) = \int_0^1 (z'^2 + (y'^2 - 1)^2 + z^2 + yz)dx$$

Exercise 4.47 *An elastic beam has vertical displacement $y(x)$, $x \in [0, l]$. (The x-axis is horizontal and the y-axis is vertical and directed upwards.) Let ρ be the load per unit length on the beam. The ends of the beam are supported, that is, $y(0) = y(l) = 0$. Then the displacement y minimizes the energy functional*

$$L(y) = \int_0^l [\frac{1}{2}D(y''(x))^2 + \rho g y(x)]dx,$$

where D, ρ and g are positive constants. Write down the differential equation and the rest of the boundary conditions that $y(x)$ must satisfy and then show that the solution is

$$y(x) = -\frac{\rho g}{24D}x(l-x)[l^2 + x(l-x)].$$

Exercise 4.48 *Find the extremal $y = y(x)$, $z = z(x)$, for the variational with boundary conditions*

$$L(y,z) = \int_0^{\pi/2} (y'^2 + z'^2 + 2yz)dx$$

$$y(0) = 1, \quad y(\pi/2) = 1, \quad z(0) = 0, \quad z(\pi/2) = -1.$$

Answer: $y(x) = \sin(x)$, $z(x) = -\sin(x)$.

Exercise 4.49 *Find the extremal $y = y(x)$, $z = z(x)$ for fixed end points of the variational*

$$L(y,z) = \int_a^b (2yz - 2y^2 - (y')^2 + (z')^2)dx.$$

Exercise 4.50 *Find the extremal $y = y(x)$, $z = z(x)$ for fixed end points of the variational*

$$L(y,z) = \int_a^b (y'z' + y^2 + z^2)dx.$$

Exercise 4.51 *Find the extremal $y = y(x)$, $z = z(x)$ for the variational with boundary conditions*

$$L(y,z) = \int_0^1 (2y + (y')^2 + (z')^2)dx,$$

$$y(0) = 1, \quad y(1) = \frac{3}{2}, \quad z(0) = 1, \quad z(1) = 1.$$

Answer: $y(x) = 1 + x^2/2$, $z(x) = 1$.

4.7   Natural Boundary Conditions

In the preceding sections, we only considered variational with fixed endpoints. In this section, our aim is to redevelop the theory that either one or both endpoints are free to move. We begin by letting $y = y(x) \in C^2([a,b])$ be an extremal of the variational

$$L(y) = \int_a^b F(x,y,y')dx, \quad y(a) = A, \, y(b) \quad \text{is unspecified.} \qquad (4.50)$$

Such a problem is called *free endpoints problem*. Note that $y(b)$ takes values at the vertical line $x = b$, as illustrated in Fig. 4.7. It seems that if y is an extremal, additional condition(s) must be imposed at the second boundary point $x = b$. Most of the next derivations are similar to those in Theorem 4.3.

Let $\eta = \eta(x) \in C^2([a,b])$ with $\eta(a) = 0$. In the functional $L(y)$ replace y by $y + \varepsilon \eta$. Setting

$$\frac{dL(\varepsilon)}{d\varepsilon}\Big|_{\varepsilon=0}$$

we arrive at

$$\int_a^b \left[\frac{\partial F}{\partial y}(x, y + \varepsilon\eta, y' + \varepsilon\eta')\eta + \frac{\partial F}{\partial y'}(x, y + \varepsilon\eta, y' + \varepsilon\eta')\eta' \right]dx \Big|_{\varepsilon=0}.$$

Therefore, we obtain the necessary condition

$$\int_a^b \left[\frac{\partial F}{\partial y}(x,y,y')\eta(x) + \frac{\partial F}{\partial y'}(x,y,y')\eta'(x) \right]dx = 0.$$

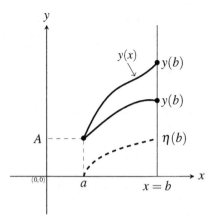

FIGURE 4.7
Free boundary condition at $x = b$.

We perform an integration by parts on the second term in the integrand of the above integral.

$$\int_a^b \frac{\partial F}{\partial y'} \eta' dx = \frac{\partial F}{\partial y'} \eta(x)\Big|_a^b - \int_a^b \frac{d}{dx}\left(\frac{\partial F}{\partial y'}\right) \eta(x) dx$$

$$= \frac{\partial F(b, y(b), y'(b))}{\partial y'} \eta(b) - \frac{\partial F}{\partial y'} \eta(a) - \int_a^b \frac{d}{dx}\left(\frac{\partial F}{\partial y'}\right) \eta(x) dx$$

$$= \frac{\partial F(b, y(b), y'(b))}{\partial y'} \eta(b) - \int_a^b \frac{d}{dx}\left(\frac{\partial F}{\partial y'}\right) \eta(x) dx,$$

since $\eta(a) = 0$. Substituting back into the integral we arrive at

$$\int_a^b \left[\frac{\partial F}{\partial y} - \frac{d}{dx}\left(\frac{\partial F}{\partial y'}\right)\right] \eta(x) dx + F_{y'}(b, y(b), y'(b)) \eta(b) = 0. \qquad (4.51)$$

Since (4.51) holds for all values of η, it must hold for η also satisfying the condition $\eta(b) = 0$. Hence

$$\int_a^b \left[\frac{\partial F}{\partial y} - \frac{d}{dx}\left(\frac{\partial F}{\partial y'}\right)\right] \eta(x) dx = 0,$$

and by Lemma 10 , it follows that

$$\frac{d}{dx}\left(\frac{\partial F}{\partial y'}\right) - \frac{\partial F}{\partial y} = 0, \qquad (4.52)$$

for all functions $\eta(x)$. A substitution of (4.52) into (4.51) gives

$$F_{y'}(b, y(b), y'(b)) := F_{y'}\big|_{x=b} = 0. \qquad (4.53)$$

Similar results can be easily obtained for cases when $y(a)$ is unspecified or both $y(a)$ and $y(b)$ are unspecified. We summarize the results in the next theorem but first we state

$$F_{y'}(a,y(a),y'(a)) := F_{y'}\big|_{x=a} = 0. \tag{4.54}$$

Theorem 4.15 *Let* $y = y(x) \in C^2[a,b]$ *be an extremal for the variational*

$$L(y) = \int_a^b F(x,y,y')dx \tag{4.55}$$

with boundary conditions specified or unspecified at $x = a$ *and* $x = b$.

1) *If both boundary conditions are specified,* $(y(a) = A,\ y(b) = B)$ *then a necessary condition for* $y(x)$ *to be an extremal of* (4.55) *is the Euler-Lagrange equation given by* (4.52).

2) *If* $y(a)$ *is not specified and* $y(b)$ *is specified* $(y(b) = B)$, *then the necessary conditions for* $y(x)$ *to be an extremal of* (4.55) *are the Euler-Lagrange equation given by* (4.52) *and* (4.54).

3) *If* $y(a)$ *is specified* $(y(a) = A)$ *and* $y(b)$ *is unspecified then the necessary conditions for* $y(x)$ *to be an extremal of* (4.55) *are the Euler-Lagrange equation given by* (4.52) *and* (4.53).

4) *If neither* $y(a)$ *nor* $y(b)$ *is specified then the necessary conditions for* $y(x)$ *to be an extremal of* (4.55) *are the Euler-Lagrange equation given by* (4.52), *plus* (4.53) *and* (4.54) .

Example 4.16 Let $y = y(x)$ be an extrema of the variational

$$L(y) = \int_0^1 (y'^2 + y^2)dx, \quad y(0) = 1,\ y(1) \quad \text{is unspecified.}$$

Then $f_{y'} = 2y'$, and hence $F_{y'}\big|_{x=1} = 2y'(1) = 0$. Moreover, the corresponding Euler-Lagrange equation is $y'' - y = 0$. Thus, we are left with solving the second-order differential equation

$$y'' - y = 0, \quad y(0) = 1, \quad y'(1) = 0,$$

which has the solution

$$y(x) = \frac{1}{1+e^{-2}}e^{-x} + \frac{e^{-2}}{1+e^{-2}}e^{x}.$$

$\square$

Example 4.17 *[River crossing]* A boat wants to cross a river with two parallel banks at a distance b apart. One of the banks coincides with the y-axis. The other bank is line $x = b$ as depicted in Fig. 4.8. The water is assumed to be moving parallel to

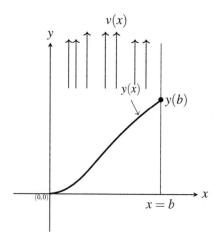

FIGURE 4.8
Boat route.

the banks with speed $v(x)$. The boat's constant speed is c such that $c^2 > v^2$. Assume $(0,0)$ is the departure point. We are interested in finding the route that the boat should take to reach the opposite bank in the shortest possible time.

To do so, we assume the boat moves along a path $y = y(x)$. Let α be the angle at which the boat is steered. Then the velocity of the boat in the river is

$$y' = \frac{dy}{dx} = \frac{dy/dt}{dx/dt} = \frac{v + c\sin(\alpha)}{c\cos(\alpha)} = \frac{v}{c}\sec(\alpha) + \tan(\alpha).$$

On the other hand, the time T required to cross the river is

$$T = \int_0^b t'(x)dx = \int_0^b \frac{dt}{dx}dx = \int_0^b \frac{1}{\frac{dx}{dt}}dx = \int_0^b \frac{1}{c}\sec(\alpha)dx.$$

From the preceding equation of y' we have

$$cy' = v\sec(\alpha) + c\tan(\alpha).$$

Or

$$(cy' - v\sec(\alpha))^2 = c^2\tan^2(\alpha) = c^2(\sec^2(\alpha) - 1).$$

After rearranging the terms we arrive at the quadratic equation in $\sec(\alpha)$,

$$(c^2 - v^2)\sec^2(\alpha) + 2cvy'\sec(\alpha) - c^2(1 + y'^2) = 0,$$

that we need to solve. Since $\sec(\alpha) > 0$ in the first quadrant we have that

$$\begin{aligned}
\sec(\alpha) &= \frac{-cvy' + \sqrt{c^2v^2y'^2 + c^2(c^2 - v^2)(1 + y'^2)}}{c^2 - v^2} \\
&= \frac{-cvy' + c\sqrt{c^2(1 + y'^2) - v^2}}{c^2 - v^2}.
\end{aligned}$$

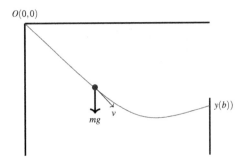

FIGURE 4.9
Brachistochrone free end point.

A substitution of $\sec(\alpha)$ in the integrand of T yields the variational

$$L(y) = \int_0^b \frac{\sqrt{c^2(1+y'^2(x)) - v^2(x)} - v(x)y'(x)}{c^2 - v^2(x)}, \quad y(0) = 0, \ y(b) \quad \text{is unspecified.}$$
(4.56)

Minimizing expression (4.56) yields finding the shortest trajectory or path $y = y(x)$ that the boat follows in order to cross to the other bank of the river, which is equivalent to finding the shortest possible time. □

Example 4.18 *[Brachistochrone problem revisited]* We revisit the Brachistochrone problem that was considered in Section 4.5. For simplicity, we assume the bead's starting point is the origin. That is, $(x_1, y_1) = (0, 0)$. As before, we let $x_2 = b$ so that we have compatible notation as in Section 4.5. In this problem, we are seeking the shape of the wire $y(x)$ that enables the bead to get from the origin to a point on the line $x = b > 0$ in the shortest time. In other words, $y(b)$ is unspecified; see Fig. 4.9. Since we are starting at the origin, the conservation of energy equation (4.43) takes the form

$$\frac{1}{2}mv^2 = mgy.$$

Solving for v and setting $y_1 = 0$, the variational in (4.44) is reduced to

$$L(y) = \int_0^b \frac{\sqrt{1+y'^2(x)}}{\sqrt{2gy(x)}}\,dx, \quad y(0 = 0, \ y(b) \quad \text{is free.}$$
(4.57)

□

Natural boundary conditions of higher-orders.

We follow the same set up as in Section 4.6. Let $y \in C^4([a,b])$ and consider the variational

$$L(y) = \int_a^b F(x,y,y',y'')\,dx,$$
(4.58)

with boundary conditions

$$y(a) = A_1, \quad y'(a) = A_2, \quad y(b) = B_1, \quad y'(b) = B_2.$$

Setting $\dfrac{dL(\varepsilon)}{d\varepsilon}\Big|_{\varepsilon=0}=0$,

$$
\begin{aligned}
\frac{dL(\varepsilon)}{d\varepsilon} &= \frac{d}{d\varepsilon}\int_a^b F(x,y+\varepsilon\eta,y'+\varepsilon\eta',y''+\varepsilon\eta'')dx\Big|_{\varepsilon=0} \\
&= \int_a^b \Big[F_y\eta+F_{y'}\eta'+F_{y''}\eta''\Big]dx. \quad (4.59)
\end{aligned}
$$

since $\dfrac{dx}{d\varepsilon}=0$. We perform an integration by parts on the second and third terms in the integrand. After some work we end up with

$$
\begin{aligned}
\int_a^b \Big[F_y\eta+F_{y'}\eta'+F_{y''}\eta''\Big]dx &= F_{y'}\eta\Big|_{x=a}^{x=b}+F_{y''}\eta'\Big|_{x=a}^{x=b}-\frac{d}{dx}F_{y''}\eta\Big|_{x=a}^{x=b} \\
&\quad + \int_a^b\Big[\frac{d^2}{dx^2}F_{y''}-\frac{d}{dx}F_{y'}+F_y\Big]\eta\,dx \\
&= \Big(F_{y''}\eta'-(\frac{d}{dx}F_{y''}-F_{y'})\eta\Big)\Big|_{x=a}^{x=b} \\
&\quad + \int_a^b\Big[\frac{d^2}{dx^2}F_{y''}-\frac{d}{dx}F_{y'}+F_y\Big]\eta\,dx.
\end{aligned}
$$

Thus the natural boundary conditions depend on the relation

$$
\Big[F_{y''}\eta'-(\frac{d}{dx}F_{y''}-F_{y'})\eta\Big]\Big|_{x=a}^{x=b}=0. \quad (4.60)
$$

Then a combination of the following natural boundary conditions are needed when one or more boundary condition is unprescribed or unspecified. To be specific, we may deduce from (4.60) the following:

$$
F_{y''}\big|_{x=a}=0, \quad \text{if } y'(a) \text{ is unspecified,} \quad (4.61)
$$
$$
F_{y''}\big|_{x=b}=0, \quad \text{if } y'(b) \text{ is unspecified,} \quad (4.62)
$$
$$
(\frac{d}{dx}F_{y''}-F_{y'})\Big|_{x=a}=0, \quad \text{if } y(a) \text{ is unspecified,} \quad (4.63)
$$

and

$$
(\frac{d}{dx}F_{y''}-F_{y'})\Big|_{x=b}=0, \quad \text{if } y(b) \text{ is unspecified.} \quad (4.64)
$$

Recall that in order for $y(x)$ to be an extremal of (4.58) it must satisfy the Euler-Lagrange equation given by

$$
\frac{d^2}{dx^2}F_{y''}-\frac{d}{dx}F_{y'}+F_y=0,
$$

that readily follows from (4.59). We have the following example.

Example 4.19 Find the extremal $y = y(x)$ for the functional

$$L(y) = \int_0^{\pi/2} \left(-y^2 + (y'')^2 \right) dx$$

subject to

$$y(0) = 1, \quad y'(0) = 2, \quad y(\pi/2), \quad \text{and} \quad y'(\pi/2) \text{ are unspecified.}$$

The corresponding necessary Euler-Legrange equation is

$$y^{(4)} - y = 0.$$

Using the method of Section 1.8 we obtain the general solution

$$y(x) = c_1 e^{-x} + c_2 e^x + c_3 \sin(x) + c_4 \cos(x).$$

The two natural boundary conditions that we need are (4.62) and (4.64). Condition (4.62) yields

$$y''(\pi/2) = 0.$$

Similarly, from condition (4.64) we get

$$y'''(\pi/2) = 0.$$

Hence, by applying all four boundary conditions, we arrive at the system of equations

$$c_1 + c_2 + c_4 = 1,$$
$$-c_1 + c_2 + c_3 = 2,$$
$$c_1 e^{-\pi/2} + c_2 e^{\pi/2} - c_3 = 0,$$
$$-c_1 e^{-\pi/2} + c_2 e^{\pi/2} + c_4 = 0,$$

with solution

$$c_1 = 3.35786, \quad c_2 = 0.80197, c_3 = 4.55589, \quad c_4 = -3.15983.$$

$\square$

Next, we provide an application for reducing a cantilever beam's potential energy. A more general case of the study of beam will be considered in Chapter 5. As a result of an underlying force that pulls a body toward its source, a system has a propensity to reduce potential energy. Or shoving a body away if the force is repellent. As a result, the distance is reduced, which reduces potential energy. Hence, potential energy is a measure of potential movement; potential energy is a measure of potential motion. Clearly, if the two attractive bodies are already together, there is no movement and no potential energy. Now, this justification holds true for both elastic and electric potential energy. There is an underlying force that moves the material in each of these instances. While these forces can produce movement, if their nature is attracting, the corresponding potential energy increases with distance. In conclusion, the support situation, profile (form of the cross-section), geometry, equilibrium situation, and material of a beam are its defining characteristics.

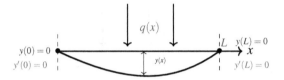

FIGURE 4.10
Clamped Beam at both end points.

Example 4.20 Suppose we have a beam of length L with small transverse displacement $y(x)$ under transverse load $q(x)$. The beam is subject to infinitesimal deflections only. According to the force and moment balance approach, the displacement is governed by the fourth-order differential equation

$$eI\frac{d^4y}{dx^4} = q(x), \tag{4.65}$$

where e is the modulus of elasticity of the beam's material and $I(x)$ is the moment of inertia of the beam's cross-sectional area about a point x. We are interested in minimizing the potential energy. It is thought that applying the minimal total potential energy approach will make future extensions of the beam equation into large deflections, nonlinear materials, and accurate modeling of shear forces between the cable elements simpler than using force- and moment balances. The potential energy is a combination of the *strain energy*,

$$\frac{1}{2}eI\left(\frac{d^2y}{dx^2}\right)^2,$$

or the deformed energy stored in the elastic plus the *work potential*. The work potential is the negative work done by external forces, which is $-qy$. Thus, the total potential energy is given by the variational

$$L(y) = \int_0^L \left[\frac{1}{2}eI\left(\frac{d^2y}{dx^2}\right)^2 - q(x)y(x)\right]dx, \tag{4.66}$$

where e, q, and I are known quantities. Note that then Euler-Lagrange equation of (4.66) is (4.65). In what to follows, we will consider different cases of conditions corresponding to support systems for the beam, and we assume that e and I are constants.

(I). The beam is clamped at each end, as Fig. 4.10 shows. In this case, we have the four boundary conditions

$$y(0) = y'(0) = 0, \quad y(L) = y'(L) = 0,$$

and hence no natural boundary conditions are in play.

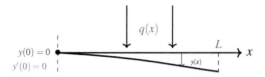

FIGURE 4.11
Clamped Beam at $x = 0$.

(II). The beam is only clamped at $x = 0$ as Fig. 4.11 shows. A beam that is fixed at one end and free at the other end is known as a cantilever beam. A cantilever beam is one that is free-hanging at one end and fixed at the other. This type of beam is capable of carrying loads with both bending moment and sheer stress and is typically used when building bridge trusses or similar structures. The end that is fixed is typically attached to a column or wall. The tension zone of a cantilever beam, is found at the top of the beam with the compression zone at the bottom of the beam. In such a case, we are considering a cantilever beam, which is a rigid structure supported at one end and free at the other. We are assuming small deflection of the beam since the end point $x = L$ is unclamped. In this case we need the natural boundary conditions (4.62) and (4.64). Conditions (4.62) and (4.64) yields

$$eIy'''(L) = 0, \quad \text{and} \quad eIy''(L) = 0.$$

The condition $y'''(L) = 0$ means that the reaction force at $x = L$ is zero. Similarly, the condition $y''(L) = 0$ means that the reaction moment force at $x = L$ is zero.

(III). We assume the beam is simply supported at the end points as depicted in Fig. 4.12. Simply supported beams are those that have supports at both ends of the beam. These are most frequently utilized in general construction and are very versatile in terms of the types of structures that they can be used with. A simply supported beam has no moment resistance at the support area and is placed in a way that allows for free rotation at the ends on columns or walls. In other words, the beam is pinned at both ends, and no restrictions are imposed on y' at $x = 0$ and $x = L$. The relevant natural boundary conditions in this instance are (4.61) and (4.62) and as a consequence, we obtain $y''(0) = 0$ and $y''(L) = 0$.

(IV). Double overhanging: This is a simple beam with both ends extending beyond its supports on both ends. Then all four natural boundary conditions (4.61)–(4.64) are in play. Consequently, they yield

$$y''(0) = y''(L) = 0, \quad \text{and} \quad y'''(0) = y'''(L) = 0.$$

Physically, this means that the reaction force and moment at each end of the beam must be zero under these circumstances.

□

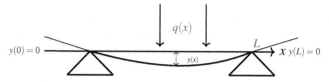

FIGURE 4.12
Simply supported beam.

4.8 Impact of y'' on Euler-Lagrange Equation

In this brief section, we examine variational in which y'' enters linearly. Thus, we are interested in variational of the form

$$L(y) = \int_a^b \Big(N(x,y)y'' + M(x,y) \Big) dx, \qquad (4.67)$$

with boundary conditions

$$y(a) = A_1, \quad y'(a) = A_2, \quad y(b) = B_1, \quad y'(b) = B_2.$$

Assume N and M are continuous with continuous partial derivatives on some subset of $\mathbb{R}^2$. Let

$$F(x,y) = N(x,y)y'' + M(x,y).$$

Then, $F_{y''} = N$, and therefore

$$\frac{d}{dx}F_{y''} = \frac{d}{dx}N = N_x + N_y y'.$$

Moreover,

$$\frac{d^2}{dx^2}F_{y''} = N_{xx} + N_{xy}y' + N_y y'' + N_{yx}y' + N_{yy}y'^2.$$

In addition, $F_{y'} = 0$, and $F_y = N_y y'' + M_y$. Thus, the Euler-Lagrange equation

$$\frac{d^2}{dx^2}F_{y''} - \frac{d}{dx}F_{y'} + F_y = 2N_y y'' + N_{yy}y'^2 + N_{xy}y' + M_y = 0,$$

which is a second-order differential equation, and hence not all four boundary conditions can be satisfied in most cases.

4.8.1 Exercises

Exercise 4.52 *Let $y = y(x)$ be an extremal of the variational*

$$L(y) = \int_0^1 (y'^2 + y^2)dx.$$

(a) Find $y(x)$ when $y(0)$ is unspecified and $y(1) = 1$.

(b) Find $y(x)$ when both end points are unspecified, and argue that it minimizes L.

Exercise 4.53 *Find the extremal $y = y(x)$ of the variational*

$$L(y) = \int_0^{\pi/4} (-y'^2 + y^2) dx,$$

when $y(0) = 1$ and $y(\pi/4)$ is unspecified.

Exercise 4.54 *Find the extremal $y = y(x)$ of the variational*

$$L(y) = \int_1^e (\frac{x^2}{2} y'^2 - \frac{y^2}{8}) dx,$$

when $y(1) = 1$ and $y(e)$ is unspecified.

Exercise 4.55 *Find the extremal $y = y(x)$ of the variational*

$$L(y) = \int_0^1 \frac{1}{2} [y'^2 + yy' + y] dx,$$

when $y(0)$ and $y(1)$ are unspecified.

Exercise 4.56 *[River crossing] Compute $F_{y'}|_{x=b}$ for the variational (4.56).*

Answer: $y'(b) = \dfrac{v(b)}{c}$.

Exercise 4.57 *Suppose $p(x)$ and $q(x)$ are continuous and positive functions on $[0,1]$. Find the Euler-Lagrange equation and the natural boundary condition for the variational*

$$L(y) = \int_0^1 [p(x)y'^2 - q(x)y^2] dx,$$

$y(0) = 0$, $y(1)$ free.

Exercise 4.58 *Solve the variational problem (4.57) that describes the shortest path for Brachistochrone with free end point.*

Exercise 4.59 *Assume $g(x,y) \neq 0$ for all (x,y). Find the natural boundary condition for the variational*

$$L(y) = \int_a^b g(x,y)\sqrt{1 + (y')^2} \, dx,$$

$y(0)$ is free and $y(1) = 0$.

Exercise 4.60 *Show that the extremal for*

$$L(y) = \int_a^b \frac{1}{y}\sqrt{1 + (y')^2} \, dx, \quad y > 0$$

is

$$(x-B)^2 + y^2 = R^2,$$

for appropriate constants B and R.

Exercise 4.61 *Find the extremals $y = y(x)$, $z = z(x)$ for the variational*

$$L(y,z) = \int_0^\pi (4y^2 + z^2 - y'^2 - z'^2)dx;$$

$y(0) = 1 = z(0)$, $y(\pi)$ *and* $z(\pi)$ *are unspecified.*

Exercise 4.62 *Find $y = y(x)$ the extremal of the variational*

$$L(y) = \int_0^1 (1 + (y'')^2)dx,$$

when $y(0) = 0$, $y'(0) = 1$ and $y(1)$, $y'(1)$ are unspecified.

Exercise 4.63 *Find the extremal $y = y(x)$ for the functional*

$$L(y) = \int_0^{\pi/2} (-y^2 + 2yx^3 + (y'')^2)dx;$$

$y(0)$, $y'(0)$ *are unspecified and* $y(\pi/2) = 1$, $y'(\pi/2) = 2$.

Exercise 4.64 *Compute*

$$\frac{d^2}{dx^2}F_{y''} - \frac{d}{dx}F_{y'} + F_y$$

for the functional

$$L(y) = \int_a^b \left(N(x,y)y'' + P(x,y)y' + M(x,y) \right)dx$$

4.9 Discontinuity in Euler-Lagrange Equation

Consider the variational

$$L(y) = \int_a^b F(x,y,y')dx, \quad y(a) = A, \ y(b) = B \tag{4.68}$$

and suppose one or both of the terms F_y and $\frac{d}{dx}F_{y'}$ are discontinuous at one or more points in (a,b). For illustrative purpose, we assume there is one point of discontinuity, $c \in (a,b)$. Divide the interval $[a,b]$ into two subintervals such that $[a,b] = [a,c^-] \cup (c^+, b]$. We are searching for a continuous extremal $y(x)$ of (4.68) and as a consequence the following condition must hold.

$$\lim_{x \to c^-} y(x) = \lim_{x \to c^+} y(x). \tag{4.69}$$

Our η function is assumed to be continuous in the sense that

$$\eta(c^+) = \lim_{x \to c^+} \eta(x) = \eta(c^-) = \lim_{x \to c^-} \eta(x) = \eta(c).$$

By a similar arguments as in Section 4.7, one has

$$\int_a^b \left[\frac{\partial F}{\partial y}(x, y + \varepsilon\eta, y' + \varepsilon\eta')\eta + \frac{\partial F}{\partial y'}(x, y + \varepsilon\eta, y' + \varepsilon\eta')\eta' \right] dx \Big|_{\varepsilon=0}$$

$$= \int_a^{c^-} \left[\frac{\partial F}{\partial y}(x, y + \varepsilon\eta, y' + \varepsilon\eta')\eta + \frac{\partial F}{\partial y'}(x, y + \varepsilon\eta, y' + \varepsilon\eta')\eta' \right] dx \Big|_{\varepsilon=0}$$

$$+ \int_{c^+}^b \left[\frac{\partial F}{\partial y}(x, y + \varepsilon\eta, y' + \varepsilon\eta')\eta + \frac{\partial F}{\partial y'}(x, y + \varepsilon\eta, y' + \varepsilon\eta')\eta' \right] dx \Big|_{\varepsilon=0}$$

Set $\frac{dL(\varepsilon)}{d\varepsilon}\big|_{\varepsilon=0} = 0$ and integrate by parts to obtain,

$$\int_a^{c^-} \left[\frac{\partial F}{\partial y}\eta + \frac{\partial F}{\partial y'}\eta' \right] dx + \int_{c^+}^b \left[\frac{\partial F}{\partial y}\eta + \frac{\partial F}{\partial y'}\eta' \right] dx$$

$$= \int_a^{c^-} \left(F_y - \frac{d}{dx}F_{y'} \right) \eta \, dx + \int_{c^+}^b \left(F_y - \frac{d}{dx}F_{y'} \right) \eta \, dx$$

$$+ F_{y'}(c^-, y(c^-), y'(c^-))\eta(c^-) - F_{y'}(a, y(a), y'(a))\eta(a)$$

$$+ F_{y'}(b, y(b), y'(b))\eta(b) - F_{y'}(c^+, y(c^+), y'(c^+))\eta(c^-) = 0. \qquad (4.70)$$

As a consequence of (4.70) we obtain the following conditions.

$$F_y - \frac{d}{dx}F_{y'} = 0, \quad a < x < c^-; \quad F_y - \frac{d}{dx}F_{y'} = 0, \quad c^+ < x < b, \qquad (4.71)$$

$$F_{y'}(a, y(a), y'(a)) = 0 \qquad (4.72)$$

$$F_{y'}(b, y(b), y'(b)) = 0, \qquad (4.73)$$

and

$$\lim_{x \to c^-} F_{y'} = \lim_{x \to c^+} F_{y'}. \qquad (4.74)$$

Theorem 4.16 *Let $y = y(x) \in C^2[a, b]$ be an extremal for the variational*

$$L(y) = \int_a^b F(x, y, y') dx \qquad (4.75)$$

with boundary conditions specified or unspecified at $x = a$ and $x = b$. Assume there is a discontinuity at at point $c \in (a, b)$.

1) *If both boundary conditions are specified, $(y(a) = A, y(b) = B)$ then conditions (4.69), (4.71), and (4.74) are needed.*

2) *If $y(a)$ is not specified and $y(b)$ is specified $(y(b) = B)$, then the necessary conditions (4.69), (4.71), (4.72), and (4.74) are needed.*

3) *If $y(a)$ is specified ($y(a) = A$) and $y(b)$ is unspecified then the necessary conditions (4.69), (4.71), (4.73), and (4.74) are needed.*

4) *If neither $y(a)$ nor $y(b)$ is specified then the necessary conditions (4.69), (4.71)–(4.74) are needed.*

Example 4 Consider the functional

$$L(y) = \int_{-1}^{1} (f(x)y'^2 + y)dx, \quad y(-1) = y(1) = 0$$

with

$$f(x) = \begin{cases} 1, & -1 \leq x < 0 \\ 2, & 0 < x \leq 1. \end{cases}$$

Obviously, we have discontinuity at $c = 0$. Regardless of the discontinuity, the Euler-Lagrange equation is

$$\frac{d}{dx}(2f(x)y') - 1 = 0. \tag{4.76}$$

For $-1 \leq x < 0$, we have $2y'' - 1 = 0$, with the general solution

$$y(x) = \frac{x^2}{4} + c_1 x + c_2. \tag{4.77}$$

Similarly, for $0 < x \leq 1$, we have $4y'' - 1 = 0$, with the general solution

$$y(x) = \frac{x^2}{8} + d_1 x + d_2. \tag{4.78}$$

An application of $0 = y(-1)$ to (4.77) gives

$$c_2 - c_1 = -\frac{1}{4}.$$

Next apply $0 = y(1)$ to (4.77) and get

$$d_1 + d_2 = -\frac{1}{8}.$$

An application of (4.69)

$$\lim_{x \to 0^-} y(x) = \lim_{x \to 0^+} y(x),$$

yields

$$c_2 = d_2.$$

Finally, condition (4.74) yields

$$c_1 = 2d_1.$$

Next we substitute $d_1 = c_1/2$, and $d_2 = c_2$ into $d_1 + d_2 = -\frac{1}{8}$ to obtain $c_1 + 2c_2 = -\frac{1}{4}$.
Finally, solving

$$c_2 - c_1 = -\frac{1}{4}; \quad c_1 + 2c_2 = -\frac{1}{4},$$

one obtains

$$c_1 = \frac{1}{12}, \quad c_2 = -\frac{1}{6}.$$

Also, It follows that

$$d_1 = \frac{1}{24}, \quad d_2 = -\frac{1}{6}.$$

In conclusion, the solution over the whole interval is given piecewise

$$y(x) = \begin{cases} \frac{x^2}{4} + \frac{1}{12}x - \frac{1}{6}, & -1 \leq x < 0 \\\\ \frac{x^2}{8} + \frac{1}{24}x - \frac{1}{6}, & 0 < x \leq 1. \end{cases}$$

which is continuous at $x = 0$.   □

4.9.1   Exercises

Exercise 4.65 *Find the extrema $y = y(x)$ for the functional*

$$L(y) = \int_{-1}^{1} (f(x)y'^2 + 8y^2)dx, \quad y(-1) = 0, \ y(1) = 1,$$

with

$$f(x) = \begin{cases} \frac{1}{2}, & -1 \leq x < 0 \\ 2, & 0 < x \leq 1. \end{cases}$$

Exercise 4.66 *Find the extrema $y = y(x)$ for the functional*

$$L(y) = \int_{0}^{\pi/2} (f(x)y'^2 - 2y^2)dx, \quad y(0) = 0, \ y(\pi/2) = 1,$$

with

$$f(x) = \begin{cases} 2, & 0 \leq x < \pi/4 \\ \frac{1}{2}, & \pi/4 < x \leq \pi/2. \end{cases}$$

Exercise 4.67 *Let $f(x) = \begin{cases} -1, & 0 \leq x < 1/4 \\ 1, & 1/4 < x \leq 1 \end{cases}$ and consider the functional*

$$L(y) = \int_{0}^{1} (y'^2 f(x))dx.$$

Find the extremal of $L(y)$ when

(a) $y(0) = 0$, $y(1) = 1$,

(b) $y(0) = 0$ and $y(1)$ is unassigned,

(c) $y(0)$ is unassigned and $y(1) = 1$,

(d) both $y(0)$ and $y(1)$ are unassigned.

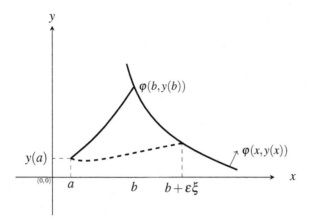

FIGURE 4.13
Transversality condition.

4.10 Transversality Condition

Let $y = y(x) \in C^2([a,b])$ and consider the variational,

$$L(y) = \int_a^b F(x,y,y')dx, \quad y(a) = A. \tag{4.79}$$

So far, we have investigated specified boundaries and unspecified boundaries that take values along vertical lines. In this section, we are interested in exploring the scenario when the free point moves along a specified curve. Without loss of generality, we assume $y(a)$ is fixed and $y(b)$ slides or lies on a curve defined by the equation $\varphi(x,y) = 0$. Assume that φ_x and φ_y don't vanish simultaneously on the domain of interest. As before, we assume a function $\eta \in C^2([a,b])$ with $\eta(a) = 0$ so that the function $y(x) + \eta(x)$ is in the admissible space of functions. See Fig. 4.13. Let $(b, \varphi(b))$ be the terminal point of the extremal of y on $\varphi(x,y)$. From Fig. 4.13, the terminal point of the varied path $y(x) + \varepsilon\eta(x)$ is

$$y(b+\varepsilon\xi) + \varepsilon\eta(b+\varepsilon\xi) = y(b) + \varepsilon\big(\xi y'(b) + \eta(b)\big) + O(\varepsilon^2),$$

for $\xi > 0$.

Since the same point lies on the curve $\varphi(x,y) = 0$ we have that

$$\varphi\Big(b+\varepsilon\xi, y(b) + \varepsilon(\xi y'(b) + \eta(b))\Big) = 0. \tag{4.80}$$

Expanding expression (4.80) to the first-order ε, yields

$$\varphi(b,y(b)) + \xi\varphi_x + \big(\xi y'(b) + \eta(b)\big)\varphi_y = 0.$$

Use the fact that $\varphi\big(b,y(b)\big) = 0$ and rearrange the terms to get

$$\xi\big(\varphi_x + \varphi_y\, y'(b)\big) + \eta(b)\varphi_y = 0. \tag{4.81}$$

Now, we are ready to compute $\frac{dL(\varepsilon)}{d\varepsilon}$. Let

$$L(y + \varepsilon\eta) = \int_a^{b+\varepsilon\xi} F(x, y + \varepsilon\eta, y' + \varepsilon\eta')\,dx.$$

Then by Leibniz rule, which says,

$$\frac{d}{dx}\int_{g(x)}^{f(x)} B(x,t)\,dt = B(x,f(x))f'(x) - B(x,g(x))g'(x) + \int_{g(x)}^{f(x)} \frac{\partial}{\partial x} B(x,t)\,dt,$$

we see that

$$
\begin{aligned}
\frac{dL(\varepsilon)}{d\varepsilon} &= \frac{d}{d\varepsilon}\int_a^{b+\varepsilon\xi} F(x, y + \varepsilon\eta, y' + \varepsilon\eta')\,dx\Big|_{\varepsilon=0} \\
&= F\big(b + \varepsilon\xi, y(b + \varepsilon\xi) + \varepsilon\eta(b + \varepsilon\xi), y'(b + \varepsilon\xi) + \varepsilon\eta'(b + \varepsilon\xi)\big)\xi\Big|_{\varepsilon=0} \\
&\quad + \int_a^b \Big[\frac{\partial F}{\partial y}(x, y + \varepsilon\eta, y' + \varepsilon\eta')\eta + \frac{\partial F}{\partial y'}(x, y + \varepsilon\eta, y' + \varepsilon\eta')\eta'\Big]\,dx\Big|_{\varepsilon=0}.
\end{aligned}
$$

Setting $\varepsilon = 0$, $\frac{dL(\varepsilon)}{d\varepsilon} = 0$ and integrating by parts, the above expression yields,

$$F\big(b, y(b), y'(b)\big)\xi + F_{y'}\big(b, y(b), y'(b)\big)\eta(b) + \int_a^b \big(F_y - \frac{d}{dx}F_{y'}\big)\eta(x)\,dx = 0. \tag{4.82}$$

Solving for ξ in (4.81) yields

$$\xi = -\frac{\eta(b)\varphi_y}{\varphi_x + \varphi_y\, y'(b)}.$$

Substituting into (4.82) and factoring $\eta(b)$ give

$$\Big[-\frac{F\varphi_y}{\varphi_x + \varphi_y\, y'(b)} + F_{y'}\Big]\Big|_{x=b}\eta(b) + \int_a^b \big(F_y - \frac{d}{dx}F_{y'}\big)\eta(x)\,dx = 0. \tag{4.83}$$

The above relation (4.83) holds for all $\eta(x)$, $a \le x \le b$ and in particular it must hold when $\eta(b) = 0$. Thus (4.83) implies

$$\int_a^b \big(F_y - \frac{d}{dx}F_{y'}\big)\eta(x)\,dx = 0,$$

and so by Lemma 10 we arrive at

$$F_y - \frac{d}{dx}F_{y'} = 0. \tag{4.84}$$

Substituting (4.84) into (4.83) yields

$$\left[-\frac{F\varphi_y}{\varphi_x+\varphi_y\,y'(b)}+F_{y'}\right]\Big|_{x=b}=0,$$

or

$$\left[F_{y'}\left(\varphi_x+y'(b)\varphi_y\right)-\varphi_y F\right]\Big|_{x=b}=0. \tag{4.85}$$

Condition (4.85) is called the *transversality condition*. A similar work can be performed to obtain

$$\left[F_{y'}\left(\psi_x+y'(a)\psi_y\right)-\psi_y F\right]\Big|_{x=a}=0 \tag{4.86}$$

when $y(a)$ varies along the curve $\psi(x,y)=0$ and $y(b)$ is fixed.

Let us take a closer look at the transversality condition given by (4.85). Suppose we can solve for y in terms of x in $\varphi(x,y)=0$. If so, then we set $y=g(x)$. Now

$$\frac{d}{dx}\varphi(x,y)=\varphi_x+\varphi_y\,y'=0.$$

This implies that

$$y'=-\frac{\varphi_x}{\varphi_y}=g'(x).$$

We may solve for φ_x and obtain $\varphi_x=-g'(x)\varphi_y$. Substituting φ_x into (4.85) yields

$$\left[F+(g'(x)-y'(b))F_{y'}\right]\Big|_{x=b}=0. \tag{4.87}$$

Reminder:

$$F\Big|_{x=b}=F(b,y(b),y'(b)),\quad\text{and}\quad F_{y'}\Big|_{x=b}=F_{y'}(b,y(b),y'(b)).$$

Along the lines of the preceding discussion, if $y(b)$ is fixed and the left end point $y(a)$ varies along a curve $y=h(x)$, then the corresponding transversality condition is

$$\left[F+(h'(x)-y'(a))F_{y'}\right]\Big|_{x=a}=0. \tag{4.88}$$

Thus, we proved the following theorem.

Theorem 4.17 *Let $y=y(x)\in C^2[a,b]$ be an extremal for the variational (4.79) with boundary conditions specified or unspecified at $x=a$ and $x=b$.*

1) *If $y(a)$ moves along the curve $y=h(x)$ and $y(b)$ is specified ($y(b)=B$), then the necessary conditions for $y(x)$ to be an extremal of (4.79) are the Euler-Lagrange equation given by (4.84) and (4.88).*

2) *If $y(a)$ is specified ($y(a)=A$) and $y(b)$ moves along the curve $y=g(x)$, then the necessary conditions for $y(x)$ to be an extremal of (4.79) are the Euler-Lagrange equation given by (4.84) and (4.87).*

3) *If both endpoints are allowed to move freely along the curves h and g, then the necessary conditions for y(x) to be an extremal of (4.79) are the Euler-Lagrange equation given by (4.84), plus (4.87) and (4.88).*

Natural boundary conditions can be easily derived from this discussion. For example, if $y(a)$ is fixed and $y(b)$ varies along the line $x = b$, then $\varphi(x,y) = x - b$. This implies that $\varphi_x = 1$, and $\varphi_y = 0$. Substituting into (4.85), we obtain $F_{y'}(b,y(b),y'(b)) = 0$.

Example 4.21 Find the shortest distance from the point $(0,0)$ to the nearest point on the curve $xy = 1$, $x,y > 0$. Basically, by Example 4.5 we are to minimize

$$L(y) = \int_0^b \sqrt{1+(y')^2}dx, \quad y(0) = 0$$

and $y(b)$ lies on the curve $g(x) = \frac{1}{x}$. Then,

$$F = \sqrt{1+(y')^2},$$

which is independent of x, and y, and hence we make use of $F_{y'} = C$ that is given in Corollary 4.12. It follows that

$$\frac{y'}{\sqrt{1+(y')^2}} = C.$$

Solving for y' we end up with

$$y' = \text{constant} = K,$$

where K is some function of C (another constant). Hence, $y(x) = Kx + D$. Applying $0 = y(0)$ we get $D = 0$. We are in need of another boundary condition to solve for K. We make use of the transversality condition (4.87), which requires that

$$\left[F + (g'(x) - y'(b))F_{y'}\right]\Big|_{x=b} = 0.$$

Or

$$\sqrt{1+y'^2(b)} + \frac{y'(b)}{\sqrt{1+y'^2(b)}}\left(-\frac{1}{b^2} - y'(b)\right) = 0.$$

Since $y' = k$, the above expression reduces to

$$\sqrt{1+K^2} + \frac{K}{\sqrt{1+K^2}}\left(-\frac{1}{b^2} - K\right) = 0.$$

Multiply by $\sqrt{1+K^2}$ to arrive at $K = b^2$. Consequently, the shortest distance from the point $(0,0)$ to the nearest point $(b,y(b))$ on the curve $xy = 1$ is

$$y(x) = b^2x.$$

For example if $b = 1$, then $y(x) = x$ is a straight line with the shortest distance between the origin and the point $(1,1)$ that lies on the parabola $y = 1/x$ as depicted in Fig. 4.14. □

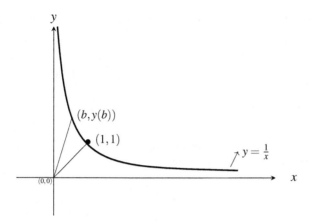

FIGURE 4.14
Shortest distance to a parabola.

4.10.1 Problem of Bolza

Now we extend the results of Section 4.10 to the *Bolza Problem*. For $y = y(x) \in C^2([a,b])$ we are interested in finding the extremal of the functional

$$L(y) = h(b, y(b)) + \int_a^b F(x, y, y')dx, \quad y(a) = A. \tag{4.89}$$

Without loss of generality, we assume $y(a)$ is fixed and $y(b)$ slides or lies on a curve defined by the equation $\varphi(x, y) = 0$. The set up is very identical to the one in Section 4.10. Thus, following the same derivations, we have, with slight modification due to the presence of the function h that

$$
\begin{aligned}
L(y + \varepsilon \eta) &= h\big(b + \varepsilon \xi, y(b + \varepsilon \xi) + \varepsilon \eta(b + \varepsilon \xi)\big) \\
&\quad + \int_a^{b+\varepsilon \xi} F(x, y + \varepsilon \eta, y' + \varepsilon \eta')dx.
\end{aligned}
$$

Then by Leibniz rule, we have that

$$
\begin{aligned}
\frac{dL(\varepsilon)}{d\varepsilon} &= \frac{d}{d\varepsilon} h\big(b + \varepsilon \xi, y(b + \varepsilon \xi) + \varepsilon \eta(b + \varepsilon \xi)\big) \\
&\quad + \frac{d}{d\varepsilon} \int_a^{b+\varepsilon \xi} F(x, y + \varepsilon \eta, y' + \varepsilon \eta')dx \bigg|_{\varepsilon = 0} \\
&= h_x(b, y(b))\xi + h_y(b, y(b))\Big[y'(b)\xi + \eta(b)\Big] \\
&\quad + F(b, y(b), y'(b))\xi + \int_a^b \Big[F_y(x, y, y')\eta + F_y'(x, y, y')\eta'\Big]dx.
\end{aligned}
$$

After integrating by parts and rearranging the terms, the above expression simplifies to

$$\left[h_x(b,y(b)) + h_y(b,y(b))y'(b) + F(b,y(b),y'(b))\right]\xi + \int_a^b \left(F_y - \frac{d}{dx}F_{y'}\right)\eta(x)dx$$

$$+ \left[h_y(b,y(b)) + F_{y'}(b,y(b),y'(b))\right]\eta(b). \tag{4.90}$$

The value of ξ is not affected by the presence of the function h and hence, using the results of the previous section we see that

$$\xi = -\frac{\eta(b)\varphi_y}{\varphi_x + \varphi_y\, y'(b)}.$$

Suppose we can solve for y in terms of x in $\varphi(x,y) = 0$. If so, then we set $y = g(x)$. Now

$$\frac{d}{dx}\varphi(x,y) = \varphi_x + \varphi_y\, y' = 0.$$

This implies that

$$y' = -\frac{\varphi_x}{\varphi_y} = g'(x).$$

We may solve for φ_x and obtain $\varphi_x = -g'(x)\varphi_y$. As a consequence, we will have

$$\xi = -\frac{\eta(b)\varphi_y}{\varphi_x + \varphi_y\, y'(b)} = \frac{\eta(b)}{g'(b) - y'(b)}.$$

Substituting into (4.90) and factoring $\eta(b)$ give

$$\left[\frac{h_x + h_y y' + F}{g' - y'} + h_y + F_{y'}\right]\eta\Big|_{x=b} + \int_a^b \left(F_y - \frac{d}{dx}F_{y'}\right)\eta(x)dx = 0. \tag{4.91}$$

Arguing as before one obtains from (4.91) that

$$F_y - \frac{d}{dx}F_{y'} = 0 \tag{4.92}$$

and the transversality condition

$$\frac{h_x + h_y y' + F}{g' - y'} + h_y + F_{y'},$$

which simplifies to

$$\left[h_x + F + g'h_y + (g' - y')F_{y'}\right]\Big|_{x=b} = 0. \tag{4.93}$$

Note that the term y in (4.93) is the solution of the Euler-Lagrange equation given by (4.92). Along the lines of the preceding discussion, if $y(b)$ is fixed and the left end point $y(a)$ varies along a curve $y = l(x)$, then the corresponding transversality condition is

$$\left[h_x + l'h_y - F - (l' - y')F_{y'} \right]\Big|_{x=a} = 0 \qquad (4.94)$$

Example 4.22 Find the extremal of the functional

$$L(y) = (\pi/2)^2 + \int_0^{\pi/2} \left((y')^2 - y^2 \right) dx, \quad y(0) = 0,$$

and $y(\pi/2)$ varies along the curve $y + 2 - x^2 = 0$.

Here we have

$$F = (y')^2 - y^2, \quad h(x) = x^2.$$

Thus, (4.92) yields

$$y''(x) + y(x) = 0,$$

which has the general solution

$$y(x) = c_1 \cos(x) + c_2 \sin(x).$$

Applying the first boundary condition, we arrive at $c_1 = 0$. To obtain c_2 we make use of (4.93). Let $y(x) = c_2 \sin(x)$. By computing all necessary terms, condition (4.93) yields,

$$\left[2b + \left((y')^2(b) - y^2(b) \right) + \left(-2b - y'(b) \right)(2y'(b)) \right]\Big|_{b=\frac{\pi}{2}} = 0.$$

Since $y(\pi/2) = c_2$ and $y'(\pi/2) = 0$, the above expression yields

$$2(\pi/2) + (0 - c_2^2) + (-2(\pi/2) - 0)(0) = 0.$$

Solving for c_2 we obtain $c_2 = \pm\sqrt{\pi}$, and so the extremal is

$$y(x) = \pm\sqrt{\pi}\sin(x).$$

$\square$

4.10.2   Exercises

Exercise 4.68 *Find the shortest distance from the point (a,A) to the nearest point $(b,y(b))$ on the line with slope m, $y = mx + c$.*

Exercise 4.69 *Find the extremal $y = y(x)$ for the functional*

$$J(y) = \int_0^b \frac{\sqrt{1+y'^2}}{y}\, dx, \quad y(0) = 0$$

and $y(b)$ varies along the circle

$$(x-9)^2 + y^2 = 9.$$

Exercise 4.70 *Find the extremal* $y = y(x)$ *for the functional*

$$J(y) = \int_0^b \frac{\sqrt{1+y'^2}}{y} \, dx, \quad y(0) = 0$$

and $y(b)$ *varies along the line* $y = x - 5$.

Exercise 4.71 *Consider the variational*

$$L(y) = \int_a^b xy\sqrt{1+y'^2} \, dx, \quad y(a) = A$$

and $y(b)$ *varies along the curve* $y = g(x)$. *Show that at the point* $x = b$,

$$g'(b)y'(b) = -1.$$

Of course same results hold if we interchange the boundary conditions.

Exercise 4.72 *Find the extremal* $y = y(x)$ *for the functional*

$$J(y) = \int_1^b x^3 y'^2 \, dx, \quad y(1) = 0$$

and $y(b)$ *varies along the curve* $x^2(y+2) - 2 = 0$.

Exercise 4.73 *Find the extremal* $y = y(x)$ *for the functional*

$$J(y) = \int_0^b y'^2 \, dx$$

(a) $y(0) = 1$ *and* $y(b)$ *varies along the curve* $y - 2x + 3 = 0$.

(b) $y(0) = 2$ *and* $y(b)$ *varies along the curve* $y - \sin(x) = 0$.

Exercise 4.74 *Derive* (4.94) *for the functional*

$$L(y) = h(a, y(a)) + \int_a^b F(x, y, y') \, dx, \quad y(b) = B$$

and y *at* a *varies along the curve* $l(x)$.

Exercise 4.75 *Find the extremal of the functional*

$$L(y) = (\pi/2)^2 + y^2(\pi/2) + \int_0^{\pi/2} ((y')^2 - y^2) \, dx, \quad y(0) = 0,$$

and $y(\pi/2)$ *varies along the curve* $y + 1 - x^2 = 0$.

4.11   Corners and Broken Extremal

In Example 4.6 of Section 4.2 we touched on broken extremal. In this section we want to make the concept formal and more precise. So far, we have looked at extremal $y(x) \in C^2([a,b])$, which is not always the case. Let's begin with the following example.

Example 4.23   Consider the variational

$$L(y) = \int_{-2}^{2} y^2(2-y')^2 dx, \quad y(-2) = 0, \, y(2) = 2.$$

Then the second-order differential equation corresponding to the Euler-Lagrange equation

$$F - y'F_{y'} = c_1$$

is

$$y^2(2-y')(2+y') = c_1,$$

or

$$y^2(4-y'^2) = c_1.$$

If $c_1 = 0$, then we obtain the two solutions

$$y = 0, \quad \text{or} \quad y = \pm 2x + B.$$

Easy to see that neither solution satisfy both boundary conditions. Thus, we suspect at least for now that $c_1 \neq 0$. So we assume $c_1 \neq 0$ and obtain

$$y'^2 = \frac{y^2 - c_1}{y^2}.$$

After separating the variables we arrive at

$$dx = \pm \frac{y}{\sqrt{y^2 - c_1}} dy.$$

An integration of both sides yields

$$x = \pm\sqrt{y^2 - c_1} + c_2.$$

Rearrange the terms to obtain the solution

$$(x - c_2)^2 = y^2 - c_1,$$

which is hyperbola. Next, we make use of both boundary conditions to evaluate c_1 and c_2. With that being said, the following two equations are obtained.

$$(-2 - c_2)^2 = -c_1$$

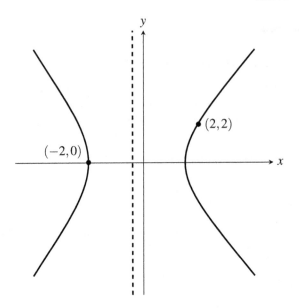

FIGURE 4.15
There is no smooth path that connects boundary conditions.

$$(2 - c_2)^2 = 4 - c_1.$$

Solving for c_1 in the first equation and substituting it into the second equation yields

$$(2 - c_2)^2 = 4 + (2 + c_2)^2.$$

After expanding the terms we obtain $c_2 = -\frac{1}{2}$. Consequently, using $c_1 = -(2 + c_2)^2$ we see that $c_1 = -\frac{9}{4}$. Finally, the solution is

$$y^2 = (x + 1/2)^2 - \frac{9}{4},$$

which is a hyperbola as depicted in Fig. 4.15.

It is clear from Fig. 4.15 that the endpoints are on opposite branches of the hyperbola and hence there is no smooth extremal curve that connects $(-2, 0)$ and $(2, 2)$. Therefore, we must seek a broken curve or curve with corners to connect the endpoints. A broken extremal is a continuous extremal whose derivative has jump discontinuities at a finite number of points. We will revisit this example once we develop the needed conditions to obtain a piecewise continuous extremal. □

In what to follow, we assume there is one corner point and obtain necessary conditions for the continuity of the broken extremal. Assume we have a corner point at $x_* \in (a, b)$ and let y be an extremal of

$$L(y) = \int_a^b F(x, y, y')dx, \quad y(a) = A, \; y(b) = B \tag{4.95}$$

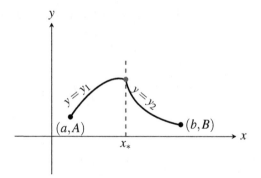

FIGURE 4.16
Broken path with one corner point.

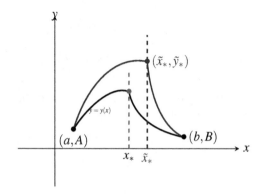

FIGURE 4.17
Perturbing corner point x_* with $\tilde{x}_*$.

where

$$y(x) = \begin{cases} y_1(x), & a \leq x \leq x_* \\ y_2(x), & x_* \leq x \leq b. \end{cases}$$

See Fig. 4.16

Now we perturb the corner point along with the broken extremal. See Fig. 4.17. Let ξ_1, ξ_2 be functions of x and positive. Then the perturbed point $(\tilde{x}_*, \tilde{y}_*)$ must satisfy, for the purpose of compatibility, the relations

$$\tilde{x}_* = x_* + \varepsilon \xi_1,$$

$$\tilde{y}_* = y_* + \varepsilon \xi_2. \tag{4.96}$$

As before, notice that

$$\tilde{y}_* = y(x_* + \varepsilon \xi_1) + \varepsilon \eta(x_* + \varepsilon \xi_1).$$

We follow the same procedure as in Section 4.10. We write our variational as the sum of two variations in the sense that

$$
\begin{aligned}
L(y) &= L_1(y_1) + L_2(y_2) \\
&= \int_a^{x_*} F(x, y_1, y_1') dx + \int_{x_*}^b F(x, y_2, y_2') dx.
\end{aligned}
$$

First, we consider $L(y_1)$.

$$
\begin{aligned}
0 = \frac{dL_1(\varepsilon)}{d\varepsilon} &= \frac{d}{d\varepsilon} \int_a^{x_* + \varepsilon \xi_1} F(x, y_1 + \varepsilon \eta, y_1' + \varepsilon \eta') dx \Big|_{\varepsilon=0} \\
&= F(x_*, y_1(x_*), y_1'(x_*)) \xi_1 \\
&\quad + \int_a^{x_*} \left[\frac{\partial F}{\partial y_1}(x, y_1 + \varepsilon \eta, y_1' + \varepsilon \eta') \eta + \frac{\partial F}{\partial y_1'}(x, y_1 + \varepsilon \eta, y_1' + \varepsilon \eta') \eta' \right] dx \Big|_{\varepsilon=0}.
\end{aligned}
$$

After integrating by parts, the above expression yields,

$$
F(x_*, y_1(x_*), y_1'(x_*)) \xi_1 + F_{y_1'}(x_*, y_1(x_*), y'(x_*)) \eta(x_*) + \int_a^{x_*} \left(F_{y_1} - \frac{d}{dx} F_{y_1'} \right) \eta(x) dx = 0.
$$
(4.97)

Next we compute $\eta(x_*)$. By Taylor's theorem and for small ε, the first term from the right of third equation in (4.96) yields

$$
\begin{aligned}
y(x_* + \varepsilon \xi_1) &= y(x_*) + \varepsilon \xi_1 y'(x_*) + O(\varepsilon^2) \\
&= y_* + \varepsilon \xi_1 y'(x_*) + O(\varepsilon^2).
\end{aligned}
$$

Similarly,

$$
\varepsilon \eta(x_* + \varepsilon \xi_1) = \varepsilon \eta(x_*) + O(\varepsilon^2).
$$

Substituting the two expressions into the right-side of the third equation of (4.96) and then using the second equation of (4.96) yield

$$
\begin{aligned}
y_* + \varepsilon \xi_2 &= y(x_* + \varepsilon \xi_1) + \varepsilon \eta(x_* + \varepsilon \xi_1) \\
&= y_* + \varepsilon \xi_1 y'(x_*) + \varepsilon \eta(x_*) + O(\varepsilon^2).
\end{aligned}
$$

This provides us with

$$
\varepsilon \xi_2 = \varepsilon \xi_1 y'(x_*) + \varepsilon \eta(x_* + O(\varepsilon^2).
$$

Solving for $\eta(x_*)$ gives

$$
\eta(x_*) = \xi_2 - \xi_1 y'(x_*) + O(\varepsilon).
$$
(4.98)

Substituting (4.98) into (4.97) and using simplified notations we arrive at

$$
\xi_1 [F - y_1' F_{y_1'}] \Big|_{x=x_*} + \xi_2 F_{y_1'} \Big|_{x=x_*} + \int_a^{x_*} \left(F_{y_1} - \frac{d}{dx} F_{y_1'} \right) \eta(x) dx = 0.
$$

From the above expression we obtain the familiar Euler-Lagrange equation

$$F_{y_1} - \frac{d}{dx}F_{y_1'} = 0, \tag{4.99}$$

plus the additional condition

$$\left\{ \xi_1[F - y_1'F_{y_1'}] + \xi_2 F_{y_1'} \right\}\Big|_{x=x_*} = 0. \tag{4.100}$$

By doing similar work we obtain from $L_2(y_2)$, equation (4.98) and the additional condition

$$\left\{ -\xi_1[F - y_2'F_{y_2'}] - \xi_2 F_{y_2'} \right\}\Big|_{x=x_*} = 0. \tag{4.101}$$

Combining conditions (4.100) and (4.101) we arrive at

$$\left\{ \xi_1\left[\left(F(x,y_1,y_1') - y_1'F_{y_1'} \right) - \left(F(x,y_2,y_2') - y_2'F_{y_2'} \right) \right] + \xi_2\left(F_{y_1'} - F_{y_2'} \right) \right\}\Big|_{x=x_*} = 0.$$

In light of the fact that the point of discontinuity is free to change, we can independently change both ξ_1 and ξ_2 or set them both to zero. We can therefore divide the condition into two conditions.

$$\left\{ [F(x,y_1,y_1') - y_1'F_{y_1'}] - [F(x,y_2,y_2') - y_2'F_{y_2'}] \right\}\Big|_{x=x_*} = 0,$$

$$[F_{y_1'} - F_{y_2'}]\Big|_{x=x_*} = 0.$$

The above corner conditions can be expressed in terms of limits from the left and right rather than dividing y into y_1 and y_2. That is

$$\lim_{x\to x_*^-} [F(x,y,y') - y'F_{y'}] = \lim_{x\to x_*^+} [F(x,y,y') - y'F_{y'}], \tag{4.102}$$

$$\lim_{x\to x_*^-} F_{y'} = \lim_{x\to x_*^+} F_{y'} \tag{4.103}$$

must hold at very corner point. The corners conditions given by (4.102) and (4.103) are called *Weirstrass-Erdmann corner conditions*. We proved the following theorem.

Theorem 4.18 *For the functional* (4.95) *with one corner point* $x_* \in (a,b)$ *conditions* (4.102) *and* (4.103) *must hold.*

We note that (4.102) and (4.103) hold everywhere in (a,b) since if we are not at a corner point $y'(x)$ is continuous as is $F_{y'}$.

Back to Example 4.23. We saw that for $c_1 \neq 0$, then there is no smooth extremal that connect both endpoints. Thus, we must look for an extremal that is piecewise defined or has a corner. We are left with the choice of $c_1 = 0$. In this case the Euler-Lagrange equation has the two solutions

$$y = 0, \text{ or } y = 2x + B.$$

The branch of the solution $y = 0$, satisfies the first boundary condition $y(-2) = 0$. In addition, the second part of the solution $y = 2x + B$ satisfies $y(2) = 2$, for $B = -2$. The corner conditions (4.102) and (4.103) are satisfied independently of the location of the corner point in $(-2, 2)$ since

$$F - y'F_{y'} = 0 \quad \text{and} \quad F_{y'} = 0.$$

Thus, to have a continuous extremal, we may take $x_* = 1$, (corner point at 1) and then the solution is defined by

$$y(x) = \begin{cases} 0, & -2 \leq x \leq 1 \\ 2x - 2, & 1 \leq x \leq 2. \end{cases}$$

Corollary 8 *If $F_{y'y'} \neq 0$, then an extremal for the functional (4.95) must be smooth. That is it can not have corners.*

Proof *Let y_0 be an extremal of (4.95) with a corner point at $x_* \in (a,b)$. Then from the corner condition (4.103), we must have the continuity condition*

$$\lim_{x \to x_*^-} F_{y'} = \lim_{x \to x_*^+} F_{y'}.$$

That is

$$F_{y'}\left(x_*^-, y(x_*^-), y'(x_*^-)\right) - F_{y'}\left(x_*^+, y(x_*^+), y'(x_*^+)\right) = 0. \tag{4.104}$$

Let $p = y'(x_*^-)$ and $q = y'(x_*^+)$. Then by the Mean value theorem, there exists an $\alpha \in (0,1)$ such that

$$F_{y'}\left(x_*, y(x_*), p)\right) - F_{y'}\left(x_*, y(x_*), q)\right) = (p - q)F_{y'y'}\left(x_*, y(x_*), q + \alpha(p - q)\right).$$

But then from (4.104), this implies that

$$(p - q)F_{y'y'}\left(x_*, y(x_*), q + \alpha(p - q)\right) = 0.$$

Or,

$$F_{y'y'}\left(x_*, y(x_*), q + \alpha(p - q)\right) = 0,$$

which is a contradiction to the fact that $F_{y'y'} \neq 0$. This completes the proof.

Example 4.24 According to Corollary 8 the extremal of the variational

$$L(y) = \int_a^b \left(\alpha y'^2 + \varphi(y) + \phi(x)\right) dx, y(a) = A, \ y(b) = B$$

where φ and ϕ are continuous functions of y and x, respectively, has no corner points when $\alpha \neq 0$, since $F_{y'y'} = 2\alpha$. $\qquad\square$

The next example shows that $F_{y'y'} \neq 0$, is only a necessary condition.

Example 4.25 The variational

$$L(y) = \int_1^2 \left[(x^3 + y^2)y' + 3x^2 y \right] dx, \quad y(1) = 1, \ y(2) = -1$$

was considered in Example 4.8, and it was shown that the functional was path independent. In addition, the corresponding Euler-Lagrange equation is $-2yy' = 0$, from which we obtain either $y(x) = 0$ or $y(x) = constant$. Hence, neither one satisfies both boundary conditions. Notice that $F_{y'y'} = 0$. Clearly, the path $y_0(x) = 2x - 1$ connects both endpoints and it can be easily computed and verified that $L(y_0(x)) = -\frac{29}{3}$. (See Example 4.8). Next, we construct another path with a corner point that will piecewise connect both endpoints. Note that since the functional is path independent we may assume a corner point anywhere in $(1,2)$. Thus we may take the corner point to be at $(3/2, 2)$, and we wish to construct a piecewise continuous and linear path in the form of

$$y(x) = \begin{cases} A_1 x + B_1, & 1 \le x \le 3/2 \\ A_2 x + B_2, & 3/2 \le x \le 2. \end{cases}$$

Applying $y(1) = 1$, $y(2) = -1$, we obtain $B_1 = 1 - A_1$ and $B_2 = -1 - 2A_2$. Therefore,

$$y(x) = \begin{cases} A_1 x + 1 - A_1, & 1 \le x \le 3/2 \\ A_2 x - 1 - 2A_2, & 3/2 \le x \le 2. \end{cases} \tag{4.105}$$

Applying the corner condition

$$\lim_{x \to x_*^-} F_{y'} = \lim_{x \to x_*^+} F_{y'}$$

at $x_* = 3/2$ we arrive at

$$\lim_{x \to (3/2)^-} y(x) = \lim_{x \to (3/2)^+} y(x),$$

or

$$A_1 \frac{3}{2} + 1 - A_1 = A_2 \frac{3}{2} - 1 - 2A_2.$$

This results into

$$A_1 + A_2 = -4. \tag{4.106}$$

Making use of the other corner condition

$$\lim_{x \to (3/2)^-} \left[F - y' F_{y'} \right] = \lim_{x \to (3/2)^+} \left[F - y' F_{y'} \right],$$

yields to

$$\lim_{x \to (3/2)^-} (3x^2 y) = \lim_{x \to (3/2)^+} (3x^2 y).$$

Simplifying $3x^2$ from both sides, we arrive at the same expression (4.106). Due to the continuity requirement at $3/2$, we must have the solution match at $3/2$. That is, $y(3/2) = 2$. Applying this to the first branch of the solution we obtain

$$A_1 \frac{3}{2} + 1 - A_1 = 2,$$

or $A_1 = 2$. Using (4.106), we arrive at $A_2 = -6$. Substituting A_1 and A_2 with their values into (4.105), leads to the continuous and linear path

$$y_*(x) = \begin{cases} 2x - 1, & 1 \le x \le 3/2 \\ -6x + 11, & 3/2 \le x \le 2. \end{cases}$$

One may check that $L(y_*(x)) = -\frac{29}{3}$, also, since the variational is path independent.

□

4.11.1 Exercises

Exercise 4.76 *In the spirit of Example 4.23 discuss the variational*

$$L(y) = \int_{-1}^{1} y^2(1 - y')^2 dx, \quad y(-1) = 0, \ y(1) = 1.$$

Exercise 4.77 *Provide all details for obtaining (4.101).*

Exercise 4.78 *Show the solution of the variational*

$$L(y) = \int_{0}^{3} (y')^3 dx, \quad y(0) = 0, \ y(3) = 1,$$

has no corner point and find its extremal.

Hint: Check the corner conditions.

Exercise 4.79 *In the spirit of Example 4.23 discuss the variational*

$$L(y) = \int_{0}^{1} (1 - y'^2)^2 dx, \quad y(0) = 0, \ y(1) = 1/4.$$

Exercise 4.80 *Find the broken extremal of the variational*

$$L(y) = \int_{0}^{4} (y'^2 - 1)^2 (y' + 1)^2 dx, \quad y(0) = 0, \ y(4) = 2.$$

Exercise 4.81 *Redo Example 4.24, with corner point at $(4/3, 5)$.*

4.12 Variational Problems with Constraints

So far, we have dealt with functionals where the boundary points are fixed or allowed to freely move along a well-defined curve. In this section, we generalize that result to situations where equality constraints are imposed on the admissible curves. It might be helpful to review the finite-dimensional problem with constraints. Recall from calculus that such problems can be optimized using the concept of the Lagrange

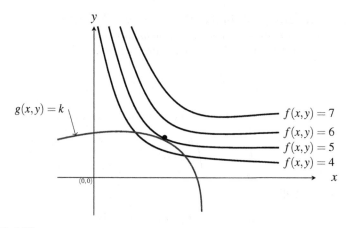

FIGURE 4.18
Level curves and Lagrange multiplier.

multiplier. The same concept will be used to deal with variational problems with constraints. First, we begin with a short review of Lagrange multipliers for finite-dimensional optimization problems.

Suppose we want to find the extreme value of the function $f(x,y)$ subject to the constraint $g(x,y) = k$, for a fixed constant k. In other words, if $f(x.y)$ has an extrema at (x^*, y^*), then (x^*, y^*) must lie on the level $g(x,y) = k$. To maximize $f(x,y)$ subject to $g(x,y), = k$ is to find the largest value of c such that the level curve $f(x,y) = c$ intersects $g(x,y) = k$. In Fig 4.18, $c = 4, 5, 6, 7$. Also, it appears from Fig. 4.18 that this happens when the curves touch each other; that is, when they have a common tangent line. (Otherwise, the values of c could be increased further.) This can only mean that the normal lines at (x_0, y_0) where they touch are identical. This implies that the gradient vectors are parallel. In other words,

$$\nabla f(x_0, y_0) = \lambda \nabla g(x_0, y_0),$$

for some scalar λ. The number λ is called a *Lagrange multiplier*. The next theorem can be found in any advanced calculus textbook.

Theorem 4.19 (Lagrange Multiplier Rule) *Let f and g be differentiable functions with $g_x(x_0, y_0)$ and $g_y(x_0, y_0)$ not both zero. If (x_0, y_0) provides an extreme value to $f(x,y) = 0$ subject to the constraint $g(x, y) = k$, then there exists a constant λ such that*

$$f_x^*(x_0, y_0) = 0, \quad f_y^*(x_0, y_0) = 0,$$

and $g^*(x_0, y_0) = k$, where $f^* = f + \lambda g$.

The above theorem is valid for functions in $\mathbb{R}^n$. Recall that fort $x \in \mathbb{R}^n$ and $f : \mathbb{R}^n \to \mathbb{R}$ is a smooth function, then the gradient of f, denoted by ∇f is the vector

$$\nabla f = < \frac{\partial f}{\partial x_1}, \frac{\partial f}{\partial x_2}, \dots, \frac{\partial f}{\partial x_n} >.$$

Theorem 4.20 (Lagrange Multiplier Rule) *Let $\omega \subset \mathbb{R}^n$ and let $f, g : \omega \to \mathbb{R}$, be smooth functions. Suppose f has a local extremum at $x^* \in \mathbb{R}^n$ subject to the constraint $g(x) = 0$. If $\nabla f(x^*) \neq 0$, then there is a number λ such that*

$$\nabla f(x^*) = \lambda \nabla g(x^*).$$

Let $y, z \in C^2[a, b]$, and consider the variational with two variables y and z

$$L(y, z) = \int_a^b F(x, y, y', z, z') dx, \tag{4.107}$$

with boundary conditions

$$y(a) = A_1, \ y(b) = B_1 \ z(a) = A_2, \ z(b) = B_2,$$

and subject to the constraint

$$\varphi(x, y, z) = 0. \tag{4.108}$$

Let $\eta_1 = \eta_1(x) \in C([a, b])$ and $\eta_2 = \eta_2(x) \in C([a, b])$, such that

$$\eta_1(a) = \eta_1(b) = \eta_2(a) = \eta_2(b) = 0.$$

From Section 4.6 we have that

$$\int_a^b \left[\left(F_y - \frac{d}{dx} F_{y'} \right) \eta_1(x) dx + \left(F_z - \frac{d}{dx} F_{z'} \right) \eta_2(x) \right] dx = 0. \tag{4.109}$$

For the same $\eta_1(x), \eta_2(x)$ we see that

$$
\begin{aligned}
\delta \varphi(x, y, z) &= \frac{d\varphi}{d\varepsilon}(y + \varepsilon \eta_1, z + \varepsilon \eta_2)\big|_{\varepsilon=0} \\
&= \left[\varphi_y(y + \varepsilon \eta_1, z + \varepsilon \eta_1) \eta_1 + \varphi_z(y + \varepsilon \eta_2, z + \varepsilon \eta_2) \eta_2 \right]\big|_{\varepsilon=0} \\
&= \varphi_y(y, z) \eta_1 + \varphi_z(y, z) \eta_2.
\end{aligned}
$$

Setting $\delta \varphi(x, y, z) = 0$, we obtain

$$\varphi_y(y, z) \eta_1 + \varphi_z(y, z) \eta_2 = 0.$$

Multiply the above expression with Lagrange multiplier λ and then integrate the resulting equation from a to b to obtain

$$\int_a^b \lambda \left(\varphi_y(y, z) \eta_1 + \varphi_z(y, z) \eta_2 \right) dx = 0. \tag{4.110}$$

Subtracting (4.110) from (4.109) yields the following expression,

$$\int_a^b \left[\left(F_y - \frac{d}{dx} F_{y'} - \lambda \varphi_y \right) \eta_1(x) dx + \left(F_z - \frac{d}{dx} F_{z'} - \lambda \varphi_z \right) \eta_2(x) \right] dx = 0. \tag{4.111}$$

Applying Lemma 10 one obtains from (4.111) the Euler-Lagrange equations,

$$F_y - \frac{d}{dx}F_{y'} - \lambda\varphi_y = 0 \tag{4.112}$$

and

$$F_z - \frac{d}{dx}F_{z'} - \lambda\varphi_z = 0. \tag{4.113}$$

Theorem 4.21 *Let* $y, z \in C^2[a, b]$ *be extremals for the variational* (4.107) *with boundary conditions specified at* $x = a$ *and* $x = b$, *subject to the constraint function* (4.108). *Then* $y(x)$ *and* $z(x)$ *must satisfy the Euler-Lagrange equations given by* (4.112) *and* (4.113).

The next theorem easily generalizes Theorem 4.21 to n constraints functions and its proof is Exercise 4.82.

Theorem 4.22 *Let* $y, z \in C^2[a, b]$ *be extremals for the variational* (4.107) *with boundary conditions specified at* $x = a$ *and* $x = b$, *subject to the n constraints*

$$\varphi_i(y, z) = 0, \quad i = 1, 2, \dots n.$$

Then $y(x)$ *and* $z(x)$ *must satisfy the Euler-Lagrange equations*

$$F_y - \frac{d}{dx}F_{y'} - \sum_{i=1}^{n}\lambda_i\frac{\partial\varphi_i}{\partial y} = 0, \quad F_z - \frac{d}{dx}F_{z'} - \sum_{i=1}^{n}\lambda_i\frac{\partial\varphi_i}{\partial z} = 0.$$

Example 4.26 Find the extremals y and z that minimize the functional

$$L(y, z) = \int_0^{\pi/2}\left(1 + y'^2 + z'^2\right)dx, \quad y(0) = z(0) = y(\pi/2) = z(\pi/2) = 0,$$

subject to the constraint $y^2 + z^2 = 5$. Here $\varphi(y, z) = y^2 + z^2 - 5$. Thus (4.112) and (4.113) generate the two second-order differential equations

$$y'' + \lambda y = 0, \quad z'' + \lambda z = 0.$$

Remember $\lambda \in \mathbb{R}$ and so, special care must be applied. We will do this in three separate cases.

case 1 $\lambda = 0$. In this case the general solution for the first differential equation is

$$y(x) = c_1 x + c_2.$$

Applying the boundary conditions, we get $c_1 = c_2 = 0$. This results in the trivial solution $y(x) = 0$, which has to be rejected since it does not satisfy the constraint.

case 2. $\lambda < 0$. Say $\lambda = -\alpha^2$, where $\alpha > 0$. Then the general solution is

$$y(x) = c_1 e^{\alpha x} + c_2 e^{-\alpha x}$$

Applying the boundary condition, we get $c_1 = c_2 = 0$. Again, this results in the trivial solution $y(x) = 0$, which has to be rejected since it does not satisfy the constraint.

case 3. $\lambda > 0$. Say $\lambda = \alpha^2$, where $\alpha > 0$. Then the general solution is

$$y(x) = c_1 \cos(\alpha x) + c_2 \sin(\alpha x)$$

Applying the boundary conditions $y(0) = 0$, yields $c_1 = 0$. Similarly, $0 = y(\pi/2)$ implies that

$$c_2 \sin(\alpha \frac{\pi}{2}).$$

So we have either $c_2 = 0$, which results in the trivial solution again, or we set

$$\sin(\alpha \frac{\pi}{2}) = 0,$$

which holds when $\alpha \frac{\pi}{2} = n\pi$, or when $\alpha = 2n$, $n = 1, 2, \ldots$ A similar argument can be applied to the differential equation in z and obtain

$$y(x) = c \sin(2nx), \quad z(x) = d \sin(2nx), \, n = 1, 2, \ldots$$

Next we evaluate L at the obtained y and z to see if they minimize L since the integrand of L is positive for all functions y and z.

$$
\begin{aligned}
L\big(c \sin(2nx), d \sin(2nx)\big) &= \int_0^{\pi/2} \Big(1 + 4n^2(c^2 + d^2)\cos^2(2nx)\Big) dx \\
&= \frac{\pi}{2} + 4n^2(c^2 + d^2) \int_0^{\pi/2} \cos^2(2nx) dx \\
&= \frac{\pi}{2} + \pi n^2(c^2 + d^2). \qquad\qquad (4.114)
\end{aligned}
$$

Note that expression (4.114) is increasing in n and therefore its minimum is achieved when $n = 1$. That is y and z minimize L for $n = 1$. Therefore, the extremals are

$$y(x) = c \sin(2x), \quad z(x) = d \sin(2x),$$

where

$$c^2 \sin^2(2x) + d^2 \sin^2(2x) = 5, \, 0 < x < \pi/2.$$

$\square$

4.12.1 Exercises

Exercise 4.82 *Prove Theorem 4.22.*

Exercise 4.83 *Show that if $y, z, w \in C^2[a, b]$ are extremals for the variational*

$$L(y, z) = \int_a^b F(x, y, y', z, z') dx,$$

with boundary conditions specified at $x = a$ and $x = b$, subject to the constraint function

$$\varphi(y, z, w) = 0,$$

then $y(x), z(x)$ and $w(x)$ must satisfy the Euler-Lagrange equations

$$F_y - \frac{d}{dx} F_{y'} - \lambda \varphi_y = 0,$$

$$F_z - \frac{d}{dx} F_{z'} - \lambda \varphi_z = 0,$$

$$F_w - \frac{d}{dx} F_{w'} - \lambda \varphi_w = 0.$$

Exercise 4.84 *Use Exercise 4.83 to find the extremals y, z and w that minimizes the functional*

$$L(y, z, w) = \int_0^b \frac{1}{2} (y'^2 + z'^2 + w'^2) dx,$$

with boundary conditions

$$y(0) = z(0) = 0, \ w(0) = 0, \ y(b) = z(b) = 0, \ w(b) = 0,$$

subject to the constraint $y^2 + z^2 + w^2 = 1$.

4.13   Isoperimetric Problems

Let $y = y(x) \in C^2([a, b])$ and consider the *isoperimetric problem*

$$L(y) = \int_a^b F(x, y, y') dx, \quad y(a) = A, \ y(b) = B, \tag{4.115}$$

subject to the integral constraint

$$W(y) = \int_a^b G(x, y, y') dx = d, \tag{4.116}$$

where d is a fixed constant. The fixed functions F and G are assumed to be twice continuously differentiable. The subsidiary condition (4.116) is called *isoperimetric constraint*. Before, we assumed a local extremal $y(x)$ in a family of admissible functions with respect to which we carry out the extremization. A one parameter family $y(x) + \varepsilon \eta(x)$ is not , however, a suitable choice since those curves may not maintain the consistency of W. Therefore, we introduce a two parameters family

$$z = y(x) + \varepsilon_1 \eta_1(x) + \varepsilon_2 \eta_2(x),$$

where $\eta_1, \eta_2 \in C^2([a,b])$ such that $\eta_1(a) = \eta_1(b) = \eta_2(a) = \eta_2(b) = 0$, and ε_1 and ε_2 are real parameters ranging over the intervals containing the origin. We make the assumption that y is not an extremal of W. Therefore, for any choice of η_1 and η_2 there will be values ε_1 and ε_2 in the neighborhood of $(0,0)$, for which $W(z) = d$. Let

$$S_1(\varepsilon_1, \varepsilon_2) = \int_a^b F(x,z,z')dx,$$

and

$$S_2(\varepsilon_1, \varepsilon_2) = \int_a^b G(x,z,z')dx = C.$$

Since y is a local extremal of (4.115), subject to the constraint (4.116), the point $(\varepsilon_1, \varepsilon_2) = (0,0)$ must be a local extremal for $S_1(\varepsilon_1, \varepsilon_2)$ subject to the constraint $S_2(\varepsilon_1, \varepsilon_2) = C$. This is just a differential calculus problem and so the Lagrange multiplier rule might be applied. That is, there must be a constant λ such that

$$\frac{\partial S^*}{\partial \varepsilon_1} = \frac{\partial S^*}{\partial \varepsilon_2} = 0, \quad \text{at } (\varepsilon_1, \varepsilon_2) = (0,0), \tag{4.117}$$

where

$$S^* = S_1 + \lambda S_2 = \int_a^b F^*(x,z,z')dx,$$

with

$$F^* = F + \lambda G.$$

Substituting $z = y(x) + \varepsilon_1 \eta_1(x) + \varepsilon_2 \eta_2(x)$ into S^* and then calculating partial derivatives with respect to ε_1 and ε_2 we arrive at

$$\frac{\partial S^*}{\varepsilon_i}(\varepsilon_1, \varepsilon_2) = \int_a^b \left[F_y^*(x,y,y')\eta_i(x) + F_{y'}^*(x,y,y')\eta_i'(x) \right] dx, \quad i = 1,2.$$

Setting

$$\frac{\partial S^*}{\varepsilon_i}(\varepsilon_1, \varepsilon_2)\Big|_{(\varepsilon_1, \varepsilon_2)=(0,0)} = 0,$$

followed by an integration by parts on the term that involves η' and then applying Lemma 10, we arrive at the Euler-Lagrange equation

$$F_y^*(x,y,y') - \frac{d}{dx}F_{y'}^*(x,y,y') = 0, \tag{4.118}$$

which is a necessary condition for an extremal. We proved the following theorem.

Theorem 4.23 *Let $y \in C^2[a,b]$. If is y not an extremal of (4.116) but an extremal for the variational (4.115) with boundary conditions specified at $x = a$ and $x = b$, subject to the isoperimetric constraint (4.116), then $y(x)$ satisfies the Euler-Lagrange equation (4.118), or*

$$\frac{\partial}{\partial y}\left(F + \lambda G\right) - \frac{d}{dx}\left[\frac{\partial}{\partial y'}\left(F + \lambda G\right)\right] = 0.$$

We furnish an example.

Example 4.27 In this example, we show that the sphere is the solid figure of revolution that, for a given surface area l, has the maximum volume. Consider a curve $y(x) \geq 0$ with $y(0) = 0$, and $y(a) = 0, a > 0$. Revolve $y(x)$ along the x-axis. Then, any short circular strip with a radius y and a height ds has a surface area of $2\pi y \, ds$. Consequently, the total surface area of revolution is

$$l = \int_0^a 2\pi y \, ds = \int_0^a 2\pi y \sqrt{1+y'^2} \, dx.$$

On the other hand, the volume of the solid of revolution is $\int_0^a \pi y^2 \, dx$. Thus, the problem can be formulated as a variational problem with constraints. In other words, we want to maximize

$$L(y) = \int_0^a \pi y^2 \, dx,$$

subject to the constraint

$$\int_0^a 2\pi y \sqrt{1+y'^2} \, dx = l \quad \text{(constant)}.$$

Set

$$F^* = \pi y^2 + \lambda 2\pi y \sqrt{1+y'^2}.$$

Since x does not enter in F^*, we will use the Euler-Lagrange equation

$$F^* - y' F_{y'}^* = c.$$

Or,

$$\pi y^2 + 2\pi \lambda y \sqrt{1+y'^2} - y' \frac{2\pi \lambda y y'}{\sqrt{1+y'^2}} = c,$$

which simplifies to

$$\pi y^2 + \frac{2\pi \lambda y}{\sqrt{1+y'^2}} = c.$$

Now $y = 0$, at $x = 0$ and at $x = a$, which can be true if $c = 0$, and wherefore we have

$$y = -\frac{2\lambda}{\sqrt{1+y'^2}}. \tag{4.119}$$

By squaring both sides and then solving for y' we arrive at

$$y' = \frac{\sqrt{4\lambda^2 - y^2}}{y}.$$

Setting $y' = \frac{dy}{dx}$, separating the variables followed by an integration give

$$\int \frac{y}{\sqrt{4\lambda^2 - y^2}} dy = \int dx.$$

This yields the expression

$$-\sqrt{4\lambda^2 - y^2} = x + k, \text{ for constant of integration } k. \tag{4.120}$$

Using $y(0) = 0$, in (4.120) we get $k = \pm 2\lambda$. Substituting k into (4.120) gives

$$\sqrt{4\lambda^2 - y^2} = x \pm 2\lambda.$$

By squaring both sides and rearranging the terms we obtain the solution

$$(x \pm 2\lambda)^2 + y^2 = 4\lambda^2.$$

Hence the obtained curve is a circle centered at $(\pm 2\lambda, 0)$ and radius 2λ. This shows that the solid of revolution is a sphere. To find λ, we make use of (4.119) and obtain $y\sqrt{1+y'^2} = -2\lambda$. Substituting this into the integral constraint we arrive at

$$l = \int_0^a 2\pi y \sqrt{1+y'^2}\, dx = \int_0^a 2\pi(-2\lambda)\, dx = -4\pi\lambda a.$$

This gives $\lambda = \frac{-l}{4\pi a}$.

$\square$

The next theorem easily generalizes Theorem 4.23 and its proof is left as an exercise.

Theorem 4.24 *Let $y, z \in C^2[a,b]$ be extremals for the variational*

$$L(y,z) = \int_a^b F(x,y,z,y',z')dx,$$

with fixed end points, and subject to the isoperimetric constraint

$$W(y,z) = \int_a^b G(x,y,z,y',z')dx = d, \tag{4.121}$$

where d is a fixed constant. If y and z are not extremals to (4.121), then they must satisfy the Euler-Lagrange equations

$$\frac{\partial}{\partial y}\left(F + \lambda G\right) - \frac{d}{dx}\left[\frac{\partial}{\partial y'}\left(F + \lambda G\right)\right] = 0, \tag{4.122}$$

and

$$\frac{\partial}{\partial z}\left(F + \lambda G\right) - \frac{d}{dx}\left[\frac{\partial}{\partial z'}\left(F + \lambda G\right)\right] = 0. \tag{4.123}$$

We provide the following example.

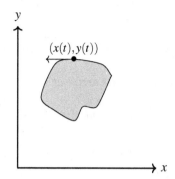

FIGURE 4.19
Dido's area.

Example 4.28 *[Dido's problem]* Most traditions identify Dido as the Phoenician city-state of Tyre's queen, who fled oppression to create her own city in northwest Africa. Tyre is now in Lebanon. The legendary Dido requested a plot of land to farm when she landed in Carthage (Tunisia) in 814 BC. Her request was accepted with the provision that an *n* oxhide should encircle the area. She divided the oxhide into incredibly tiny pieces and arranged them to completely enclose the available land. View Figure 4.19. The problem comes down to finding *the closed curve with a fixed perimeter that encloses the maximum area.* Let's describe the curve by the parametric equations $(x(t), y(t))$ with velocity $(x'(t), y'(t))$ and $x(0) = x(1)$ and $y(0) = y(1)$. Then the length of the curve, or its perimeter is

$$\oint \sqrt{x'^2(t) + y'^2(t)}\, dt = d,$$

where d is the allowed perimeter. To find a formula for the enclosed area, we make use of Green's Theorem, which states that over a region D in the plane with boundary ∂D we have

$$\oint_{\partial D} f\,dx + g\,dy = \int_D \left(\frac{\partial g}{\partial x} - \frac{\partial f}{\partial y}\right) dx\,dy.$$

If we set $f = -\frac{y}{2}$ and $g = \frac{x}{2}$, we get

$$\frac{1}{2}\int_D x\,dy - y\,dx = \int_D dx\,dy.$$

Thus the enclosed area is given by

$$\frac{1}{2}\oint (xy' - yx')\,dt.$$

So the problem comes down to maximizing

$$L(x, y) = \frac{1}{2}\oint (xy' - yx')\,dt,$$

subject to the constraint

$$W(x,y) = \oint \sqrt{x'^2(t) + y'^2(t)}\,dt = d,$$

Using equations (4.122) and (4.123) we arrive at

$$y' = \frac{d}{dt}\frac{\lambda x'}{\sqrt{x'^2 + y'^2}}, \quad and \quad x' = -\frac{d}{dt}\frac{\lambda y'}{\sqrt{x'^2 + y'^2}}.$$

An integration of both equations yield

$$y = \frac{\lambda x'}{\sqrt{x'^2 + y'^2}} + c_1, \quad and \quad x = \frac{\lambda y'}{\sqrt{x'^2 + y'^2}} + c_2.$$

Rearrange the terms and then square both sides of the two equations and get

$$(y - c_1)^2 = \frac{\lambda^2 x'^2}{x'^2 + y'^2}, \quad and \quad (x - c_2)^2 = \frac{\lambda^2 y'^2}{x'^2 + y'^2}.$$

Adding both equations gives

$$(x - c_2)^2 + (y - c_1)^2 = \lambda^2$$

which is a circle centered at (c_2, c_1) of radius λ. To find the radius λ, we substitute the solutions into the isoperimetric constraint. Note that the solutions may be written as

$$x(t) = -\lambda \cos(2\pi t) + c_2, \quad y(t) = \lambda \sin(2\pi t) + c_1.$$

W.l.o.g, assume the circle is centered at $(0,0)$. Then

$$\int_0^1 \sqrt{x'^2(t) + y'^2(t)}\,dt = \int_0^1 2\pi\lambda\,dt = d,$$

implies that $\lambda = \frac{d}{2\pi}$. Another way to find λ is to set the perimeter of a circle with radius λ equal the given length of the circle. That is $2\pi d = \lambda$. $\square$

The next theorem addresses variationals with higher-order derivatives subject to constraints.

Theorem 4.25 *[Euler-Lagrange equation] If a function $y = y(x) \in C^4([a,b])$ is an extremal to the variational problem in (4.45), subject to the constraint*

$$W(y) = \int_a^b G(x, y, y', y'')\,dx = d,$$

then $y(x)$ must satisfy the Euler-Lagrange equation

$$F_y^* - \frac{d}{dx}F_{y'}^* + \frac{d^2}{dx^2}F_{y''} = 0,$$

where

$$F^* = F + \lambda G.$$

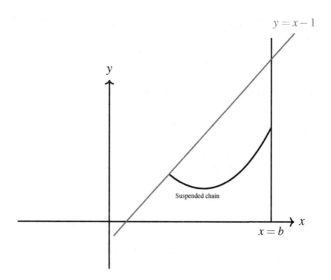

FIGURE 4.20
Catenary; transversality and natural conditions.

The next theorem addresses natural boundary conditions and transversality conditions of variational that are subject to given constraints.

Theorem 4.26 *Let $y = y(x) \in C^2[a,b]$ be an extremal for the variational (4.79) with boundary conditions specified or unspecified at $x = a$ and $x = b$, and subject to the constraint*

$$W(y) = \int_a^b G(x,y,y')dx = d.$$

For $\lambda \in \mathbb{R}$, set

$$F^* = F(x,y,y') + \lambda G(x,y,y'). \tag{4.124}$$

Then 1)-3) of Theorem 4.17 hold when F is replaced with F^*.

Example 4.29 (Catenary revisited) In Exercise 4.39 we presented the problem of the hanging of heavy chain from two fixed points. Now, we are considering the length of the chain to be $l > 1$. In addition, unlike the situation in Exercise 4.39 we let the chain slides freely along the vertical line $x = b$. The left end of the cable, that is at $x = a$ the chain slides along a tilted pole or skewed line. Again, for mathematical convenience we assume the chain density and gravity are both one. See Fig. 4.20. Let $y - x + 1 = 0$ be the tilted pole. Then, we must have $1 \le a < b$. Then the problem is to minimize to potential energy

$$L(y) = \int_a^b y\sqrt{1+y'^2}dx,$$

with $y(b)$ being unspecified and $y(a)$ moves along the curve $y = h(x) = x - 1$, subject to the constraint

$$W(y) = \int_a^b \sqrt{1+y'^2}\,dx = l.$$

Equivalently, we are to find the path $y(x)$ that minimizes L, where

$$F^* = y\sqrt{1+y'^2} + \lambda\sqrt{1+y'^2}.$$

Since F^* does not explicitly depend on the variable x, by Corollary 7 the Euler-Lagrange equation that y must satisfy is

$$y'F_{y'}^* - F^* = D.$$

Or,

$$\frac{y'^2(y+\lambda)}{\sqrt{1+y'^2}} - (y+\lambda)\sqrt{1+y'^2} = D,$$

which simplifies to

$$\frac{y+\lambda}{\sqrt{1+y'^2}} = D.$$

Solving for y' we arrive at

$$y' = \frac{1}{D}\sqrt{(y+\lambda)^2 - D^2}.$$

By letting

$$y+\lambda = D\cosh(t),$$

and then imitating the work of Section 4.5 on minimal surface we arrive at the solution

$$y+\lambda = c_1\cosh(\frac{x+c_2}{c_1}), \tag{4.125}$$

where the constants c_1 and c_2 are to be found. We have a natural boundary condition at $x = b$ which implies that

$$F_{y'}^*\big|_{x=b} = \frac{y'(y+\lambda)}{\sqrt{1+y'^2}}\bigg|_{x=b}.$$

This yields that $y'(b) = 0$, or $y(b) = -\lambda$. Now, if $y(b) = -\lambda$, then (4.125) implies that $\cosh(\frac{b+c_2}{c_1}) = 0$, which can not be. Therefore, we must take

$$y'(b) = 0.$$

From (4.125) we get $y'(x) = \sinh(\frac{x+c_2}{c_1})$. Apply $y'(b) = 0$ to get $\frac{b+c_2}{c_1} = 0$, or $c_2 = -b$. Thus,

$$y(x) = -\lambda + c_1\cosh(\frac{x-b}{c_1}).$$

The transversality condition

$$F^* + \left(h'(x) - y'(a)\right)F_{y'}^*\Big|_{x=a} = 0,$$

yields

$$\frac{(y+\lambda)}{\sqrt{1+y'^2}}(1+y') = 0,$$

or $y'(a) = -1$. Combining this with $y'(x) = \sinh(\frac{x-b}{c_1})$ we arrive at

$$\sinh(\frac{b-a}{c_1}) = 1.$$

Using the isoperimetric constraint we get

$$\int_a^b \sqrt{1+\sinh^2(\frac{x-b}{c_1})}\,dx = c_1\left(0+\sinh(\frac{b-a}{c_1})\right) = l.$$

Combining

$$\sinh(\frac{b-a}{c_1}) = 1 \text{ and } c_1\sinh(\frac{b-a}{c_1}) = l,$$

gives $c_1 = l$. Thus, $\sinh(\frac{b-a}{l}) = 1$, from which we obtain $b-a = l\sinh^{-1}(1)$. Using (4.125) we obtain the solution

$$y(x) = -\lambda + l\cosh(\frac{x-b}{l}).$$

Left to determine λ. Since $y(a)$ lies on the line $h(x) = x - 1$, we have $y(a) = a - 1$. In addition,

$$\begin{aligned} y(a) &= -\lambda + l\cosh(\frac{b-a}{l}) \\ &= -\lambda + l\cosh\left(\frac{1}{l}\sinh^{-1}(1)l\right) \\ &= -\lambda + l\sqrt{2}. \end{aligned}$$

Setting $y(a) = y(a)$, yields $\lambda = 1 - a + l\sqrt{2}$. Thus, the solution is

$$y(x) = a - 1 - l\sqrt{2} + l\cosh(\frac{x-b}{l}).$$

$\square$

4.13.1 Exercises

Exercise 4.85 *Prove Theorem 4.24.*

Exercise 4.86 *Find the extremal that minimizes*

$$L(y) = \int_0^\pi y'^2 dx, \quad y(0) = y(\pi) = 0,$$

subject to the constraint $W(y) = \int_0^\pi y^2 dx = 1.$

Exercise 4.87 *Find the extremals for*

$$L(y) = \int_0^\pi (y'^2 + x^2) dx, \quad y(0) = y(\pi) = 0,$$

subject to the constraint $W(y) = \int_0^\pi y^2 dx = 8.$

Exercise 4.88 *Maximize the surface area*

$$S(y) = 2\pi \int_0^\pi y\sqrt{1+y'^2} dx, \quad y(0) = 0, \ y(\pi) = 0,$$

subject to the constraint $V(y) = \pi \int_0^\pi y^2 dx = l$ *which is the volume of the given surface, for positive constant l.*

Exercise 4.89 *Find the extremal that minimizes*

$$L(y) = \int_{-1}^1 \sqrt{1+y'^2} dx, \quad y(-1) = y(1) = 0,$$

subject to the constraint $W(y) = \int_{-1}^1 y^2 dx = l, \quad l > 2.$

Exercise 4.90 *Find the extremal for*

$$L(y) = \int_{-2}^2 y dx, \quad y(-2) = y(2) = 0,$$

subject to the constraint $W(y) = \int_{-2}^2 \sqrt{1+y'^2} dx = 2\pi.$

Exercise 4.91 *Find the extremal for*

$$L(y) = \int_0^2 y'^2 dx, \quad y(0) = 0, \ y(2) = 1,$$

subject to the constraint $W(y) = \int_0^2 y dx = l, \quad l > 0.$

Exercise 4.92 *[Dido's problem in polar coordinates] Suppose in Dido's problem we require the enclosed land to be along a straight river, say the x-axis. We want to find the curve that encloses the maximum area. Assume the curve is bounded by the river and under the graph* $y = f(x), \ -1 \le x \le 1,$ *with given perimeter* $L > 2.$ *Set up of the problem in polar coordinates to maximize*

$$L(r) = \int_0^\pi r^2 dr, \quad r(0) = 0, \ r(\pi) = 0,$$

where $r = r(\theta)$ subject to the constraint

$$W(y) = \int_0^\pi \sqrt{r^2 + r'^2}\, d\theta = L.$$

Exercise 4.93 *Find the curve for which the functional*

$$J(y) = \int_0^b \frac{\sqrt{1+y'^2}}{y}\, dx, \quad y(0) = 0$$

and $y(b)$ *varies along the circle*

$$(x-9)^2 + y^2 = 9.$$

Exercise 4.94 *Minimize the functional*

$$J(y,z) = \int_0^{\pi/2} (1 + y'^2 + z'^2)\, dx, \quad y(0) = z(0) = y(\pi/2) = z(\pi/2) = 0.$$

subject to $y^2 + 2z = 2$.

Exercise 4.95 *Find an extremal corresponding to* $J(y) = \int_{-1}^1 y\, dx$ *when subject to* $y(-1) = y(1) = 0$ *and* $\int_{-1}^1 (y^2 + y'^2)\, dx = 1$.

Exercise 4.96 *Find an extremal corresponding to* $J(y) = \int_1^e x^2 y'^2\, dx$ *when subject to* $y(1) = y(e) = 0$ *and* $\int_1^e y^2\, dx = 1$.

Exercise 4.97 *Find an extremal corresponding to*

$$J(y) = \int_0^\pi y'^2\, dx, \quad y'(0) = y'(\pi) = 0,$$

when subject to $\int_0^\pi y^2\, dx = 1$.

Exercise 4.98 *Find an extremal corresponding to*

$$L(y) = \int_0^1 (y''^2 + x^2)\, dx, \quad y(0) = y(1) = y'(0) = y'(1) = 0$$

and

$$W(y) = \int_0^1 (y^2 + 1)\, dx = 2.$$

Exercise 4.99 *Find the curve of fixed length* πa *joining the two points* $(-a,0)$ *and* $(a,0)$ *and situated above the x-axis such that the area below it and above the x-axis is maximum.*

Exercise 4.100 *Consider Example 4.29, but this time the chain in freely sliding on the line* $x = 0$ *(y-axis). Also, the right end of the chain is left free to slide on a tilted pole, given by the equation* $cx + dy = cd$, *where* $c, d > 0$. *Find the equation of the chain that minimizes the potential energy.*

4.14   Sturm-Liouville Problem

Consider the second-order differential equation with parameter $\lambda \in \mathbb{R}$,

$$y''(x) + P(x)y'(x) - Q(x)y(x) - \lambda R(x)y(x) = 0, \ a \le x \le b,$$

where P, Q, R are continuous, and R is positive on $[a,b]$. Multiply both sides of the above equation with

$$r(x) = e^{\int P(x)dx},$$

and then by observing that

$$\left(r(x)y'\right)' = r'(x)y' + r(x)y'',$$

the above equation may take the form

$$\left(r(x)y'\right)' - q(x)y - \lambda p(x)y = 0, \tag{4.126}$$

where $q(x) = r(x)Q(x)$, and $p(x) = r(x)R(x) > 0$ for all $x \in [a,b]$. For constants α_1, α_2, β_1, β_2, we impose the boundary conditions

$$\alpha_1 y(a) + \beta_1 y'(a) = 0, \quad \alpha_2 y(b) + \beta_2 y'(b) = 0, \tag{4.127}$$

with

$$\alpha_1^2 + \beta_1^2 \ne 0; \quad \alpha_2^2 + \beta_2^2 \ne 0.$$

The differential equation given by (4.126) along with (4.127) is called *Sturm-Liouville problem* (SLP). There is a habitual relation between variational with isoperimetric constraint and Sturm-Liouville problem. To see this, let $y = y(x) \in C^2([a,b])$ be an extremal for the variational

$$L(y) = \int_a^b \left(r(x)y'^2 + q(x)y^2\right)dx,$$

subject to

$$W(y) = \int_a^b p(x)y^2 dx.$$

Then

$$F^* = r(x)y'^2 + q(x)y^2 + \lambda p(x)y^2,$$

and

$$F_y^* - \frac{d}{dx}F_{y'}^* = 0,$$

implies that

$$r(x)y'' + r'(x)y' - q(x)y - \lambda p(x)y = 0,$$

which is equivalent to

$$\left(r(x)y'\right)' - q(x)y - \lambda p(x)y = 0.$$

Notice that the nontrivial y does not satisfy the Euler-Lagrange equation for the constraint $W(y)$, since

$$-2p(x)y(x) = 0,$$

is not possible due to the fact that $p(x) > 0$ for all $x \in [a,b]$. Thus, we have shown that the (SLP) can be recasted as variational problem with isoperimetric constraint.

The (SLP) has a wide range of applications. The boundary conditions make it conveniently suitable for standing wave. In addition, (SLP) models the one dimensional time dependent Schrödinger equation

$$-\left(\psi'(x)\right)' + \frac{2m}{\hbar^2}V(x)\Psi(x) - \lambda\Psi(x) = 0.$$

We make the following definitions.

Definition 4.10 *If for a certain value of λ, the (SLP) given by (4.126) and (4.127) has a nontrivial solution $y(x)$, then λ is called an eigenvalue and $y(x)$ the corresponding eigenfunction.*

Definition 4.11 *If for any two functions f and g we have*

$$\int_0^l f(x)g(x)p(x)dx = 0, \tag{4.128}$$

then we say f and g are orthogonal on $[0,l]$ with respect to the weight function $p(x) > 0$.

The integral on the left side of (4.128) is called the *inner product* of f and g and is denoted by (f,g). Thus,

$$(f,g) = \int_0^l f(x)g(x)p(x)dx.$$

The number $\|f\|$ defined by

$$\|f\| = \left(\int_0^l f^2(x)p(x)dx\right)^{\frac{1}{2}}$$

is called the *norm* of f. Clearly,

$$\|f\| = \left(f,f\right)^{\frac{1}{2}}.$$

Now we are able to make the following definition.

Definition 4.12 *A set of functions $f_1(x), f_2(x), \ldots, f_n(x)$ is orthonormal if*

$$(f_n, f_m) = \delta_{nm} = \left\{ \begin{array}{ll} 0, & n \neq m \\ 1, & n = m \end{array} \right.$$

It is then clear that an orthogonal set of functions can be made into an orthonormal set by dividing each function in the set by its norm. The next theorem characterizes the eigenvalues and eigenfunctions solutions of (SLP). Its proof can be found in various places such as [1] of Chapter 2.

Theorem 4.27 *For the Sturm-Liouville propblem given by (4.126) and (4.127) the following statements hold.*

i) *The eigenvalues are real and to each eigenvalue there corresponds a single eigenfunction up to a constant multiple.*

ii) *The eigenvalues form an infinite sequence $-\lambda_1, -\lambda_2, \ldots, -\lambda_n, \ldots$, and can be ordered in a manner that $0 < -\lambda_1 < -\lambda_2 < -\lambda_3 < \ldots$ with*

$$\lim_{n \to \infty} (-\lambda_n) = \infty.$$

iii) *If $-\lambda_m$ and $-\lambda_n$ are two distinct eigenvalues, then the corresponding eigenfunctions $y_m(x)$ and $y_n(x)$ are orthogonal on the interval $[0, l]$.*

Example 4.30 Consider the variational

$$L(y) = \int_1^e x^2 y'^2 dx, \quad y(1) = y(e) = 0,$$

subject to

$$W(y) = \int_1^e y^2 dx = 1.$$

Then the corresponding Euler-Lagrange equation is

$$(x^2 y')' - \lambda y = 0,$$

or

$$x^2 y'' + 2xy' - \lambda y = 0, \tag{4.129}$$

which is a (SLP) with $r(x) = x^2$, $q(x) = 0$, and $p(x) = 1$. Using the method of Section 1.11, we arrive at

$$y'' + y' - \lambda y = 0,$$

where the independent variable x has been changed to the independent variable t under the transformation $x = e^t$. If we assume solutions of the form $y = e^{mt}$, then we have

$$m = -\frac{1}{2} \pm \frac{\sqrt{1 + 4\lambda}}{2}.$$

For $\lambda = 0$, we get the solution $y(t) = c_1 + c_2 e^{-t}$, or $y(x) = c_1 + \frac{c_2}{x}$. Applying the given boundary conditions we obtain $c_1 = c_2 = 0$, which corresponds to the trivial solution $(y = 0)$.

If $1 + 4\lambda > 0$, then the solution is given by

$$y(x) = c_1 x^{m_1} + c_2 x^{m_2},$$

where $m_1 = -\frac{1}{2} + \frac{\sqrt{1+4\lambda}}{2}$, and $m_2 = -\frac{1}{2} - \frac{\sqrt{1+4\lambda}}{2}$. Again, applying the given boundary conditions we obtain $c_1 = c_2 = 0$, which corresponds to the trivial solution $(y = 0)$.

If $1 + 4\lambda < 0$, then we have complex roots and the solution is (see Section 1.11)

$$y(x) = x^{-\frac{1}{2}} \left(c_1 \cos(\frac{\sqrt{-(1+4\lambda)}}{2} \ln(x)) + c_2 \sin(\frac{\sqrt{-(1+4\lambda)}}{2} \ln(x)) \right).$$

Using $0 = y(1)$, automatically gives $c_1 = 0$ and we are left with the solution

$$y(x) = x^{-\frac{1}{2}} c_2 \sin(\frac{\sqrt{-(1+4\lambda)}}{2} \ln(x)).$$

If we apply $y(e) = 0$, we get

$$0 = e^{-\frac{1}{2}} c_2 \sin(\frac{\sqrt{-(1+4\lambda)}}{2}).$$

We have the trivial solution for $c_2 = 0$, since $c_1 = 0$. Thus, to avoid the trivial solution, we set

$$\sin(\frac{\sqrt{-(1+4\lambda)}}{2}) = 0,$$

which holds for

$$\frac{\sqrt{-(1+4\lambda)}}{2} = n\pi, \; n = 1, 2, \ldots .$$

Or,

$$\lambda = -n^2 \pi^2 - \frac{1}{4}, \; n = 1, 2, \ldots .$$

Thus, the eigenvalues and the corresponding eigenfunctions are given by

$$-\lambda_n = n^2 \pi^2 + \frac{1}{4}, \; y_n(x) = c_2 x^{-\frac{1}{2}} \sin(n\pi \ln(x)), \; n = 1, 2, \ldots . \tag{4.130}$$

To find the constant c_2, we make use of the isoperimetric constraint $W(y)$.

$$\int_1^e y^2 dx = c_2^2 \int_1^e \frac{\sin^2(n\pi \ln(x))}{x} dx \tag{4.131}$$

$$= \frac{c_2^2}{2} \int_1^e \frac{1 - \cos(2n\pi \ln(x))}{x} dx$$

$$= \frac{c_2^2}{2} \Big[\ln(x) \Big|_1^e - \int_1^e \frac{\cos\left(2n\pi \ln(x)\right)}{x} dx \Big].$$

To evaluate the integral we make the substitution $u = 2n\pi \ln(x)$. Then $\frac{du}{2n\pi} = \frac{dx}{x}$ and

$$\int \frac{\cos\left(2n\pi \ln(x)\right)}{x} dx = \int \frac{\cos(u)}{2n\pi} du = \frac{\sin(2n\pi \ln(x))}{2n\pi}.$$

So, (4.131) yields,

$$W(y) = \int_1^e y^2 dx \quad = \quad \frac{c_2^2}{2} \Big[\ln(x) - \frac{\sin(2n\pi \ln(x))}{2n\pi} \Big]\Big|_1^e$$

$$= \quad \frac{c_2^2}{2} [1 - 0] = 1.$$

Or,

$$c_2 = \pm\sqrt{2}.$$

Then, by (4.130) the eigenfunctions are

$$y_n(x) = \pm\sqrt{2} x^{-\frac{1}{2}} \sin(n\pi \ln(x)), \ n = 1, 2, \ldots.$$

$\square$

4.14.1 The First Eigenvalue

In Example 4.30, we saw that the corresponding variational problem with isoperimetric constraint has infinitely many eigenvalues, and to each eigenvalue there corresponds an eigenfunction. Let $y_n(x)$ be given by (4.130) with $c_2 = 1$. Due to the Principle of superposition; p. 51, and the linearity of the Euler-Lagrange equation, any function of the form

$$z(x) = \sum_{n=1}^{\infty} b_n y_n(x),$$

is an extremal to the variational provided that z satisfies the isoperimetric constraint. We have naturally presumed that the series converges and that each of its terms is twice differentiable. We analyze the constraint at z to further clarify the concept. In the next argument, we make use of the orthogonality of the eigenfunctions found in Example 4.30 (see Exercise 4.103)

$$W(z) = \int_1^e z^2 dx \quad = \quad \int_1^e \Big(\sum_{n=1}^{\infty} b_n \frac{\sin(n\pi \ln(x))}{\sqrt{x}} \Big)^2 dx$$

$$= \quad \int_1^e \sum_{n=1}^{\infty} \sum_{m=1}^{\infty} b_m b_n \frac{\sin(n\pi \ln(x))}{\sqrt{x}} \frac{\sin(m\pi \ln(x))}{\sqrt{x}} dx$$

$$= \quad \sum_{n=1}^{\infty} \sum_{m=1}^{\infty} \int_1^e b_m b_n \frac{\sin(n\pi \ln(x))}{\sqrt{x}} \frac{\sin(m\pi \ln(x))}{\sqrt{x}} dx$$

$$= \sum_{n=1}^{\infty} b_n^2 \int_1^e \frac{\sin^2(n\pi \ln(x))}{x} dx \text{ due to orthogonality}$$

$$= \frac{1}{2} \sum_{n=1}^{\infty} b_n^2 = 1. \tag{4.132}$$

Thus,

$$\sum_{n=1}^{\infty} b_n^2 = 2. \tag{4.133}$$

Let $y_n(x)$ be given by (4.130) with $c_2 = 1$. We have shown $z(x) = \sum_{n=1}^{\infty} b_n y_n(x)$, with b_n satisfying (4.133) is an extremal of the variational problem given in Example 4.30. However, if we want that same extremal to minimize the variational problem, then we need to look deeper into the sequence of the eigenvalues λ_n, $n = 1, 2, \ldots$. Let's evaluate the variational L at z. In the coming calculations we make use of the following:

Let

$$f_n(x) = \frac{\sin(n\pi \ln(x))}{\sqrt{x}}, \quad g_n(x) = \frac{\cos(n\pi \ln(x))}{\sqrt{x}}, \quad x \in [1, e].$$

Then,

$$(f_n(x), f_m(x)) = (g_n(x), g_m(x)) = 0, \text{ for all } n \neq m; \; n, m = 1, 2, \ldots, \tag{4.134}$$

and

$$(f_n(x), g_m(x)) = 0, \quad \text{for all} \quad n, m = 1, 2, \ldots, \tag{4.135}$$

The next argument is similar to the preceding one, so we skip some of the details.

$$L(z) = \int_1^e x^2 z'^2 dx$$

$$= \sum_{n=1}^{\infty} b_n^2 \left(\int_1^e \frac{\sin^2(n\pi \ln(x))}{4x} + \int_1^e \frac{\cos^2(n\pi \ln(x))}{x} \right) dx \text{ (by (4.134) and (4.135))}$$

$$= \sum_{n=1}^{\infty} b_n^2 \left(\frac{1}{8} + \frac{n^2 \pi^2}{2} \right). \tag{4.136}$$

Since, the variational L is positive for non trivial solution, and the right side of (4.136) is increasing in n, then it is likely that the minimum of L is achieved at $n = 1$, or at the *first eigenvalue* $-\lambda_1 = \pi^2 + \frac{1}{4}$. Recall, $-\lambda_n = n^2 \pi^2 + \frac{1}{4}$, $n = 1, 2, \ldots$. The extremal eigenfunction that corresponds to the first eigenvalue is

$$y_1(x) = \pm\sqrt{2} x^{-\frac{1}{2}} \sin(\pi \ln(x)).$$

Of course at this eigenfunction, $W(y) = 1$. Note that the number 1 in $W(y) = 1$, is "symbolic". As a matter of fact, the above discussion should hold for any number $l > 0$, such that $W(y) = l$. Additionaly, (4.136) implies for $n = 1$ that

$$L(y_1(x)) = \frac{1}{4} + \pi^2 = -\lambda_1,$$

where we have used $b_1^2 = 2$ that was obtained from (4.133). Next, we make it clear that $y_1(x)$ minimizes L. Suppose there is another function f that minimizes L such that f is different from y_1. Due to the completeness property of Fourier series, (see Appendix A) f must be of the form $\sum_{n=1}^{\infty} b_n y_n(x)$. Since y_1 and f differ, there is an integer $k \geq 2$ such that $b_K \neq 0$. Thus, from (4.136) we get

$$
\begin{aligned}
L(f) &= \sum_{n=1}^{\infty} b_n^2 \left(\frac{1}{8} + \frac{n^2 \pi^2}{2} \right) \\
&= \frac{1}{8} \sum_{n=1}^{\infty} b_n^2 + \frac{\pi^2}{2} \sum_{n=1}^{\infty} n^2 b_n^2 \\
&= \frac{1}{8} \sum_{n=1}^{\infty} b_n^2 + \frac{\pi^2}{2} \left(\sum_{n=1}^{K-1} n^2 b_n^2 + K^2 b_K^2 + \sum_{n=K+1}^{\infty} n^2 b_n^2 \right) \\
&\geq \frac{1}{8} \sum_{n=1}^{\infty} b_n^2 + \frac{\pi^2}{2} \left(K^2 b_K^2 + \sum_{n=1}^{K-1} b_n^2 + \sum_{n=K+1}^{\infty} n^2 b_n^2 \right) \\
&> \frac{1}{8} \sum_{n=1}^{\infty} b_n^2 + \frac{\pi^2}{2} \left(K^2 b_K^2 - b_K^2 + \sum_{n=1}^{\infty} b_n^2 \right) \\
&= \frac{\pi^2}{2} (K^2 - 1) b_K^2 + \left(\frac{1}{8} + \frac{\pi^2}{2} \right) \sum_{n=1}^{\infty} b_n^2 \\
&> \left(\frac{1}{8} + \frac{\pi^2}{2} \right) \sum_{n=1}^{\infty} b_n^2 \\
&= \left(\frac{1}{4} + \pi^2 \right) \frac{1}{2} \sum_{n=1}^{\infty} b_n^2 \\
&= \frac{1}{4} + \pi^2, \quad (by(4.132)).
\end{aligned}
$$

This shows

$$
L(f) > \frac{1}{4} + \pi^2 = L(y_1).
$$

This proves that y_1 minimizes L subject to the constraint W.

The next theorem asserts that, in general, the corresponding eigenfunction to the first eigenvalue of (SLP), does indeed minimize the variational subject to its isoperimetric constraint. For a quick reference we restate the (SLP). Consider the variational

$$
L(y) = \int_a^b \left(r(x) y'^2 + q(x) y^2 \right) dx, \quad y(a) = y(b) = 0, \tag{4.137}
$$

and subject to

$$
W(y) = \int_a^b p(x) y^2 dx = 1. \tag{4.138}
$$

Note that, (4.138) holds when the corresponding eigenfunctions are normalized with respect to the weight function p.

Theorem 4.28 *Suppose* $-\lambda_1$ *is the first eigenvalue of* (4.137) *and* (4.138) *with corresponding normalized eigenfunction* $y_1(x) \in C^2([a,b])$. *Then among all admissible normalized eigenfunctions* $y \in C^2([a,b])$, *the function* $y = y_1(x)$ *minimizes L, subject to* (4.138). *Moreover,* $L(y_1) = -\lambda_1$.

Proof *We mention that the presence of the number* $-\lambda_1$ *and not* λ_1, *depends solely on the way we decided to consider* (4.126). *We begin by multiplying* (4.126) *with y and then integrating by parts the first term in the resulting equation from* $x = a$, *to* $x = b$.

$$\int_a^b \left[y(r(x)y')' - q(x)y^2 - \lambda p(x)y^2 \right] dx = 0.$$

Letting $u = y$, $dv = (r(x)y'(x))' dx$ *and making use of* $y(a) = y(b) = 0$, *we arrive at*

$$\int_a^b y(x)(r(x)y'(x))' dx = r(x)y(x)y'(x)\big|_{x=a}^b - \int_a^b r(x)y'^2(x)dx = -\int_a^b r(x)y'^2(x)dx.$$

Substituting into the previous equations and rearranging terms we arrive at

$$\int_a^b \left(r(x)y'^2 + q(x)y^2 \right) dx = -\lambda \int_a^b p(x)y^2 dx. \qquad (4.139)$$

As $W(y) = 1$, (4.139) *implies that*

$$\int_a^b \left(r(x)y'^2 + q(x)y^2 \right) dx = -\lambda.$$

Since y is nontrivial, the number $-\lambda$ *is an eigenvalue. By ii) of Theorem 5.4, the first eigenvalue is* $-\lambda_1$ *and hence it has the corresponding normalized eigenfunction* y_1. *This shows L is minimized at* $-\lambda_1$; *that is*

$$L(y_1) = -\lambda_1 > 0.$$

This completes the proof.

It is crucial that we examine the ratio of $L(y)$ and $W(y)$ in more detail. Expression (4.139) gives

$$L(y) = -\lambda_1 W(y), \quad \text{or} \quad \frac{L(y)}{W(y)} = -\lambda_1.$$

We define the *Rayleigh quotient*

$$R(y) = \frac{L(y)}{W(y)}. \qquad (4.140)$$

It is evident from (4.140) that for any nontrivial solution $\phi_n(x)$, that corresponds to eigenvalues $-\lambda_n$, we have that

$$R(\phi_n) = -\lambda_n, \quad n = 1, 2, \dots. \qquad (4.141)$$

It is important to remark that (4.141) holds for all nontrivial eigenfunctions y whether they are normalized or not, since the same excess factor will appear in the numerator and denominator, and hence it cancels out.

Also, (4.141) is handy when the eigenvalues can be computed, which is not the case in some situations. The Rayleigh quotient can be easily generalized to the (SLP) with general boundary conditions. Let $y = y(x) \in C^2([a,b])$ be an extremal for

$$L(y) = \int_a^b \left(r(x)y'^2 + q(x)y^2 \right) dx,$$

with boundary conditions (4.127) and subject to the constraint

$$W(y) = \int_a^b p(x)y^2 dx.$$

Then

$$-\lambda = \frac{-r(x)y(x)y'(x)\big|_{x=a}^b + \int_a^b \left(r(x)y'^2 + q(x)y^2 \right) dx}{\int_a^b p(x)y^2 dx}. \qquad (4.142)$$

The verification of (4.142) comes from integrating by parts the first term in

$$\int_a^b \left(y(r(x)y')' - q(x)y^2 - \lambda p(x)y^2 \right) dx = 0,$$

and then solving for $-\lambda$. Under the special boundary conditions

$$y(a) = y(b) = 0, \quad \text{(Dirichlet boundary conditions)}$$

$$y'(a) = y'(b) = 0, \quad \text{(Neumann boundary conditions),}$$

the first term in (4.142)

$$r(x)y(x)y'(x)\big|_{x=a}^b = 0.$$

In such cases, we have from (4.140) and (4.142) that

$$\frac{L(y)}{W(y)} = -\lambda_1 = R(y).$$

The next theorem says that the Rayleigh quotient yields an upper bound to the true value of the lowest eigenvalue $-\lambda_1$.

Theorem 4.29 *Suppose $-\lambda_1$ is the first eigenvalue of (4.137) and (4.138). Let*

$$\sigma = \{u : u \in C^2([a,b]), \, u(a) = 0, \, u(b) = 0\}.$$

Let the Rayleigh quotient be given by (4.140). Then

$$\min_{u \in \sigma} R(u) = -\lambda_1. \qquad (4.143)$$

Note that Theorem 4.29 does not requires the function u to be an extremal of (4.137) subject to the constraint (4.138). You may think of the set σ as the set of "trial functions".

Proof

For $y \in \sigma$, we let $\hat{y} = y + \varepsilon\eta$, with $\eta(a) = \eta(b) = 0$. Set

$$\min_{y \in \sigma} R(y) = M.$$

By Taylor's expansion about ε we have

$$
\begin{aligned}
L(\hat{y}) &= \int_a^b \left(r(x)(y' + \varepsilon\eta')^2 + q(x)(y + \varepsilon\eta)^2 \right) dx \\
&= \int_a^b \left(r(x)y'^2 + q(x)y^2 \right) dx \\
&\quad + 2\varepsilon \int_a^b \left(r(x)y'\eta'(x) + q(x)y\eta(x) \right) dx + O(\varepsilon^2).
\end{aligned}
$$

An integration by parts yields,

$$
\begin{aligned}
\int_a^b r(x)y'\eta'(x)dx &= r(x)y'(x)\eta(x)\big|_{x=a}^b - \int_a^b \eta(x)\left(r(x)y'(x)\right)'dx \\
&= -\int_a^b \eta(x)\left(r(x)y'(x)\right)'dx.
\end{aligned}
$$

Thus,

$$L(\hat{y}) = L(y) + 2\varepsilon \int_a^b \eta\left(-(r(x)y')' + q(x)y \right) dx + O(\varepsilon^2).$$

Similarly, but without the integration by parts,

$$W(\hat{y}) = W(y) + 2\varepsilon \int_a^b \eta p(x)y\,dx + O(\varepsilon^2).$$

Then by (4.140) we see that

$$L(\hat{y}) = MW(y) + 2\varepsilon \int_a^b \eta\left(-(r(x)y')' + q(x)y \right) dx + O(\varepsilon^2).$$

Rearranging the terms we get

$$L(\hat{y}) - MW(\hat{y}) = 2\varepsilon \int_a^b \eta\left(-(r(x)y')' + q(x)y - Mpy \right) dx + O(\varepsilon^2).$$

Since $W(y) \neq 0$ for nontrivial y, it follows that

$$R(\hat{y}) - R(y) = \frac{L(\hat{y})}{W(\hat{y})} - \frac{L(y)}{W(y)}$$

$$= \frac{W(y)L(\hat{y}) - W(\hat{y})L(y)}{W(y)W(\hat{y})}$$

$$= \frac{L(\hat{y}) - MW(\hat{y})}{W(\hat{y})} \quad (\text{using the fact that } M = \frac{L(y)}{W(y)})$$

$$= \frac{2\varepsilon \int_a^b \eta\left(-(r(x)y')' + q(x)y - Mpy\right)dx + O(\varepsilon^2)}{W(\hat{y})}.$$

We claim that

$$-(r(x)y')' + q(x)y - Mpy = 0.$$

Otherwise, we may choose η so that, for small ε the integral is negative, which violates the fact that $R(\hat{y}) - R(y) > 0$. We conclude that $(r(x)y')' - q(x)y - Mpy = 0$, which is the (SLP), and therefore M must be an eigenvalue. From Theorem 4.28, we have

$$R(\phi_n) = -\lambda_n \geq \min_{y \in \sigma} R(y) = M = -\lambda_n, \quad \text{for all } n.$$

Therefore,

$$\min_{y \in \sigma} R(y) = -\lambda_1.$$

This completes the proof.

4.14.2   Exercises

Exercise 4.101 *Put the second-order differential equation in the form of (4.126),*

$$y'' + 4y' - 3y - \lambda y = 0, \quad y(0) = 0, \ y(1) = 0.$$

Exercise 4.102 *Put the second-order differential in the form of (4.126),*

$$x^2 y'' + \frac{1}{x}y' - \lambda y = 0, \quad y(1) = 0, \ y(2) = 0.$$

Exercise 4.103 *Show that the eignefunctions*

$$y_n(x) = \pm\sqrt{2}x^{-\frac{1}{2}}\sin(n\pi \ln(x))$$

that were found in Example 4.30 are orthogonal and normalize the eigenfunctions.

Exercise 4.104 *Prove (4.134) and (4.135).*

Exercise 4.105 *Redo Example 4.30 for the variational problem*

$$L(y) = \int_0^\pi y'^2 dx, \quad y(0) = y(\pi) = 0,$$

subject to $W(y) = \int_0^\pi y^2 dx = 3$. Then evaluate L at the eigenfunctions $y_n(x) = \sum_{n=1}^\infty a_n \sin(nx)$, to find a formula for $\sum_{n=1}^\infty a_n^2$. Finally, argue or refer to other statements, that the eigenfunction corresponding to the first eigenvalue minimizes L.

Exercise 4.106 *Put the (SLP)*

$$(xy')' - \frac{\lambda}{x} = 0, \quad y(1) = y(b) = 0$$

in the form of (4.137) and (4.138). Find the eigenvalues and normalized eigenfunctions and show that the normalized eigenfunction corresponding to the first eigenvalue minimizes L.

Exercise 4.107 *Consider the variational problem*

$$L(y) = \int_0^l y'^2 dx, \quad y(0) = y(l) = 0,$$

subject to $W(y) = \int_0^l y^2 dx$. *Find the eigenvalues* $-\lambda_n$, *and corresponding eigenfunctions* $y_n(x)$, $n = 1,2,...$*(No need to normalize them). We already know that* $R(y_1) = -\lambda_1$.

For the next parts, take $l = 1$ *and use*(4.140)

(a) Compute $R(y_T)$ *at the trial function*

$$y_T = \begin{cases} x, & 0 \le x \le 1/2 \\ 1-x, & 1/2 \le x \le 1 \end{cases}$$

(b) Redo part (a) for

$$y_T = x(1-x).$$

(c) Which function is a better estimate of $-\lambda_1$?

(d) Use the trial function

$$y_T = x(x - \frac{1}{2})(x - 1)$$

to estimate $-\lambda_2$. *How close is the estimate of* y_T *to* $-\lambda_2$?

Exercise 4.108 *Consider the boundary value problem*

$$y'' - \lambda(1 + \frac{1}{10}x)y = 0, \quad y(0) = y(\pi) = 0.$$

Compute $R(y)$ *for the following trial functions:*

(a)

$$y_T = \begin{cases} x, & 0 \le x \le \pi/2 \\ \pi-x, & 1/2 \le x \le \pi \end{cases}$$

(b)

$$y_T = x(\pi - x).$$

4.15   Rayleigh Ritz Method

The *Rayleigh-Ritz method* is a numerical procedure to obtain approximate solutions to problems that can be recast as variational problems. The method takes us from an infinite-dimensional problem to a finite-dimensional problem. We seek a function $y = y(x)$ that is an extremal to the variational problem.

$$L(y) = \int_a^b F(x,y,y')dx, \quad y(a) = A, \; y(b) = B, \qquad (4.144)$$

in the form

$$y(x) \approx \phi_0(x) + c_1\phi_1(x) + c_2\phi_1(x) + \ldots + c_N\phi_N(x) = \phi_0(x) + \sum_{n=1}^N c_n\phi_n(x), \quad (4.145)$$

where the constants c_i, $i = 1, 2, \ldots N$ are to be found. The mystery is to find or select the functions ϕ_i, $i = 0, 1, \ldots N$, that will efficiently do the job. Below we list couple observations:

1) If the boundary conditions are given, then chose $\phi_0(x)$ so that it satisfies all the problem's boundary conditions and the others ϕ_i $i = 1, 2, \ldots N$, vanish at the boundary conditions.

2) The functions ϕ_i $i = 1, 2, \ldots N$, can be chosen in problems when one is aware of the form of the answer in order for the equation (4.145) to take that form.

For example if we have a boundary value problem in which the solutions are of the form $y = c + dx$, and boundary conditions $y(0) = 0, y(1) = 1$, then we may take $\phi_0(x) = x$, and $\phi_1(x) = x(x-1)$, and hence

$$y(x) \approx x + c_1 x(1-x).$$

Note that $\phi_0(0) = 0$, $\phi_0(1) = 1$, and $\phi_1(0) = \phi_1(1) = 0$. Here we only decided to select ϕ_0 and ϕ_1. If we were to write down all of them, then we would set

$$\phi_2(x) = x^2(1-x), \phi_3(x) = x^3(1-x), \ldots, \phi_N(x) = x^N(1-x),$$

which corresponds to an approximate of the form

$$y(x) \approx x + x(x-1)\left(c_1 + c_2 x + \ldots + c_N x^{N-1}\right).$$

Next we substitute (4.145) into the variational in (4.144) and suppose we want to

$$\text{Minimize } L(y) = \int_a^b F\left(x, \phi_0(x) + \sum_{n=1}^N c_n\phi_n(x), \phi_0'(x) + \sum_{n=1}^N c_n\phi_n'(x)\right)dx.$$

The independent variable x will integrate out and we are left with a function of the unknown constants say, $L(c_1, c_2, \ldots, c_N)$. The problem reduces to

$$\min L(y) = \min_{c_1, c_2, \ldots, c_N} L(c_1, c_2, \ldots, c_N).$$

Using our knowledge of calculus, we require

$$\frac{\partial L}{\partial c_i} = 0, \quad i = 1, 2, \ldots, N.$$

We are left with solving N linear equations in N unknown variables $c_1, c_2, \ldots, c_N$.

This procedure generates the best estimate when we begin with the initial estimate $y(x) \approx \phi_0(x) + c_1\phi_1(x)$ and c_1 are determined by the substitution of y into the variational. Then, set $y(x) \approx \phi_0(x) + c_1\phi_1(x) + c_2\phi_2(x)$, and *redetermine* c_1 and determine c_2 by substituting y into the variational. Repeat the same process, and at each stage, the following is true: At the Nth stage, the terms $c_1, c_2, \ldots, c_{N-1}$ that have been previously determined are predetermined. This guarantees a better estimate at the Nth stage

$$y(x) \approx \phi_0(x) + c_1\phi_1(x) + \ldots c_N\phi_N(x)$$

than the approximation at the $N - 1^{st}$ stage

$$y(x) \approx \phi_0(x) + c_1\phi_1(x) + \ldots c_{N-1}\phi_{N-1}(x).$$

The above process should lead to a convergence of the approximations to the real extremizer. That is

$$\lim_{N \to \infty} \left(\phi_0(x) + \sum_{n=1}^{N} c_n\phi_n(x) \right) = y_0(x),$$

where $y_0(x)$ is the *extremizing* function. We furnish the following example.

Example 4.31 In Example 4.2 we found the true solution for the extremal of

$$L(y) = \int_0^1 \left((y')^2 - xy - y^2 \right) dx, \quad y(0) = 1, \ y(1) = 2$$

to be

$$y(x) = \frac{2e^{-1} - 5}{2(e^{-1} - e)} e^x + \frac{5 - 2e}{2(e^{-1} - e)} e^{-x} - \frac{1}{2}x.$$

Next we apply the Rayleigh-Ritz Method. Set $\phi_0(x) = 1 + x$. then $\phi_0(0) = 1$, and $\phi_0(1) = 2$. Thus $\phi_0(x)$ satisfies the boundary conditions as required by the method. Choose, $\phi_1(x) = x(1 - x)$. Clearly, $\phi_1(x)$ vanishes at the boundaries and it has no zeroes in $(0,1)$. Set

$$y_1(x) = \phi_0(x) + c_1\phi_1(x) = 1 + x + c_1 x(1 - x).$$

Next we substitute y_1 into the variational and obtain

$$L(y_1) = \int_0^1 \left((y_1')^2 - xy_1 - y_1^2 \right) dx$$

$$= \int_0^1 \left\{ [1+c_1(1-2x)]^2 - x - x^2 - c_1 x^2(1-x) - [1+x+c_1 x(1-x)]^2 \right\} dx$$

$$= \int_0^1 \left\{ (-3x - 2x^2) + c_1(2 - 6x - x^2 + 3x^3) + c_1^2(1 - 4x + 3x^2 + 2x^3 - x^4) \right\} dx$$

$$= -\frac{13}{6} - \frac{7}{12}c_1 + \frac{3}{10}c_1^2.$$

Solving $\dfrac{\partial L}{\partial c_1} = 0$ yields $c_1 = \dfrac{35}{4}$. Thus

$$y_1(x) = 1 + x + \frac{35}{4}x(1-x),$$

as the first approximate solution. We remark that, since $\frac{d^2 L}{dc_1^2} > 0$ at $c_1 = \frac{35}{4}$, $y_1(x)$ is a minimizer candidate. The relation

$$y(x) \approx 1 + x + x(1-x)[c_1 + c_2 x + \ldots + c_N x^{N-1}]$$

offers higher-order approximations. For example, when $N = 1$, we get $y_1(x)$, and when $N = 2$, we get $y_2(x) = 1 + x + x(1-x)[c_1 + c_2 x]$. If we substitute $y_2(x)$ into L, then we would set

$$\frac{\partial L}{\partial c_1} = 0, \quad \frac{\partial L}{\partial c_2} = 0.$$

□

Remark 20 *According to our guidelines on how to choose $\phi_i(x)$, $i = 0, 1, \ldots, N$ if the boundary conditions are $y(0) = y(1) = 0$, then you could set $\phi_0(x) = 0$. In this case y_1 may take the form $y_1(x) = c_1 \phi_1(x) = c_1 x(1-x)$.*

4.15.1   Exercises

Exercise 4.109 *Compute $y_2(x) = 1 + x + x(1-x)[c_1 + c_2 x]$ in Example 4.31.*

Exercise 4.110 *Compute the second-order approximation $y_2(x)$ for the variational*

$$L(y) = \int_0^1 \left((y')^2 - 2xy - y^2 \right) dx, \quad y(0) = 1, \ y(1) = 2,$$

and compute the true extremal. Graph both functions; that is, the true solution and $y_2(x)$ on the same graph.

Exercise 4.111 *Redo Exercise 4.110 when the boundary conditions are*

$$y(0) = 0, \ y(1) = 0.$$

Exercise 4.112 *Put the boundary value problem*

$$xy'' + y' + y - x = 0, \quad y(0) = 0, \ y(1) = 1,$$

into a variational form and use Rayleigh Ritz method to obtain an approximation in the form

$$y_2(x) \approx x + x(1-x)[c_1 + c_2 x].$$

Exercise 4.113 *Compute the second-order approximation $y_2(x)$ for the variational*

$$L(y) = \int_0^1 \left((y')^2 - 2xy - 2y\right)dx, \quad y(0) = 2, \ y(1) = 1,$$

and compute the true extremal. Graph both functions; that is, the true solution and $y_2(x)$ on the same graph.

Exercise 4.114 *Compute the second-order approximation $y_2(x)$ for the variational*

$$L(y) = \int_1^2 \left(\frac{1}{2}(x^2 y')^2 + 6xy\right)dx, \quad y(1) = y(2) = 0,$$

and compute the true extremal. Graph both functions; that is, the true solution and $y_2(x)$ on the same graph.

4.16   Multiple Integrals

Let $\mathscr{R}$ be a closed region in the xy-plane. By $C^2(\mathscr{R})$ we mean the set of all continuous functions $u = u(x,y)$ defined on $\mathscr{R}$ having continuous second partial derivatives on the interior of $\mathscr{R}$. Geometrically, $u = u(x,y)$ represents a smooth surface over the region $\mathscr{R}$. For the set of admissible functions A we take the set of functions in $C^2(\mathscr{R})$ whose values are fixed on the curve $\mathscr{C}$, which bounds the region $\mathscr{R}$ in the xy-plane. Hence $u \in A$ if $u \in C^2(\mathscr{R})$ and

$$u(x,y) = f(x,y), \ (x,y) \in \mathscr{C},$$

where $f(x,y)$ is a given function defined over $\mathscr{C}$ and whose values trace out a fixed curve Γ, which forms the boundary of the surface u. See Fig. 4.21.

The variational problem is to minimize

$$J(u) = \int \int_{\mathscr{R}} F\left(x,y,u(x,y),u_x(x,y),u_y(x,y)\right)dx\,dy, \tag{4.146}$$

where $u \in A$. We seek a necessary condition for a minimum. Let $u(x,y)$ provide a local minimum for the functional J and consider the family of admissible functions

$$u(x,y) + \varepsilon \eta(x,y)$$

where $\eta \in C^2(\mathscr{R})$ and $\eta(x,y) = 0$ for (x,y) on $\mathscr{C}$. Then

$$\delta J(u,\eta) \quad = \quad \frac{d}{d\varepsilon}J(u+\varepsilon\eta)\big|_{\varepsilon=0}$$

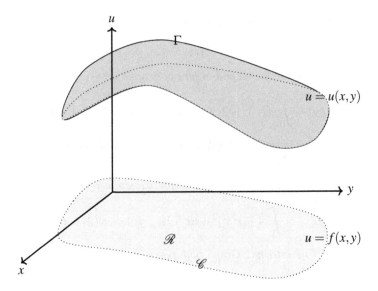

FIGURE 4.21
Surface $u = u(x,y)$ minimizing $J(u)$.

$$= \frac{d}{d\varepsilon} \int\!\!\int_{\mathcal{R}} F\left(x, y, u + \varepsilon\eta, u_x + \varepsilon\eta_x, u_y + \varepsilon\eta_y\right) dx\, dy$$

$$= \int\!\!\int_{\mathcal{R}} [F_u \eta + F_{u_x} \eta_x + F_{u_y} \eta_y]\, dx\, dy$$

$$= \int\!\!\int_{\mathcal{R}} [F_u - \frac{\partial}{\partial x} F_{u_x} - \frac{\partial}{\partial y} F_{u_y}] \eta\, dx\, dy$$

$$+ \int\!\!\int_{\mathcal{R}} [\frac{\partial}{\partial x}(\eta F_{u_x}) + \frac{\partial}{\partial y}(\eta F_{u_y})]\, dx\, dy.$$

The second integral can be transformed to a line integral over $\mathcal{C}$ using Green's theorem, which states that if P and Q are functions in $C^1(\mathcal{R})$, then

$$\int\!\!\int_{\mathcal{R}} (\frac{\partial Q}{\partial x} - \frac{\partial P}{\partial y})\, dx\, dy = \int_{\mathcal{C}} P\, dx + Q\, dy.$$

Consequently,

$$\delta J(u, \eta) = \int\!\!\int_{\mathcal{R}} [F_u - \frac{\partial}{\partial x} F_{u_x} - \frac{\partial}{\partial y} F_{u_y}] \eta\, dx\, dy$$

$$+ \int_{\mathcal{C}} \eta F_{u_x}\, dy - \eta F_{u_y}\, dx.$$

Since $\eta(x,y) = 0$ for (x,y) on $\mathcal{C}$, the line integral vanishes and we have

$$\delta J(u, \eta) = \int\!\!\int_{\mathcal{R}} [F_u - \frac{\partial}{\partial x} F_{u_x} - \frac{\partial}{\partial y} F_{u_y}] \eta\, dx\, dy. \qquad (4.147)$$

Since u is a local minimum if follows that $\delta J(u,\eta) = 0$ for every $\eta \in C^2(\mathcal{R})$ with $\eta(x,y) = 0$ on $\mathscr{C}$. Next, we extend the Fundamental Lemma of Calculus of Variations to two variables.

Lemma 14 *Suppose $g(x,y)$ is continuous over the region $\Omega \subset \mathbb{R}^2$. If*

$$\int\int_{\Omega} g(x,y)\eta(x,y)dxdy = 0,$$

for every continuous function $\eta(x,y)$ defined on Ω and satisfying $\eta = 0$ on $\partial\Omega$, then $g(x,y) = 0$ for all (x,y) in Ω.

Back to (4.147). Since u is a local minimum it follows that $\delta J(u,\eta) = 0$ on $\mathscr{C}$. Thus with the aid of Lemma 14, we have

$$F_u - \frac{\partial}{\partial x}F_{u_x} - \frac{\partial}{\partial y}F_{u_y} = 0, \tag{4.148}$$

which is the Euler-Lagrange equation for the problem (4.146). Equation (4.148) is a second-order partial differential equation for $u = u(x,y)$. It is straight forward to generalize (4.146) to m-integrals of the form

$$J(u) = \int \cdots \int_{\mathcal{R}_m} F\left(x_1, x_2, \ldots, u, \frac{\partial u}{\partial x_1}, \ldots, \frac{\partial u}{\partial x_m}\right) dx_1 \ldots dx_m,$$

where $u = u(x_1, x_2, \ldots, x_m)$, $\mathcal{R}$ is a closed region in m-dimensional Euclidean space. The Euler-Lagrange equation in this case is

$$F_u - \frac{\partial}{\partial x_1}\frac{\partial F}{\partial u_{x_1}} - \frac{\partial}{\partial x_2}\frac{\partial F}{\partial u_{x_2}} - \cdots - \frac{\partial}{\partial x_m}\frac{\partial F}{\partial u_{x_m}} = 0, \tag{4.149}$$

where $u_{x_i} = \dfrac{\partial u}{\partial u_{x_i}}$.

Example 4.32 (Plateau's problem) Given a fixed curve Γ in space, find the surface $u = u(x,y)$ with boundary whose surface area is least. That is we need to minimize

$$J(u) = \int\int_{\mathcal{R}} \sqrt{1 + u_x^2 + u_y^2}\, dx\, dy,$$

where $\mathcal{R}$ is the region enclosed by the curve $\mathscr{C}$ which is the projection of Γ onto the xy-plane. The function u should satisfy $u(x,y) = h(x,y)$, $(x,y) \in \mathscr{C}$, where h is the function defining Γ. We have $F = (1 + u_x^2 + u_y^2)^{\frac{1}{2}}$, with

$$F_u = 0; \quad F_{u_x} = \frac{u_x}{\sqrt{1 + u_x^2 + u_y^2}}; \quad F_{u_y} = \frac{u_y}{\sqrt{1 + u_x^2 + u_y^2}},$$

and the corresponding Euler equation is

$$-\frac{\partial}{\partial x}\frac{u_x}{\sqrt{1 + u_x^2 + u_y^2}} - \frac{\partial}{\partial y}\frac{u_y}{\sqrt{1 + u_x^2 + u_y^2}} = 0.$$

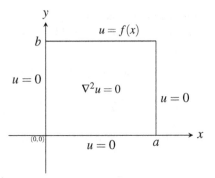

FIGURE 4.22
Dirichlet problem in rectangular coordinates.

After simplifications, the Euler equation reduces to

$$(1+u_y^2)u_{xx} - 2u_x u_y u_{xy} + (1+u_x^2)u_{yy} = 0,$$

which is the equation for the minimized surface. This is nonlinear partial differential equation and almost impossible to solve in its current form for a given boundary curve Γ. $\square$

In the next example we consider the *steady temperature in rectangular coordinates*, called the *Dirichlet problem*. For more on this, we refer to Appendix A.

Example 4.33 Consider the variational problem

$$J(u) = \int\int_{\mathscr{R}} (u_x^2 + u_y^2)\,dx\,dy,$$

where $\mathscr{R} = \{(x,y) : 0 < x < a,\ 0 < y < b\}$. Suppose on the boundary of $\mathscr{R}$, we have

$$u(0,y) = 0, \quad u(a,y) = 0 \quad (0 < y < b),$$

$$u(x,0) = 0, \quad u(x,b) = f(x) \quad (0 < x < a),$$

as depicted in Fig. 4.22. Then the corresponding Euler-Lagrange equation is

$$\nabla^2 u = u_{xx}(x,y) + u_{yy}(x,y) = 0. \tag{4.150}$$

Equation (4.150) along with the boundary conditions represent the steady temperatures $u(x,y)$ in a plates whose faces are insulated. The function $u(x,y)$ represents the *electrostatic potential* in a space formed by the planes $x = 0, x = a, y = 0$, and $y = b$ when the space is free of charges and planar surfaces are kept at potentials given by the boundary conditions. We are seeking non trivial solution and hence if we assume the solution $u(x,y)$ is the product of two functions one in x and the other in y, such that

$$u(x,y) = X(x)Y(y),$$

we obtain

$$X''(x)Y(y) + X(x)Y''(y) = 0.$$

Since $X(x) \neq 0$, and $Y(y) \neq 0$, we may divide by the term $X(x)Y(y)$, to separate the variables. That is,

$$\frac{X''(x)}{X(x)} = -\frac{Y''(y)}{Y(y)}.$$

Since the left-hand side is a function of x alone, it does not vary with y. However, it is equal to a function of y alone, and so it can not vary with x. Hence the two sides must have some constant value $-\lambda$ in common. That is,

$$\frac{X''(x)}{X(x)} = -\frac{Y''(y)}{Y(y)} = -\lambda.$$

This gives the Sturm-Liouville problems

$$X''(x) + \lambda X(x) = 0, \quad X(0) = 0, \, X(a) = 0, \tag{4.151}$$

and

$$Y''(y) - \lambda Y(y) = 0, \quad Y(0) = 0. \tag{4.152}$$

Using arguments of Section 4.14, Equations (4.151) and (4.152) have the respective eigenfunctions

$$X_n(x) = \sin(\frac{n\pi x}{a}), \quad Y_n(y) = \sinh(\frac{n\pi y}{a}), \quad n = 1, 2, \dots$$

where the eigenvalues are given by $\lambda_n = (\frac{n\pi}{a})^2$. Thus, the general solution of the Dirichlet problem

$$u(x,y) = \sum_{n=1}^{\infty} b_n \sinh(\frac{n\pi y}{a}) \sin(\frac{n\pi x}{a}).$$

For detail on computing the coefficients b_n we refer to Appendix A. Thus, b_n are given by

$$b_n = \frac{2}{a \sinh(\frac{n\pi b}{a})} \int_0^a f(x) \sin(\frac{n\pi x}{a}) dx, \quad n = 1, 2, \dots.$$

□

Next we look at a parametrized three dimensional surface. Suppose we have a surface S specified or parametrized with

$$r = r(u,v) = (x(u,v), y(u,v), z(u,v)). \tag{4.153}$$

The shortest curve lying on a surface S connecting given points on S is called geodesic. Remember, a curve lying on the surface S can be specified by $u = u(t), v = v(t)$. The arclength between the points on S corresponding to $t = t_0$ and $t = t_1$ is

$$J(u,v) = \int_{t_0}^{t_1} \sqrt{Eu'^2 + 2Ku'v' + Gv'^2} \, dt, \tag{4.154}$$

where $E, K,$ and G are called the coefficient of the *first fundamental form*

$$E = r_u \cdot r_u, \quad K = r_u \cdot r_v, \quad G = r_v \cdot r_v,$$

where $\cdot$ means the dot product. Equation (4.154) is a variational with several functions and by (4.149) we have the relevant Euler-Lagrange equations

$$F_u - \frac{d}{dt} F_{u'} = 0, \text{ and } F_v - \frac{d}{dt} F_{v'} = 0.$$

Be aware that the coefficients of the first fundamental form E, G and K depend on u and v. After some calculations the corresponding Euler-Lagrange equations are given by

$$\frac{E_u u'^2 + 2K_u u' v' + G_u v'^2}{2\sqrt{E u'^2 + 2K u' v' + G v'^2}} - \frac{d}{dt} \frac{E u' + K v'}{\sqrt{E u'^2 + 2K u' v' + G v'^2}} = 0, \qquad (4.155)$$

and

$$\frac{E_v u'^2 + 2K_v u' v' + G_v v'^2}{2\sqrt{E u'^2 + 2K u' v' + G v'^2}} - \frac{d}{dt} \frac{K u' + G v'}{\sqrt{E u'^2 + 2K u' v' + G v'^2}} = 0. \qquad (4.156)$$

In the next example we use equations (4.155) and (4.156) to find the geodesics on a circular cylinder.

Example 4.34 In this example we want to find the geodesics on a circular cylinder. Note that the circular cylinder has the parametrization

$$r = (a\cos(u), a\sin(u), v)$$

where a is the radius. Then,

$$E = r_u \cdot r_u = (-a\cos(u), a\cos(u), 0) \cdot (-a\cos(u), a\cos(u), 0) = a^2,$$

$$K = r_u \cdot r_v = (-a\cos(u), a\cos(u), 0) \cdot (0, 0, 1) = 0,$$

and

$$G = r_v \cdot r_v = (0, 0, 1) \cdot (0, 0, 1) = 1.$$

Then the corresponding equations to (4.155) and (4.156) are

$$-\frac{d}{dt} \frac{a^2 u'}{\sqrt{a^2 u'^2 + v'^2}} = 0, \quad -\frac{d}{dt} \frac{v'}{\sqrt{a^2 u'^2 + v'^2}} = 0.$$

Moreover, the corresponding solutions are given by

$$\frac{a^2 u'}{\sqrt{a^2 u'^2 + v'^2}} = c_1, \quad \frac{a^2 v'}{\sqrt{a^2 u'^2 + v'^2}} = c_2,$$

for constants c_1 and c_2. Taking the ratio we obtain

$$\frac{\dfrac{v'}{\sqrt{a^2 u'^2 + v'^2}}}{\dfrac{a^2 u'}{\sqrt{a^2 u'^2 + v'^2}}} = \frac{c_2}{c_1}.$$

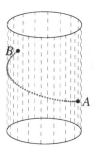

FIGURE 4.23
Geodesics on a right cylinder.

It follows that

$$\frac{v'}{a^2 u'} = \frac{c_2}{c_1} = k,$$

for another constant k. This implies that

$$\frac{v'}{u'} = a^2 k,$$

and by rewriting the derivatives we see that

$$\frac{\frac{dv}{dt}}{\frac{du}{dt}} = a^2 k.$$

Separating the variables yields the first-order ODE $dv = a^2 k\, du$, which has the solution

$$v(t) = a^2 k u(t) + c_3, \text{ for constant } c_3,$$

which is a two parameter family of helical lines on the cylinder, where the constants can be determined from the location of the two points A and B as depicted in Fig. 4.23. □

4.16.1 Exercises

Exercise 4.115 *Determine the natural boundary condition for*

$$J(u) = \int\int_{\mathscr{R}} F\big(x, y, u(x,y), u_x(x,y), u_y(x,y)\big)\, dx\, dy,$$

$u \in C^2(\mathscr{R})$ *and u unspecified on the boundary of $\partial\mathscr{R}$ of $\mathscr{R}$.*

Exercise 4.116 *Determine the Euler-Lagrange equation for*

$$J(u) = \int\int_{\mathscr{R}} \big(x^2 u_x^2 + y^2 u_y^2\big)\, dx\, dy.$$

Exercise 4.117 *Let a and b be nonzero constants. Determine the Euler-Lagrange equation for*

$$J(u) = \int\int_{\mathscr{R}} (a^2 u_x^2 + b^2 u_y^2)\, dx\, dy,$$

where $\mathscr{R} = \{(x,y) : \frac{x^2}{a^2} + \frac{y^2}{b^2} < 1\}$. *Suppose on the boundary of* $\mathscr{R}$, *we have* $u(x,y) = \frac{2x^2}{a^2} - 1$ *for all*

$$(x,y) \in \partial\mathscr{R} = \{(x,y) : \frac{x^2}{a^2} + \frac{y^2}{b^2} = 1\}.$$

Refer to Section 2.5 to find the solution.

Exercise 4.118 *Redo Example 4.33 with boundary conditions*

$$u(0,y) = 0, \quad u(a,y) = 0 \quad (0 < y < b),$$

$$u(x,0) = g(x), \quad u(x,b) = 0 \quad (0 < x < a).$$

Exercise 4.119 *Redo Example 4.33 with boundary conditions*

$$u(0,y) = 0, \quad u(a,y) = 0 \quad (0 < y < b),$$

$$u(x,0) = f(x), \quad u(x,b) = g(x) \quad (0 < x < a).$$

Exercise 4.120 *Give all details on how to obtain equations (4.155) and (4.156).*

Exercise 4.121 *Find the geodesics on the right circular cone*

$$z^2 = \frac{x^2}{4} + \frac{y^2}{9}, \quad 0 \le z \le 3.$$

Hint: Use the following parametrization for the cone

$$r = (2v\cos(u), 3v\sin(u), v), \quad 0 \le u \le 2\pi, \ 0 \le v \le 3.$$

Exercise 4.122 *Determine (4.155) and (4.156) for the surface parametrized by*

$$r = (5\sin(u)\cos(v), \sin(u)\sin(v), 2\cos(u)), \quad 0 \le u \le \pi, \ 0 \le v \le 2\pi.$$

5

Integral Equations

Integral equations are used in a wide variety of contexts, including science and engineering. Integral equations such as those derived from Volterra or Fredholm can be utilized to find solutions to a wide variety of initial and boundary value problems. Integral equations can take on a number of different forms, but in most cases they are used to model scientific procedures in which the current value of a quantity (or set of values) or its rate of change is dependent on its historical performance. This is in contrast to differential equations, which assume that the value of a quantity at any given time is the only factor that may affect the rate at which it changes. In the same way that differential equations need to be "solved," integral equations also need to be "solved" in order to describe and predict how a physical quantity will behave over a period of time. One strong argument in favor of using integral equations rather than differential equations is the fact that all of the conditions defining the initial value problems or boundary value problems for a differential equation can frequently be condensed into a single integral equation. This is one of the many reasons why integral equations are preferred over differential equations. The study of a variety of integral equations, including Fredholm first- and second-kind integral equations as well as Volterra integral equations, symmetric and separable kernels, iterative methods, the approximation of non-degenerate kernels, and the application of the Laplace transform to the solution of convoluted integral equations, will be the focus of our work. The chapter comes to a close with a discussion on integral equations that exhibit strange behavior.

5.1 Introduction and Classifications

This section focuses on integral equations when the integration is with respect to a single variable. Higher-order generalization is straightforward and unproblematic. Differential equations and integral equations differ significantly in that the former are about the local behavior of a system, while the latter are about global behavior. Local behavior is often easier to explain and grasp intuitively.

Definition 5.1 *An integral equation in the unknown function y(x) is a relation of the form*

$$y(x) = f(x) + \int K(x,\xi)y(\xi)d\xi \tag{5.1}$$

in which y(x) appears in the integrand, where K(x,ξ) is a function of two variables x and ξ and referred to as the kernel of the integral equation.

Note that we purposefully omitted the limits of integration from the formulation above because, in most circumstances, they determine the sort of integral equation we have. In (5.1) the functions f and K are given and satisfy continuity conditions and perhaps others. The following are examples of integral equations.

$$y(x) = \sin(2x) + \int_0^x (x^3 + \xi x + 1)y(\xi)d\xi,$$

and

$$e^x y(x) = \int_0^1 \sin(x\xi)y(\xi)d\xi.$$

In this chapter we discuss the Fredholm equation of the first kind

$$\alpha(x)y(x) = \int_a^b K(x,\xi)y(\xi)d\xi, \tag{5.2}$$

and the Fredholm equation of the second kind

$$y(x) = f(x) + \lambda \int_a^b K(x,\xi)y(\xi)d\xi. \tag{5.3}$$

Fredholm integral equations given by (5.2) and (5.3) have the unique property of having finite limits of integration $\xi = a$, and $\xi = b$. In addition to discussing Fredhom equations, we will discuss Volterra equations of first kind and second kind given by

$$y(x) = \int_a^x K(x,\xi)y(\xi)d\xi, \tag{5.4}$$

and

$$y(x) = f(x) + \lambda \int_a^x K(x,\xi)y(\xi)d\xi, \tag{5.5}$$

respectively. In later sections we will develop particular methods to solve integral equations with specific characteristics. Without worrying about technicality, we try to define a sequence of functions $\{y_n\}$ successively for (5.5), with $\lambda = 1$ by setting

$$
\begin{aligned}
y_0(x) &= f(x) \\
y_1(x) &= f(x) + \int_0^x K(x,\xi)y_0(\xi)d\xi \\
y_2(x) &= f(x) + \int_0^x K(x,\xi)y_1(\xi)d\xi
\end{aligned} \tag{5.6}
$$

$$\vdots$$

$$y_n(x) \quad = \quad f(x) + \int_0^x K(x, \xi) y_{n-1}(\xi) d\xi, \quad n = 1, 2, \ldots \qquad (5.7)$$

This method is referred to as the *successive approximation method*. To illustrate the above procedure we provide the following example.

Example 5.1 Consider the Volterra integral equation

$$y(x) = 1 - \int_0^x (x - \xi) y(\xi) d\xi.$$

Setting $y_0(x) = 1$, we have the recurrent formula

$$y_n(x) = 1 - \int_0^x (x - \xi) y_{n-1}(\xi) d\xi.$$

For $n = 1$ we have

$$y_1(x) = 1 - \int_0^x (x - \xi)(1) d\xi = 1 - \frac{x^2}{2}.$$

For $n = 2$, with $y_1(\xi) = 1 - \frac{\xi^2}{2}$,

$$y_2(x) = 1 - \int_0^x (x - \xi)(1 - \frac{\xi^2}{2}) d\xi = 1 - \frac{x^2}{2} + \frac{x^4}{24}.$$

Similarly, for $n = 3$ with $y_2(\xi) = 1 - \frac{\xi^2}{2} + \frac{\xi^4}{24}$, we have

$$y_3(x) = 1 - \int_0^x (x - \xi)(\frac{\xi^2}{2} + \frac{\xi^4}{24}) d\xi = 1 - \frac{x^2}{2} + \frac{x^4}{24} - \frac{x^6}{720}.$$

A continuation of this method leads to the sequence of functions

$$y_n(x) = 1 - \frac{x^2}{2} + \frac{x^4}{24} - \frac{x^6}{720} + \cdots = \sum_{k=0}^{n} \frac{(-1)^k x^{2k}}{(2k)!}.$$

Note that

$$\lim_{n \to \infty} y_n(x) = \lim_{n \to \infty} \sum_{k=0}^{n} \frac{(-1)^k x^{2k}}{(2k)!} = \cos(x).$$

We leave it to the students to verify, using either, the *Laplace transform* or by direct substitution that $y(x) = \cos(x)$ is indeed a solution of the integral equation. □

5.1.1 Exercises

Exercise 5.1 *By direct substitution, show that* $y(x) = (1 + x^2)^{-\frac{3}{2}}$ *is a solution of the Volterra integral equation*

$$y(x) = \frac{1}{1 + x^2} - \int_0^x \frac{\xi}{1 + x^2} y(\xi) d\xi.$$

Exercise 5.2 *By direct substitution, show that* $y(x) = \cos(x)$ *is a solution of the Volterra integral equation*

$$y(x) = 1 - \int_0^x (x - \xi) y(\xi) d\xi.$$

Exercise 5.3 *By direct substitution, show that* $y(x) = (x+1)^2$ *is a solution of the Volterra integral equation*

$$y(x) = e^{-x} + 2x + \int_0^x e^{\xi - x} y(\xi) d\xi.$$

Exercise 5.4 *Use the method of successive approximation and show the solution of the Volterra integral equation*

$$y(x) = 1 + \int_0^x (x - \xi) y(\xi) d\xi$$

is $y(x) = \cosh(x)$.

Exercise 5.5 *Use the method of successive approximation and show the solution of the Fredholm integral equation*

$$y(x) = x + \frac{1}{2} \int_{-1}^1 (\xi - x) y(\xi) d\xi$$

is $y(x) = \frac{3}{4}x + \frac{1}{4}$.

5.2 Connection between Ordinary Differential Equations and Integral Equations

We begin this section by addressing existence and uniqueness of initial value problem and its relationship between integral equations. Thus, we consider the (IVP)

$$x' = f(t,x), \quad x(t_0) = x_0 \tag{5.8}$$

where we assume throughout this section that $f : D \to \mathbb{R}$ is continuous and D is a subset of $\mathbb{R} \times \mathbb{R}$. In the case the differential equation given by (5.8) is linear, then a solution can be found. However, in general this approach is not feasible when the differential equation is not linear and hence another approach must be indirectly adopted that establishes the existence of a solution of (5.8). For the development of the existence theory we need a broader definition of the Lipchitz condition that we state next.

Definition 5.2 *The function* $f : D \to \mathbb{R}$ *is said to satisfy a global Lipschitz condition in x if there exists a Lipschitz constant* $k > 0$ *such that*

$$|f(t,x) - f(t,y)| \le k|x - y|, \quad \text{for} \quad (t,x), (t,y) \in D. \tag{5.9}$$

Definition 5.3 *The function $f : D \to \mathbb{R}$ is said to satisfy a local Lipschitz condition in x if for any $(t_1, x_1) \in D$ there exists a domain $(t_1, x_1) \in D_1 \subset D$ and that $f(t, x)$ satisfies a Lipschitz condition in x on D_1. That is, there exists a positive constant K_1 such that*

$$|f(t, x) - f(t, y)| \le K_1 |x - y|, \quad for \quad (t, x), (t, y) \in D_1. \tag{5.10}$$

Definition (5.4) can be easily extended to functions $f : D \to \mathbb{R}^n$, where $D \subset \mathbb{R} \times \mathbb{R}^n$ under a proper norm. Let R be any rectangle in D such that $R = \{(t, x) : |t| \le a, |x| \le b\}$. If we assume that f and $\frac{\partial f}{\partial x}$ are continuous on R, which is the case in this chapter, then f and $\frac{\partial f}{\partial x}$ are bounded on R. Therefore, there exists positive constants $\mathcal{M}$ and $\mathcal{K}$ such that

$$|f(t, x)| \le \mathcal{M} \quad and \quad \left| \frac{\partial f}{\partial x} \right| \le \mathcal{K} \tag{5.11}$$

for all points (t, x) in R. Now for any two points $(t, x_1), (t, x_2)$ in R, by the mean value theorem there exists a constant $c \in (x_1, x_2)$ such that

$$f(t, x_1) - f(t, x_2) = \frac{\partial f}{\partial x}(t, c)(x_1 - x_2),$$

form which it follows that

$$
\begin{aligned}
|f(t, x_1) - f(t, x_2)| &\le \left| \frac{\partial f}{\partial x}(t, c) \right| |x_1 - x_2| \\
&\le \mathcal{K} |x_1 - x_2|.
\end{aligned}
\tag{5.12}
$$

We have shown that if f and $\frac{\partial f}{\partial y}$ are continuous on R, then f satisfies a global Lipschitz condition on R.

We state the following definition regarding solutions of (IVP) and integral equations.

Definition 5.4 *We say x is a solution of (5.8) on an interval I, provided that $x : I \to \mathbb{R}$ is differentiable, $(t, x(t)) \in D$, for $t \in I$, $x'(t) = f(t, x(t))$ for $t \in I$, and $x(t_0) = x_0$ for $(t_0, x_0) \in D$.*

In preparation for the next theorem, we observe that the (IVP) (5.8) is related to

$$x(t) = x_0 + \int_{t_0}^{t} f(s, x(s)) ds. \tag{5.13}$$

Relation (5.13) is an *integral equation* since it contains an integral of the unknown function x. This integral is not a formula for the solution, but rather it provides another relation which is satisfied by solution of (5.8).

Definition 5.5 *We say $x : I \to \mathbb{R}$ is a solution of the integral equation given by (5.13) on an interval I, provided that $t_0 \in I$, x is continuous on I, $(t, x(t)) \in D$, for $t \in I$, and (5.13) is satisfied for $t \in I$.*

The next theorem is fundamental for the proof of the existence theorems.

Theorem 5.1 *Let D be an open subset of $\mathbb{R}^2$ and $(t_0,x_0) \in D$. Then x is a solution of (5.8) on an interval I if and only if x satisfies the integral equation given by (5.13) on I.*

Proof *Let $x(t)$ be a solution of (5.8) on an interval I. Then $t_0 \in I$, x is differentiable on I, and hence x is continuous on I. Moreover, $(t,x(t)) \in D$, for $t \in I$, with $x(t_0) = x_0$, and $x'(t) = f(t,x(t))$ for $t \in I$. Now an integration of $x'(t) = f(t,x(t))$ from t_0 to t gives (5.13) for $t \in I$. For the converse, if x satisfies (5.13) for $t \in I$, then $t_0 \in I$ and x is continuous on I. Moreover, $(t,x(t)) \in D$, for $t \in I$, and (5.13) is satisfied for $t \in I$. Thus $x(t)$ is differentiable on I. By differentiating (5.13) with respect to t we arrive at $x'(t) = f(t,x(t))$ for all $t \in I$ and $x(t_0) = x_0 + \int_{t_0}^{t_0} f(s,x(s))ds = x_0$. This completes the proof.*

For a reference, we mention Picard's local existence and uniqueness theorem, using successive approximations.

Theorem 5.2 *(Picard's Local Existence and Uniqueness) Let $D \subset \mathbb{R} \times \mathbb{R}$ be defined by*

$$D = \{(t,x) : |t - t_0| \le a, |x - x_0| \le b\},$$

where a and b are positive constants. Assume $f \in C(D,\mathbb{R})$ and f satisfies the Lipschitz condition (5.9). Let

$$M = \max_{(t,x)\in D} |f(t,x)| \tag{5.14}$$

and set

$$h = \min\{a, \frac{b}{M}\}. \tag{5.15}$$

Then the (IVP) (5.8) has a unique solution denoted by $x(t,t_0,x_0)$ on the interval $|t - t_0| \le h$ and passing through (t_0,x_0). Furthermore,

$$|x(t) - x_0| \le b, \quad for \quad |t - t_0| \le h.$$

The next lemma is convenient when converting a second-order differential equation into an integral equation of Volterra type.

Lemma 15 *Suppose $F(x)$ is continuous function on $[a,\infty)$. Then*

$$\int_a^x \int_a^\xi F(t)dtd\xi = \int_a^x (x-t)F(t)dt. \tag{5.16}$$

Proof *The proof involves changing the limits of integration. From Fig. 5.1, we define the shaded region by*

$$D = \{(t,\xi) : t \le \xi \le x, a \le t \le x\}.$$

Then

$$\int_a^x \int_a^\xi F(t)dtd\xi = \int\int_D F(t)d\xi dt$$

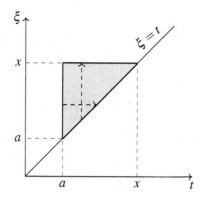

FIGURE 5.1
Shaded region of integrations.

$$= \int_a^x \int_t^x F(t) d\xi dt$$

$$= \int_a^x F(t) \int_t^x d\xi dt$$

$$= \int_a^x F(t) \left(\int_t^x d\xi \right) dt$$

$$= \int_a^x (x-t) F(t) dt.$$

This completes the proof.

We provide the following example.

Example 5.2 In this example, we show that the second-order differential equation

$$y'' + Ay' + By = 0, \quad y(0) = y(1) = 0, \tag{5.17}$$

where A and B are constants, leads to the integral equation $y(x) = \int_0^1 K(x,\xi) y(\xi)\, d\xi$, where

$$K(x,\xi) = \begin{cases} Bx(1-x) + Ax - A & \text{for } \xi < x, \\ Bx(1-\xi) + Ax & \text{for } \xi > x. \end{cases} \tag{5.18}$$

We begin by integrating all the terms in (5.17)

$$\int_0^x y''(\xi)\, d\xi + A \int_0^x y'(\xi)\, d\xi + B \int_0^x y(\xi)\, d\xi = 0,$$

which implies that

$$y'(\xi)\Big|_0^x + Ay(\xi)\Big|_0^x + B \int_0^x y(\xi)\, d\xi = 0.$$

Or,

$$y'(x) - y'(0) + Ay(x) - Ay(0) + B \int_0^x y(\xi) \, d\xi = 0.$$

We are given $y(0) = 0$ and so

$$y'(x) - y'(0) + Ay(x) + B \int_0^x y(\xi) \, d\xi = 0. \tag{5.19}$$

By integrating (5.19), we get

$$\int_0^x y'(\xi) \, d\xi - \int_0^x y'(0) \, d\xi + A \int_0^x y(\xi) \, d\xi + B \int_0^x \int_0^s y(\xi) \, d\xi \, ds = 0.$$

By Lemma 15, we see that

$$\int_0^x \int_0^s y(\xi) \, d\xi \, ds = \int_0^x (x - \xi) y(\xi) \, d\xi,$$

and so,

$$y(x) - y(0) - xy'(0) + A \int_0^x y(\xi) \, d\xi + B \int_0^x (x - \xi) y(\xi) \, d\xi = 0.$$

As $y(0) = 0$, then

$$y(x) - xy'(0) + A \int_0^x y(\xi) \, d\xi + B \int_0^x (x - \xi) y(\xi) \, d\xi = 0. \tag{5.20}$$

Now, by letting $x = 1$ and making use of $y(1) = 0$ in the above equation we obtain

$$y'(0) = A \int_0^1 y(\xi) \, d\xi + B \int_0^1 (1 - \xi) y(\xi) \, d\xi,$$

or

$$y'(0) = \int_0^1 [A + B - B\xi] y(\xi) \, d\xi. \tag{5.21}$$

Finally, substitute (5.21) into (5.20) to get

$$y(x) - x \left[\int_0^1 [A + B - B\xi] y(\xi) \, d\xi \right] + A \int_0^x y(\xi) \, d\xi + B \int_0^x (x - \xi) y(\xi) \, d\xi = 0, \tag{5.22}$$

or

$$y(x) = \int_0^1 [Ax + Bx - Bx\xi] y(\xi) \, d\xi - \int_0^x [A + Bx - B\xi] y(\xi) \, d\xi.$$

Since, $0 \leq x \leq 1$, we may use

$$\int_0^1 = \int_0^x + \int_x^1$$

and rewrite

$$
\begin{aligned}
y(x) &= \int_0^x [Ax + Bx - Bx\xi] y(\xi)\, d\xi + \int_x^1 [Ax + Bx - Bx\xi] y(\xi)\, d\xi \\
&\quad - \int_0^x [A + Bx - B\xi] y(\xi)\, d\xi \\
&= \int_0^x [Ax - A - Bx\xi + B\xi] y(\xi)\, d\xi + \int_x^1 [Ax + Bx - Bx\xi] y(\xi)\, d\xi \\
&= \int_0^x [B\xi(1-x) + Ax - A] y(\xi)\, d\xi \\
&\quad + \int_x^1 [Ax + Bx(1-\xi)] y(\xi)\, d\xi.
\end{aligned}
\tag{5.23}
$$

By observing that the first term and second term in (5.23) are valid over $0 < \xi < x$ and $x < \xi < 1$, respectively, we conclude that (5.23) can be written as

$$
y(x) = \int_0^1 K(x,\xi) y(\xi)\, d\xi,
$$

which is a Fredholm equation of first kind where

$$
K(x,\xi) = \begin{cases}
B\xi(1-x) + Ax - A & \text{for } \xi < x, \\
Bx(1-\xi) + Ax & \text{for } \xi > x.
\end{cases}
$$

$\square$

The next lemma is essential when differentiating an integral equation. It is referred to as *Leibnitz formula*

Lemma 16 *Suppose $\alpha(x)$, $\beta(x)$ are continuous such that $\frac{\partial \alpha}{\partial x}$ and $\frac{\partial \beta}{\partial x}$ exist. If F is continuous in both variables and its first partial derivatives exist, then*

$$
\begin{aligned}
\frac{d}{dx} \int_{\alpha(x)}^{\beta(x)} F(x,\xi)\, d\xi &= \int_{\alpha(x)}^{\beta(x)} \frac{\partial F}{\partial x}(x,\xi)\, d\xi \\
&\quad + F(x,\beta(x))\beta'(x) - F(x,\alpha(x))\alpha'(x)
\end{aligned}
\tag{5.24}
$$

Proof *Let*

$$
\phi(\alpha,\beta,x) = \int_{\alpha(x)}^{\beta(x)} F(x,\xi)\, d\xi,
$$

with

$$
\frac{\partial f}{\partial y} = F(x,\xi).
$$

Then

$$
\phi(\alpha,\beta,x) = f(x,\beta(x)) - f(x,\alpha(x)).
$$

Using the chain rule, we arrive at

$$\frac{d\phi}{dx} = \frac{\partial \phi}{\partial x} + \frac{\partial \phi}{\partial \beta}\frac{\partial \beta}{\partial x} + \frac{\partial \phi}{\partial \alpha}\frac{\partial \alpha}{\partial x}. \tag{5.25}$$

Moreover,

$$\frac{\partial \phi}{\partial x} = \frac{\partial}{\partial x}\int_{\alpha(x)}^{\beta(x)} F(x,\xi)d\xi = \int_{\alpha(x)}^{\beta(x)} \frac{\partial F}{\partial x}(x,\xi)d\xi,$$

since computing $\frac{\partial \phi}{\partial x}$ means that α and β are kept constants. On the other hand, using $\phi(\alpha,\beta,x) = f(x,\beta(x)) - f(x,\alpha(x))$, we get

$$\frac{\partial \phi}{\partial \alpha} = \frac{\partial f(x,\beta)}{\partial \alpha} - \frac{\partial f(x,\alpha)}{\partial \alpha} = 0 - F(x,\alpha).$$

Similarly,

$$\frac{\partial \phi}{\partial \beta} = \frac{\partial f(x,\beta)}{\partial \beta} - \frac{\partial f(x,\alpha)}{\partial \beta} = F(x,\beta) - 0.$$

Substituting the last three expressions into (5.25), we arrive at

$$\frac{d\phi}{dx} = \int_{\alpha(x)}^{\beta(x)} F(x,\xi)d\xi + F(x,\beta)\frac{\partial \beta}{\partial x} - F(x,\alpha)\frac{\partial \alpha}{\partial x},$$

which is (5.25). This completes the proof.

Example 5.3 Consider the integral equation

$$u(x) = \lambda \int_0^x (1-x)\xi u(\xi)d\xi + \lambda \int_x^1 x(1-\xi)u(\xi)d\xi. \tag{5.26}$$

Then (5.26) can be written as

$$u(x) = \lambda \int_0^1 K(x,\xi)u(\xi)d\xi$$

where $K(x,\xi)$ is defined by the relations

$$K(x,\xi) = \begin{cases} \xi(1-x) & \text{when } \xi \le x \le 1, \\ x(1-\xi) & \text{when } 0 \le x \le \xi. \end{cases}$$

It is clear that the kernel K is symmetric and from (5.26) we have that $u(0) = u(1) = 0$. Moreover, using Lemma 16, we have

$$\begin{aligned} u'(x) &= \lambda(1-x)xu(x) - \lambda \int_0^x \xi u(\xi)d\xi \\ &\quad - \lambda(1-x)xu(x) + \lambda \int_x^1 (1-\xi)u(\xi)d\xi \end{aligned}$$

$$= -\lambda \int_0^x \xi u(\xi)d\xi + \lambda \int_x^1 (1-\xi)u(\xi)d\xi.$$

Differentiating one more time gives

$$
\begin{aligned}
u''(x) &= -\lambda x u(x) - \lambda(1-x)u(x) \\
&= -\lambda u(x).
\end{aligned}
$$

Thus, the integral equation satisfies the second-order boundary value problem

$$u''(x) + \lambda u(x) = 0, \quad 0 < x < 1,$$

$$u(0) = 0, \quad u(1) = 0.$$

The boundary value problem is a Sturm-Liouville problem and we refer you to Chapter 4, Section 4.14. Note that for $\lambda \leq 0$, the problem only has the trivial solution. For $\lambda > 0$, we let $\lambda = \alpha^2, \alpha \neq 0$. Then the problem has the solution

$$u(x) = c_1 \cos(\alpha x) + c_2 \sin(\alpha x).$$

Applying the first boundary condition we immediately have $c_1 = 0$. Now, applying $u(1) = 0$, we have $c_2 \sin(\alpha) = 0$. To avoid a trivial solution, we set $\sin(\alpha) = 0$, and from which we get $\alpha = n\pi, \ n = 1, 2, \ldots$ Thus, the problem has the eigenvalues $\lambda_n = n^2 \pi^2$, with corresponding eigenfunctions

$$u_n(x) = \sin(n\pi x), \ n = 1, 2, \ldots$$

$\square$

5.2.1   Exercises

Exercise 5.6 *Show that $f(t,x) = x^{2/3}$ does not satisfy Lipschitz condition in the rectangle $R = \{(t,x) : |t| \leq 1, \ |x| \leq 1\}$.*

Exercise 5.7 *Use integration by parts to prove (5.16) of Lemma 15.*

Exercise 5.8 *(a) If $y''(x) = F(x)$, and y satisfies the initial condition $y(0) = y_0$ and $y'(0) = y_0'$, show that*

$$y(x) = \int_0^x (x-\xi)F(\xi)\, d\xi + y_0' x + y_0.$$

(b) Verify that this expression satisfies the prescribed differential equation and initial conditions.

Exercise 5.9 *(a) If $y''(x) = F(x)$, and y satisfies the end conditions $y(0) = 0$ and $y(1) = 0$, show that*

$$y(x) = \int_0^1 K(x,\xi)F(\xi)\, d\xi$$

where $K(x,\xi)$ is defined by the relations

$$K(x,\xi) = \begin{cases} \xi(x-1) & \text{when } \xi < x, \\ x(\xi-1) & \text{when } \xi > x. \end{cases}$$

(b) *Verify directly that the expression obtained satisfies the prescribed differential equation and end conditions.*

Exercise 5.10 *Verify the integral equation*

$$y(t) = 1 + t - \frac{8}{3}\int_0^t (\xi - t)^3 y(\xi)d\xi,$$

is a solution of the fourth order differential equation

$$y^{(4)}(t) - 16y(t) = 0, \quad y(0) = y'(0) = 1, \ y''(0) = y'''(0) = 0.$$

Exercise 5.11 *Reduce the integral equation*

$$y(x) = \lambda \int_0^\infty e^{|x-\xi|} y(\xi)d\xi,$$

to a differential equation.

Exercise 5.12 *Write the second-order nonhomogenous differential equation $y''(x) = \lambda y(x) + g(x)$, $x > 0$ that satisfies the initial conditions $y(0) = y'(0) = 0$ into an integral equation.*

Exercise 5.13 *Show that the second-order boundary value problem*

$$y''(x) = \lambda y(x), \quad y(a) = y(b) = 0, \ a < x < b$$

can be written in the form

$$y(x) = \lambda \int_a^b K(x,\xi)y(\xi)\,d\xi$$

where $K(x,\xi)$ is defined by the relations

$$K(x,\xi) = \begin{cases} \frac{(x-b)(\xi-a)}{b-a} & \text{when } \xi \le x \le b, \\ \frac{(x-a)(\xi-b)}{b-a} & \text{when } a \le x \le \xi. \end{cases}$$

Exercise 5.14 *Show that the second-order differential equation*

$$y'' + A(x)y' + B(x)y = g(x), \quad y(a) = a_1, \ y(b) = b_1,$$

where A and B are differentiable functions on (a,b) and g is continuous, leads to the integral equation

$$y(x) = f(x) + \int_b^a K(x,\xi)y(\xi)\,d\xi,$$

where

$$f(x) = a_1 + \int_a^x (x-\xi)g(\xi)d\xi + \frac{x-a}{b-a}\left[b_1 - a_1 - \int_a^b (b-\xi)g(\xi)d\xi\right],$$

and

$$K(x,\xi) = \begin{cases} \frac{(x-b)\left(A(\xi)-(a-\xi)(A'(\xi)-B(\xi)\right)}{b-a} & \text{when } \xi \le x \le b, \\ \frac{(x-a)\left(A(\xi)-(b-\xi)(A'(\xi)-B(\xi)\right)}{b-a} & \text{when } a \le x \le \xi. \end{cases}$$

Exercise 5.15 *Find the solution of the Volterra integral equation*

$$y(x) = 1 - x - 4\sin(x) + \int_0^x [3 - 2(x-\xi)]y(\xi)d\xi,$$

by transforming it into a nonhomogeneous second-order differential equation. In order to solve for the solution you need to compute $y(0)$ and $y'(0)$.

Exercise 5.16 *[Only if you covered Chapter 4.]*

Find the necessary Euler-Lagrange condition for a function y to be a local minimum of the functional

$$L(y) = \int_a^b \int_a^b C(t,s)y(s)y(t)dsdt + \int_a^b y^2(t)dt - 2\int_a^b y(t)f(t)dt,$$

where $y(a)$ and $y(b)$ are fixed.

Answer: $\int_a^b [C(s,t) + C(t,s)]y(s)ds + 2y(t) = 2f(t)$, which is a Fredholm integral equation.

Exercise 5.17 *For constants b and c suppose the continuous function $h(x)$ is a solution to the differential equation*

$$h''(x) + bh'(x) + ch(x) = 0, \quad h(0) = 0, \ h'(0) = 1.$$

Show that if

$$y(x) = \int_0^x h(x-\xi)g(\xi)\,d\xi,$$

then $y(x)$ solves the nonhomogeneous second-order differential equation

$$y''(x) + by'(x) + cy(x) = g(x), \quad y(0) = y'(0) = 0.$$

Exercise 5.18 *Find all continuous functions $y = y(x)$ that satisfy the relation*

$$t\int_0^x y(t)dt = (t+1)\int_0^x ty(t)dt.$$

Hint: Differentiate twice to obtain a first-order differential equation that can be solved by separation of variables.

5.3 The Green's Function

The Poisson's equation, $\nabla^2 u = f$, for the electric potential u defined inside a confined volume with certain boundary conditions on the volume's surface first appeared in George Green's work in 1828, which is when the Green's function first appeared. He introduced a function that is now known as the *Green's function*, as later defined by Riemann. We shall develop methods on finding Green's functions for initial value problems as well as boundary value problems. In the previous section, we reduced initial value problems and boundary value problems into Fredholm integral equations where the kernels $K(x, \xi)$ are actually the Green's functions. The Green's function play a major role in setting up nonlinear boundary value problems as integral equations and then some known methods are used to deduce qualitative results regarding solutions. Before we embark on finding the Green's function we prove an important result concerning *self-adjoint* second-order differential operators.

Consider the differential operator

$$\mathscr{L}_\lambda z := \frac{d}{dx}\left(p(x)\frac{dz}{dx}\right) + [q(x) + \lambda\rho(x)]z, \tag{5.27}$$

which is associated with the well-knowm Sturm-Liouville problem.

Definition 5.6 *The differential operator $\mathscr{L}$ given by (5.27) is self-adjoint if there exists a continuously differentiable function g such that*

$$\left(w\mathscr{L}_\lambda z - z\mathscr{L}_\lambda w\right)dx = dg, \tag{5.28}$$

where z and w satisfy $\mathscr{L}_\lambda z = \mathscr{L}_\lambda w = 0$. In other words, $\left(w\mathscr{L}_\lambda z - z\mathscr{L}_\lambda w\right)dx$ must be exact.

We have the following lemma.

Lemma 17 *The differential operator $\mathscr{L}$ given by (5.27) is self-adjoint.*

Proof *Let z and w satisfy $\mathscr{L}z = \mathscr{L}w = 0$. Then*

$$
\begin{aligned}
w\mathscr{L}_\lambda z - z\mathscr{L}_\lambda w &= w\frac{d}{dx}\left(p(x)\frac{dz}{dx}\right) + w[q(x) + \lambda\rho(x)]z \\
&\quad - z\frac{d}{dx}\left(p(x)\frac{dw}{dx}\right) - z[q(x) + \lambda\rho(x)]w \\
&= w\frac{d}{dx}\left(p(x)\frac{dz}{dx}\right) - z\frac{d}{dx}\left(p(x)\frac{dw}{dx}\right) \\
&= wpz'' + wp'z' - zpw'' - zp'w' \\
&= p[wz'' - zw''] + p'[wz' - zw']. \tag{5.29}
\end{aligned}
$$

On the other-hand, after simple calculation we find that

$$\frac{d}{dx}\{p(x)[wz' - zw']\} = p[wz'' - zw''] + p'[wz' - zw'].$$

This implies

$$p'[wz' - zw'] = \frac{d}{dx}\{p(x)[wz' - zw']\} - p[wz'' - zw''].$$

By substituting the above term into (5.29) we obtain

$$w\mathcal{L}_\lambda - z\mathcal{L}_\lambda = \frac{d}{dx}\{p(x)[wz' - zw']\}.$$

Or,

$$(w\mathcal{L}_\lambda z - z\mathcal{L}_\lambda w)dx = d\{p(x)[wz' - zw']\} := dg. \tag{5.30}$$

This completes the proof.

We continue by presenting methods for the construction of the Green's function.

Consider the second-order differential equation

$$\mathcal{L}y + \Phi(x) = 0, \tag{5.31}$$

where $\mathcal{L}$ is the *differential operator*

$$\mathcal{L} := \frac{d}{dx}\left(p(x)\frac{d}{dx}\right) + q(x) = p\frac{d^2}{dx^2} + \frac{dp}{dx}\frac{d}{dx} + q, \tag{5.32}$$

together with homogeneous boundary conditions of the form

$$\alpha y(a) + \beta \frac{dy}{dx}(a) = 0, \quad \alpha y(b) + \beta \frac{dy}{dx}(b) = 0, \tag{5.33}$$

for some constants α and β. It is assumed that the function $p(x)$ is continuous and that $p(x) \neq 0$ for all $x \in (a,b)$. Also $p'(x)$ and $q(x)$ are continuous on (a,b). The function $\Phi(x)$ may depend on x and $y(x)$; that is

$$\Phi(x) = \varphi(x, y(x)).$$

Note that the differential operator defined by (5.32) is the same as the one defined by (5.27) when $\lambda = 0$. We attempt to find a Green function, denoted with $G(x, \xi)$ and given by

$$G(x, \xi) = \begin{cases} G_1(x, \xi) & \text{when } x < \xi \\ G_2(x, \xi) & \text{when } x > \xi, \end{cases} \tag{5.34}$$

and satisfies the following four properties:

(i) The functions G_1 and G_2 satisfy the equation $\mathcal{L}G = 0$; that is $\mathcal{L}G_1 = 0$ when $x < \xi$ and $\mathcal{L}G_2 = 0$ when $x > \xi$.

(ii) The function G satisfies the homogeneous conditions prescribed at the end points $x = a$, and $x = b$; that is G_1 satisfies the condition prescribed at $x = a$, and G_2 satisfies the condition prescribed at $x = b$.

(iii) The function G is continuous at $x = \xi$; that is $G_1(\xi) = G_2(\xi)$.

(iv) The derivative of G has a discontinuity of magnitude $-\frac{1}{p(\xi)}$ at the point $x = \xi$; that is

$$G_2'(\xi) - G_1'(\xi) = -\frac{1}{p(\xi)}. \tag{5.35}$$

Once we determine the Green's function of (5.31) and (5.33), then the problem can be transformed to the relation

$$y(x) = \int_a^b G(x, \xi)\Phi(\xi)d\xi. \tag{5.36}$$

Note that if Φ is constant or a function of x but not $y(x)$ then (5.36) can be solved to obtain the solution. However, if Φ has $y(x)$ then (5.36) is an integral equation of the form

$$y(x) = \int_a^b G(x, \xi)\varphi(\xi, y(\xi))d\xi,$$

where y needs to be determined. We begin by determining the Green's function G. Let $y = u(x)$ be a nontrivial solution of the homogeneous equation $\mathscr{L}y = 0$ along with $\alpha y(a) + \beta \frac{dy}{dx}(a) = 0$. Similarly, we let $y = v(x)$ be a nontrivial solution of the homogeneous equation $\mathscr{L}y = 0$ and $\alpha y(b) + \beta \frac{dy}{dx}(b) = 0$. Then (i) and (ii) are satisfied if we write

$$G(x, \xi) = \begin{cases} c_1 u(x) & \text{when } x < \xi \\ c_2 v(x) & \text{when } x > \xi, \end{cases} \tag{5.37}$$

where c_1 and c_2 are constants. Condition (iii) yields

$$c_1 u(\xi) - c_2 v(\xi) = 0. \tag{5.38}$$

Whereas, (iv) implies

$$c_2 v'(\xi) - c_1 u'(\xi) = -\frac{1}{p(\xi)}. \tag{5.39}$$

Equations (5.38) and (5.39) have a unique solution if the determinant

$$W[u(x), v(x)] = \begin{vmatrix} u(\xi) & v(\xi) \\ u'(\xi) & v'(\xi) \end{vmatrix} = u(\xi)v'(\xi) - u'(\xi)v'(\xi) \neq 0. \tag{5.40}$$

Again, as in Chapter 1, W is referred to as the Wronskian of the solutions $u(x)$ and $v(x)$ of the equation $\mathscr{L}y = 0$. Since u and v are solutions to $\mathscr{L}y = 0$, we have

$$(pu')' + qu = 0 \quad \text{and} \quad (pv')' + qv = 0.$$

By multiplying the second equation by u and the first equation by v, and subtracting the results, there follows

$$u(pv')' - v(pu')' = 0.$$

But by Lemma 17, we have

$$u(pv')' - v(pu')' = [p(uv' - vu')]',$$

from which it follows that

$$[p(uv' - vu')]' = 0.$$

This results into

$$uv' - vu' = \frac{A}{p}$$

for some constant A. In other words,

$$u(\xi)v'(\xi) - v(\xi)u'(\xi) = \frac{A}{p(\xi)}. \tag{5.41}$$

Multiplying (5.39) by $-A$ and comparing the resulting expression with (5.41) we obtain

$$c_1 = -\frac{v(\xi)}{A} \quad \text{and} \quad c_2 = -\frac{u(\xi)}{A}.$$

Therefore, (5.37) takes the form

$$G(x, \xi) = \begin{cases} -\frac{v(\xi)}{A}u(x) & \text{when } x < \xi \\ -\frac{u(\xi)}{A}v(x) & \text{when } x > \xi, \end{cases} \tag{5.42}$$

where the constant A is independent of x and ξ, and is uniquely determined by (5.41). Note that in (5.42) the Green's function G is symmetric. That is

$$G(x, \xi) = G(\xi, x).$$

It turns out that the Green's function of a self-adjoint operator is symmetric. Finally substituting (5.42) into (5.36) the solution can be explicitly found to be

$$\begin{aligned} y(x) &= \int_a^b G(x, \xi)\Phi(\xi)d\xi \\ &= \int_a^x -\frac{u(\xi)}{A}v(x)\Phi(\xi)d\xi + \int_x^b -\frac{v(\xi)}{A}u(x)\Phi(\xi)d\xi \\ &= -\frac{1}{A}\left[\int_a^x u(\xi)v(x)\Phi(\xi)d\xi + \int_x^b v(\xi)u(x)\Phi(\xi)d\xi\right]. \end{aligned} \tag{5.43}$$

Remark 21 *The Green's function for (5.31) and (5.33) is independent of the function $\Phi(x)$. For example if (5.31) is replaced with*

$$\mathscr{L}y = f(x), \tag{5.44}$$

then the solution for (5.44) along with (5.33) is given by

$$y(x) = \int_a^b G(x,\xi)(-f(\xi))d\xi, \tag{5.45}$$

where $G(x,\xi)$ is given by (5.37) or (5.42).

Next we provide two examples for the purpose of better explaining the method.

Example 5.4 Solve the second-order boundary value problem

$$\mathscr{L}y + g(x) = 0, \quad y(0) = y(l) = 0,$$

where $\mathscr{L}y = y''$, using the method of Green's function.

Let $u(x)$ and $v(x)$ be solutions of $y'' = 0$, $y(0) = 0$, and $y'' = 0$, $y(l) = 0$, respectively. Then $y'' = 0$ has the solution $y(x) = ax + b$. Using $y(0) = 0$, we obtain $b = 0$. We may take $u(x) = x$ by setting $a = 1$. Similarly, applying $y(l) = 0$ gives $0 = al + b$. This implies that $b = -al$. Substituting b into $y(x) = ax + b$ gives, $y(x) = a(x - l)$. Thus, we may take $v(x) = x - l$. Note that $W[u(x), v(x)] = l \neq 0$. Set

$$G(x,\xi) = \begin{cases} c_1 x & \text{when } x < \xi \\ c_2(x - l) & \text{when } x > \xi. \end{cases}$$

Using (5.41) with $p = 1$, we arrive at $A = l$. Hence,

$$c_1 = \frac{l - \xi}{l} \text{ and } c_2 = -\frac{\xi}{l}.$$

Thus,

$$G(x,\xi) = \begin{cases} \frac{x}{l}(l - \xi) & \text{when } x < \xi \\ \frac{\xi}{l}(l - x) & \text{when } x > \xi, \end{cases}$$

and the solution to the problem is

$$y(x) = \int_0^l G(x,\xi)g(\xi)d\xi.$$

If we take $g(x) = x$, then the solution is

$$\begin{aligned} y(x) &= \int_0^l G(x,\xi)\xi d\xi \\ &= \int_0^x \frac{\xi}{l}(l - x)\xi d\xi + \int_x^l \frac{x}{l}(l - \xi)\xi d\xi \\ &= \frac{1}{6l}(xl^3 - lx^3). \end{aligned}$$

□

Here is another boundary value problem with homogeneous boundary conditions.

Example 5.5 In this example, we attempt to find the Green's function for the second-order boundary value problem with homogeneous boundary conditions

$$y''(x) = 0, \quad 0 < x < 1, \quad y(0) = 0, \quad y(1) - 3y'(1) = 0.$$

Let $u(x) = A_* + Bx$ and $v(x) = C + Dx$ be solutions of $y'' = 0$, $y(0) = 0$, and $y'' = 0$, $y(1) - 3y'(1) = 0$, respectively. Using $y(0) = 0$, we obtain $A_* = 0$ and so we may take $u(x) = x$ by setting $B = 1$. Similarly, applying $y(1) - 3y'(1) = 0$ gives $0 = C + D - 3D$. This implies that $C = 2D$. Substituting C into $y(x) = C + Dx$ gives, $y(x) = D(2+x)$. Thus, we may take $v(x) = 2 + x$. Set

$$G(x, \xi) = \begin{cases} c_1 x & \text{when } x < \xi \\ c_2(2+x) & \text{when } x > \xi. \end{cases}$$

Using (5.41) with $p = 1$, we arrive at $A = -2$. Hence,

$$c_1 = \frac{2+\xi}{2} \quad \text{and} \quad c_2 = \frac{\xi}{2}.$$

Thus,

$$G(x, \xi) = \begin{cases} \frac{1}{2}(2+\xi)x & \text{when } x < \xi \\ \frac{1}{2}\xi(2+x) & \text{when } x > \xi, \end{cases}$$

Note that $G(x, \xi) = G(\xi, x)$. □

5.3.1 Exercises

Exercise 5.19 *Consider the second-order boundary value problem*

$$y''(x) = 0, \quad a < x < b, \quad y(a) = y(b) = 0. \tag{5.46}$$

(a) Show that (5.46) has the Green's function

$$G(x, \xi) = \begin{cases} \frac{(x-a)(\xi-b)}{b-a} & \text{when } a \le x \le \xi \le b \\ \frac{(x-b)(\xi-a)}{b-a} & \text{when } a \le \xi \le x \le b. \end{cases}$$

(b) Show that for all $x, \xi \in [a, b]$,

$$\frac{a-b}{4} \le G(x, \xi) \le 0.$$

(c) Show that for all $x, \xi \in [a, b]$,

$$\int_a^b |G(x, \xi)| d\xi \le \frac{(b-a)^2}{8}.$$

(d) Show that for all $x, \xi \in [a,b]$,

$$\int_a^b |G'(x,\xi)|d\xi \leq \frac{b-a}{2},$$

where $G' = \frac{d}{dx}G$.

Exercise 5.20 *Use the Green's function to solve the second-order boundary value problem*

$$y''(x) + x^2 = 0, \quad 0 < x < 1, \ y(0) = y(1) = 0.$$

Exercise 5.21 *Use the Green's function to solve the second-order boundary value problem*

$$e^{2x}y''(x) + 2e^{2x}y'(x) = e^{3x}, \quad 0 < x < \ln(2), \ y(0) = y(\ln(2)) = 0.$$

Exercise 5.22 *Use the Green's function to solve the second-order boundary value problem*

$$y''(x) + e^x = 0, \quad a < x < b, \ y(a) = 0, \ y'(b) = 0.$$

Exercise 5.23 *(a) Show the Green's function for the second-order boundary value problem*

$$y''(x) + \alpha^2 y(x) = 0, \quad 0 < x < 1, \ \alpha \neq 0, \ y(0) = y(1) = 0$$

is given by

$$G(x,\xi) = \begin{cases} \frac{\sin[\alpha(1-\xi)]\sin(\alpha x)}{\alpha \sin(\alpha)} & \text{when } 0 \leq x \leq \xi \leq 1 \\ \frac{\sin[\alpha(1-x)]\sin(\alpha\xi)}{\alpha \sin(\alpha)} & \text{when } a \leq \xi \leq x \leq 1. \end{cases}$$

(b) Use part (a) to solve

$$y''(x) + y(x) = x, \quad y(0) = y(\pi/2) = 0.$$

Hint: After simplifying and integrating by parts, the solution is found to be

$$y(x) = x - \frac{\pi}{2}\sin(x).$$

Exercise 5.24 *Use the method of the Green's function to solve*

$$y''(x) + x = 0, \quad 0 < x < 1, \quad y(0) = 0, \ 2y(1) - y'(1) = 0.$$

Exercise 5.25 *Use the method of the Green's function to solve*

$$y''(x) + x = 0, \quad 0 < x < 1, \quad y(0) - 2y'(0) = 0, \ 2y(1) - y'(1) = 0.$$

Exercise 5.26 *Make use of (5.43) to show that y given by (5.36) satisfies the boundary value problem (5.31) and the boundary conditions given by (5.33).*

5.4 Fredholm Integral Equations and Green's Function

In Section 5.3 we discussed the Green's function of second-order differential equations of the form

$$\mathscr{L}y + \Phi(x) = 0 \tag{5.47}$$

where $\mathscr{L}$ is be defined by (5.27). Our aim is to use the concept of Section 5.3 to extend the notion of Green's function to second-order differential equations of the form

$$\mathscr{L}y + \lambda\rho(x)y(x) = f(x), \tag{5.48}$$

where $\mathscr{L}$ is given by (5.27). If we set

$$\Phi(x) = \lambda\rho(x)y(x) - f(x)$$

then by the result of Section 5.3, the solution to the boundary value problem (5.47) subject to the boundary conditions given by (5.33) is given by

$$y(x) = \lambda \int_a^b G(x,\xi)\rho(\xi)y(\xi)d\xi - \int_a^b G(x,\xi)f(\xi)d\xi, \tag{5.49}$$

where G is the Green's function of $\mathscr{L}y = 0$. If we let

$$F(x) = -\int_a^b G(x,\xi)f(\xi)d\xi, \tag{5.50}$$

then (5.49) becomes

$$y(x) = F(x) + \lambda \int_a^b G(x,\xi)\rho(\xi)y(\xi)d\xi, \tag{5.51}$$

which is Fredholm integral equation with kernel

$$K(x,\xi) = G(x,\xi)\rho(\xi).$$

Next we try to put the Fredholm integral equation given by (5.51) in symmetric form provided $\rho(x) > 0$ for $x \in (a,b)$. Assume so and multiply both sides of (5.51) by $\sqrt{\rho(x)}$ and arrive at

$$\sqrt{\rho(x)}y(x) = \sqrt{\rho(x)}F(x) + \lambda \int_a^b \sqrt{\rho(x)\rho(\xi)}G(x,\xi)\sqrt{\rho(\xi)}y(\xi)d\xi.$$

Letting $z(x) = \sqrt{\rho(x)}y(x)$ and $g(x) = \sqrt{\rho(x)}F(x)$, the preceding integral equation reduces to the symmetric Fredhom integral equation

$$z(x) = g(x) + \lambda \int_a^b \sqrt{\rho(x)\rho(\xi)}G(x,\xi)z(\xi)d\xi$$

$$= g(x) + \lambda \int_a^b K(x,\xi)z(\xi)d\xi. \tag{5.52}$$

Note that

$$K(x,\xi) = \sqrt{\rho(x)\rho(\xi)}G(x,\xi) = K(\xi,x)$$

since G is symmetric.

Example 5.6 Let Ly be given by Example 5.4 and we want to reduce the boundary value problem

$$\mathcal{L}y + \lambda y = x, \quad y(0) = y(l) = 0,$$

to a Fredholm integral equation. From the above discussion we have $\rho(x) = 1$, and hence

$$g(x) = -\int_0^l G(x,\xi)\xi d\xi,$$

where from Example 5.4 we have

$$G(x,\xi) = \begin{cases} \frac{x}{l}(l-\xi) & \text{when } x < \xi \\ \frac{\xi}{l}(l-x) & \text{when } x > \xi. \end{cases}$$

Thus,

$$\begin{aligned} g(x) &= -\int_0^l G(x,\xi)\xi d\xi \\ &= -\left[\int_0^x \frac{\xi}{l}(l-x)\xi d\xi + \int_x^l \frac{x}{l}(l-\xi)\xi d\xi\right] \\ &= \frac{x}{6}[x^2 - l^2]. \end{aligned}$$

Hence, by (5.52) the solution is given by the Fredholm integral equation

$$y(x) = \frac{x}{6}[x^2 - l^2] + \lambda \int_0^l G(x,\xi)y(\xi)d\xi,$$

since $\rho(x) = 1$.   □

5.4.1   Exercises

Exercise 5.27 *Reduce the boundary value problem*

$$y''(x) - \lambda y = x^2, \quad y(0) = y(l) = 0,$$

to a Fredholm integral equation.

Exercise 5.28 *Transform the boundary value problem*

$$y''(x) + \lambda xy = 1, \quad y(0) = y(l) = 0,$$

to the integral equation

$$y(x) = \frac{x}{l} - \frac{x}{2}(l-x) + \lambda \int_0^l G(x,\xi)\xi y(\xi)d\xi,$$

where G is given in Example 5.6.

Hint: take $\Phi(x) = \lambda xy(x) - 1$.

Exercise 5.29 *Reduce the boundary value problem*

$$y''(x) + \lambda y(x) = x, \quad 0 < x < 1, \ y(0) - 2y'(0) = 0, \ 2y(1) - y'(1) = 0$$

to a Fredholm integral equation.

Exercise 5.30 *Let*

$$\mathscr{L}y = \frac{d}{dx}\left(x\frac{dy}{dx}\right) - \frac{1}{x}y$$

(a) Show the Green's function for

$$\mathscr{L}y = 0, \quad y(0) = 0, \ y(1) = 0$$

is

$$G(x,\xi) = \begin{cases} \frac{x}{2\xi}(1-\xi^2) & \text{when } x < \xi \\ \frac{\xi}{2x}(1-x^2) & \text{when } x > \xi. \end{cases}$$

(b) Find the integral solution of

$$\mathscr{L}y + \lambda xy = 0, \quad y(0) = 0, \ y(1) = 0.$$

5.4.2   Beam problem

In this section, we briefly discuss the oscillation of thin cantilever beams. Remember, a cantilever beam is a rigid structural element supported at one end and free at the other. Cantilever construction allows overhanging structures without additional supports and bracing. This structural element is widely used in the construction of bridges, towers, and buildings and can add a unique beauty to the structure. Assume we have a long balcony that is supported by an underneath beam that extends outward from the wall. Let the origin be the point at which the beam is supported by the wall. Let the x-axis be parallel to the undeflected beam with the z-axis upward. Assume L is the length of the beam having uniform cross section under a load. Let $w(x,t)$ represent the distributed load or force acting on the beam in the negative z-direction. Let $u(x,t)$ designate the deflection of the beam from its equilibrium position (see Fig. 5.2). Assume e is the modulus of elasticity of the beam's material and $I(x)$ is the

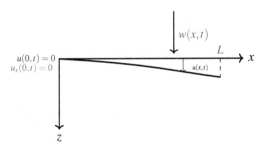

FIGURE 5.2
A thin beam undergoing a small deflections $u(x,t)$ from equilibrium.

moment of inertia of the beam's cross-sectional area about a point x. If $M(x,t)$ is the total bending moment produced by all the forces acting on the beam at point x, then the differential equation of the elastic curve of a beam is found to be

$$eI(x)\frac{\partial^2 u(x,t)}{\partial x^2} = M(x,t). \tag{5.53}$$

The bending moment is related to the applied load by the second-order partial differential equation

$$\frac{\partial^2 M(x,t)}{\partial x^2} = -w(x,t). \tag{5.54}$$

Let's decompose the applied load $w(x,t)$ into an external applied component $F(x,t)$ and an internal inertia component $\rho\frac{\partial^2 u(x,t)}{\partial x^2}$, where $\rho = \rho(x)$ is the linear mass density of the beam at the point x. Thus, if we let

$$w(x,t) = \rho\frac{\partial^2 u(x,t)}{\partial x^2} + F(x,t),$$

then (5.53) and (5.54) yield

$$\frac{\partial^2}{\partial x^2}\left[eI(x)\frac{\partial^2 u(x,t)}{\partial x^2}\right] + \rho(x)\frac{\partial^2 u(x,t)}{\partial x^2} = -F(x,t). \tag{5.55}$$

Since the beam is fixed at the end point $x = 0$, while at the end at $x = L$ is free, we have the appropriate initial and boundary conditions

$$u(0,t) = 0, \qquad \frac{\partial^2 u(L,t)}{\partial x^2} = 0,$$

$$\frac{\partial u(0,t)}{\partial x} = 0, \qquad \frac{\partial}{\partial x}\left[eI\frac{\partial^2 u}{\partial x^2}\right]\Big|_{x=L} = 0, \tag{5.56}$$

$$u(x,0) = g(x), \qquad \frac{\partial u(x,0)}{\partial t} = h(x).$$

If we assume harmonic oscillations in the sense that

$$F(x,t) = f(x)\sin(\omega t + \theta) \text{ and } u(x,t) = y(x)\sin(\omega t + \theta),$$

then (5.55) can be easily reduced to the fourth order ordinary differential equation

$$(ely'')'' - \rho\omega^2 y = -f. \tag{5.57}$$

Consequently, the first four initial and boundary conditions given by (5.56) are reduced to

$$y(0) = 0, \qquad y''(L) = 0,$$
$$y'(0) = 0, \qquad (ely'')'\big|_{x=L} = 0. \tag{5.58}$$

If G is the Green's function of

$$(ely'')'' = 0, \tag{5.59}$$

then the solution of (5.57) and (5.58) is given by the integral relation

$$y(x) = \int_0^L G(x,\xi)\big[-\omega^2\rho(\xi)y(\xi) + f(\xi)\big]d\xi.$$

Next, we briefly discuss how to compute the Green's function for the fourth-order ordinary differential operator. Consider the fourth-order differential equation

$$\mathscr{L}y + \Phi(x) = 0, \tag{5.60}$$

where $\mathscr{L}$ is the *differential operator*

$$(\mathscr{L}y)(x) := \frac{d^2}{dx^2}\left(p(x)\frac{d^2y}{dx^2}\right) + q(x)y(x) = 0, \ x \in [a,b] \tag{5.61}$$

together with homogeneous boundary conditions of the form

$$
\begin{aligned}
R_1 y &:= \alpha_1 y(a) + \alpha_2 y'(a) + \alpha_3 p(a)y''(a) + \alpha_4 (py'')'(a) = 0 \\
R_2 y &:= \beta_1 y(a) + \beta_2 y'(a) + \beta_3 p(a)y''(a) + \beta_4 (py'')'(a) = 0 \\
R_3 y &:= \gamma_1 y(b) + \gamma_2 y'(b) + \gamma_3 p(b)y''(b) + \gamma_4 (py'')'(b) = 0 \\
R_4 y &:= \eta_1 y(b) + \eta_2 y'(b) + \eta_3 p(b)y''(b) + \eta_4 (py'')'(b) = 0
\end{aligned}
\tag{5.62}
$$

for some constants $\alpha_i, \beta_i, \gamma_i$, and $\eta_i, i = 1,2,3,4$ are real constants. It is assumed that the function $p \in C^2[a,b]$ with $p(x) > 0$ on $[a,b]$ and q and is continuous on $[a,b]$. We proceed as before in finding the Green's function for (5.61) and (5.62).

We attempt to find a Green function, denoted with $G(x,\xi)$ and given by

$$G(x,\xi) = \begin{cases} G_1(x,\xi) & \text{when } x < \xi \\ G_2(x,\xi) & \text{when } x > \xi, \end{cases} \tag{5.63}$$

and satisfies the following four properties:

(i) The functions G_1 and G_2 satisfy the equation $\mathscr{L}G = 0$; that is $\mathscr{L}G_1 = 0$ when $x < \xi$ and $\mathscr{L}G_2 = 0$ when $x > \xi$.

(ii) The function G satisfies the homogeneous conditions prescribed at the end points $x = a$, and $x = b$; that is G_1 satisfies R_1y, R_2y and G_2 satisfies satisfies R_3y, R_4y.

(iii) The function G is continuous at $x = \xi$; that is

$$G_1(\xi) = G_2(\xi),$$

$$\frac{d}{dx}G_1(\xi) = \frac{d}{dx}G_2(\xi),$$

$$\frac{d^2}{dx^2}G_1(\xi) = \frac{d^2}{dx^2}G_2(\xi).$$

(iv) The third derivative of G has a discontinuity of magnitude $-\frac{1}{p(\xi)}$ at the point $x = \xi$; that is

$$\frac{d^3}{dx^3}G_2(\xi) - \frac{d^3}{dx^3}G_1(\xi) = -\frac{1}{p(\xi)}. \tag{5.64}$$

Once we determine the Green's function of (5.61) and (5.62), then the problem can be transformed to the relation

$$y(x) = \int_a^b G(x,\xi)\Phi(\xi)d\xi. \tag{5.65}$$

Our interest is to use Green's function to solve the beam problem when the inertia $I(x)$ is constant subject to the initial and boundary conditions given by (5.58). Thus, we consider the boundary value problem given by (5.59) and the initial and boundary conditions given by (5.58). Using (5.63) we obtain from

$$\left(eIy''\right)'' = 0$$

that

$$G(x,\xi) = \frac{1}{eI}\begin{cases} A_1 + A_2x + A_3x^2 + A_4x^3 & \text{when } x < \xi \\ B_1 + B_2(x-L) + B_3(x-L)^2 + B_4(x-L)^3 & \text{when } x > \xi. \end{cases} \tag{5.66}$$

Applying $y(0) = 0, y'(0) = 0$ we readily have $A_1 = A_2 = 0$. Similarly, applying the boundary conditions $y''(L) = 0$ and $\left(eIy''\right)'|_{x=L} = 0$, leads to $B_3 = B_4 = 0$. Thus, so far we have

$$G(x,\xi) = \frac{1}{eI}\begin{cases} A_3x^2 + A_4x^3 & \text{when } x < \xi \\ B_1 + B_2(x-L) & \text{when } x > \xi. \end{cases} \tag{5.67}$$

The jump condition given by (iv) yields $G_2''' - G_1''' = -\frac{1}{eI}$ or

$$0 - \frac{1}{eI}6A_4 = -\frac{1}{eI},$$

which implies that $A_4 = \frac{1}{6}$. Next we apply the continuity condition given by (iii) and obtain

$$A_3\xi^2 + A_4\xi^3 = B_1 + B_2(\xi - L),$$
$$2A_3\xi + 3A_4\xi^2 = B_2,$$
$$2A_3 + 3A_4\xi = 0.$$

From the third equation we obtain $A_3 = -\frac{1}{2}\xi$. On the other hand, the second equation yields $B_2 = -\frac{1}{2}\xi^2$. Thereupon, from the first equation we arrive at

$$B_1 = \frac{\xi^2}{2}(\xi - L).$$

The second part of the Green's function $B_1 + B_2(x - L)$ reduces to

$$B_1 + B_2(x - L) = \frac{\xi^2}{2}\left(\frac{\xi}{3} - L\right) - \frac{\xi^2}{2}(x - L) = \frac{\xi^2}{2}\left(\frac{\xi}{3} - x\right).$$

Finally, the Green's function takes the form

$$G(x,\xi) = \frac{1}{eI}\begin{cases} \frac{x^2}{2}\left(\frac{x}{3} - \xi\right) & \text{when } x < \xi \\ \frac{\xi^2}{2}\left(\frac{\xi}{3} - x\right) & \text{when } x > \xi. \end{cases} \tag{5.68}$$

Furthermore, the solution of the beam problem for constant inertia I takes the form

$$\begin{aligned} y(x) &= \int_0^L G(x,\xi)\left[-\omega^2\rho(\xi)y(\xi) + f(\xi)\right]d\xi \\ &= \frac{1}{eI}\int_0^x \frac{\xi^2}{2}\left(\frac{\xi}{3} - x\right)\left[-\omega^2\rho(\xi)y(\xi) + f(\xi)\right]d\xi \\ &+ \frac{1}{eI}\int_x^L \left(\frac{x^3}{6} - \frac{x^2\xi}{2}\right)\left[-\omega^2\rho(\xi)y(\xi) + f(\xi)\right]d\xi. \end{aligned}$$

5.4.3 Exercises

Exercise 5.31 *Find the Green's function for*

$$(eIy'')'' = 0,$$

subject to the following initial and boundary conditions:

(a)

$$\begin{aligned} y(0) &= 0, & y(L) &= 0, \\ y'(0) &= 0, & y'(L) &= 0. \end{aligned}$$

(b)

$$y(0) = 0, \qquad y'(L) = 0,$$
$$y'(0) = 0, \qquad y'''(L) = 0.$$

(c)

$$y(0) = 0, \qquad y(L) = 0,$$
$$y''(0) = 0, \qquad y''(L) = 0.$$

5.5 Fredholm Integral Equations with Separable Kernels

The ability to solve integral equations, is in most cases, depend on the type of the kernels. In this section we investigate the Fredholm integral equation of the second kind

$$y(x) = f(x) + \lambda \int_a^b K(x, \xi) y(\xi) d\xi, \tag{5.69}$$

where all functions are continuous on their respective domains. We begin with the following definition.

Definition 5.7 *The kernel K is said to be separable or degenerate if it can be written in the form*

$$K(x, \xi) = \sum_{i=1}^n \alpha_i(x) \beta_i(\xi). \tag{5.70}$$

Throughout this section we assume K is separable. For example, the kernel $K(x, \xi) = 3 + 2x\xi$ is separable since it can be written in the form $K(x, \xi) = \sum_{i=1}^2 \alpha_i(x)\beta_i(\xi)$, where $\alpha_1(x) = 3, \beta_1(\xi) = 1, \alpha_2(x) = 2x$, and $\beta_2(\xi) = \xi$. Note that $\alpha_i(x)$ and $\beta_i(\xi)$ are not unique. If we substitute (5.70) into (5.69) we arrive at the new expression

$$y(x) = f(x) + \lambda \sum_{i=1}^n \int_a^b \{\beta_i(\xi) y(\xi) d\xi\} \alpha_i(x). \tag{5.71}$$

Letting

$$c_i = \int_a^b \beta_i(\xi) y(\xi) d\xi, \tag{5.72}$$

equation (5.71) simplifies to

$$y(x) = f(x) + \lambda \sum_{i=1}^n c_i \alpha_i(x). \tag{5.73}$$

Note that the c_i given by (5.72) are unknown constants. Once they are determined the solution is given by (5.73). Multiplying (5.73) by $\beta_j(x)$ and integrating the resulting expression with respect to x from a to b gives

$$\int_a^b \beta_j(x)y(x)dx = \int_a^b \beta_j(x)f(x)dx + \lambda \sum_{i=1}^n c_i \int_a^b \beta_j(x)\alpha_i(x)dx, \ j = 1,2,\ldots n.$$
(5.74)

Interchanging i with j expression (5.74) can be written as

$$c_i = f_i + \lambda \sum_{j=1}^n c_j a_{ij}, \ i = 1,2,\ldots n$$
(5.75)

where c_i is given by (5.72) and

$$f_i = \int_a^b \beta_i(x)f(x)dx \quad \text{and} \quad a_{ij} = \int_a^b \beta_i(x)\alpha_j(x)dx.$$
(5.76)

In matrix form, equation (5.75) takes the form

$$(I - \lambda A)c = f$$
(5.77)

where I is the identity matrix,

$$A = (a_{ij}), \quad c = (c_1, c_2, \ldots, c_n)^T, \quad f = (f_1, f_2, \ldots, f_n)^T,$$

where T denotes the transpose. Thus, (5.77) represents a system of n linear algebraic equations for c. Before we embark on few examples, we recall some basic facts from Chapter 3 about linear systems. Consider the linear system

$$Bx = b$$
(5.78)

where B is an $n \times n$ matrix, b is a given n vector, and x is the unknown vector.

Lemma 18 *Suppose* $b = 0$. *Then*

(a) $\det(B) \neq 0$, *implies* (5.78) *has only the trivial solution* $x = 0$.

(b) $\det(B) = 0$, *implies* (5.78) *has infinitely many solutions. Suppose* $b \neq 0$. *Then*

(c) $\det(B) \neq 0$, *implies* (5.78) *has a unique solution.*

(d) $\det(B) = 0$, *implies* (5.78) *has no solution or infinitely many solutions.*

Based on Lemma 18 we have the following Fredholm theorem.

Theorem 5.3 (Fredholm Theorem) *Consider the Fredholm integral equation* (5.69) *with a separable kernel* K.

(i) Assume $\int_a^b \beta_i(x)f(x)dx, j = 1,\ldots n$ *are not all zero.*

    (a) If $\det(I - \lambda A) \neq 0$, *then there exists a unique solution to* (5.69) *given by* (5.73), *where* $c = (c_1, c_2, \ldots, c_n)^T$ *is the unique solution of* (5.77).

    (b) *If* $\det(I - \lambda A) = 0$, *then either no solution exists or infinitely many solutions exist.*

(ii) *Assume* $\int_a^b \beta_i(x)f(x)dx = 0, j = 1,\ldots n.$

    (a) *If* $\det(I - \lambda A) \neq 0$, *then (5.69) has the solution* $y = f(x)$.

    (b) *If* $\det(I - \lambda A) = 0$, *then (5.69) has infinitely many solutions.*

We provide two examples to illustrate the method of Fredholm equations with separable kernels.

Example 5.7 Consider the homogeneous Fredholm equation

$$y(x) = \lambda \int_0^1 (4x\xi - 5x^2\xi^2)y(\xi)d\xi. \tag{5.79}$$

So we have

$$K(x,\xi) = 4x\xi - 5x^2\xi^2 = \sum_{i=1}^2 \alpha_i(x)\beta_j(\xi).$$

We may choose

$$\alpha_1(x) = x, \quad \alpha_2(x) = x^2, \quad \beta_1(\xi) = 4\xi, \quad \beta_2(\xi) = -5\xi^2.$$

Next we use

$$a_{ij} = \int_0^1 \beta_i(x)\alpha_j(x)dx, \, i, j = 1,2$$

to compute the matrix $A = (a_{ij})$.

$$a_{11} = \int_0^1 \beta_1(x)\alpha_1(x)dx = \int_0^1 4x^2dx = \frac{4}{3},$$

$$a_{12} = \int_0^1 \beta_1(x)\alpha_2(x)dx = \int_0^1 4x^3dx = 1,$$

$$a_{21} = \int_0^1 \beta_2(x)\alpha_1(x)dx = -\int_0^1 5x^3dx = -\frac{5}{4},$$

$$a_{22} = \int_0^1 \beta_2(x)\alpha_2(x)dx = -\int_0^1 5x^4dx = -1.$$

Hence we have matrix

$$A = \begin{pmatrix} \frac{4}{3} & 1 \\ -\frac{5}{4} & -1 \end{pmatrix}$$

and

$$\det(I - \lambda A) = \begin{vmatrix} 1 - \frac{4}{3}\lambda & -\lambda \\ \frac{5}{4}\lambda & 1 + \lambda \end{vmatrix} = (1 - \frac{4}{3}\lambda)(1 + \lambda) + \frac{5}{4}\lambda^2.$$

If $\det(I - \lambda A) = 0$, then we have the simplified quadratic equation

$$\lambda^2 + 4\lambda - 12 = 0$$

that has the two roots

$$\lambda = -6, \quad 2.$$

- If $\det(I - \lambda A) \neq 0$; that is $\lambda \neq -6, 2$, then by (b) of (ii) of Theorem 5.3 equation (5.79) has only the trivial solution $y(x) = 0$, since $f(x) = 0$.

- Now we consider the case $\det(I - \lambda A) = 0$; that is $\lambda = -6, 2$. Then by (ii) of Theorem 5.3 equation (5.79) has infinitely many solutions which depend on the value of λ. Now we determine the forms of solutions by computing the vector c from the equation

$$(I - \lambda A)c = 0,$$

where $c = \begin{pmatrix} c_1 \\ c_2 \end{pmatrix}$. The corresponding system of equations is

$$(1 - \frac{4}{3}\lambda)c_1 - \lambda c_2 = 0$$

$$\frac{5}{4}\lambda c_1 + (1 + \lambda)c_2 = 0. \tag{5.80}$$

Setting $\lambda = -6$ in (5.80) we arrive at $3c_1 + 2c_2 = 0$, from either equation. Setting $c_1 = a$, implies that $c_2 = -\frac{3}{2}a$, for nonzero constant a. Thus, using (5.73) with $f = 0$ we arrive at the infinitely many solutions

$$\begin{aligned} y(x) &= 0 + \lambda \sum_{i=1}^{2} c_i \alpha_i(x) = -6(c_1\alpha_1(x) + c_2\alpha_2(x)) \\ &= -6(ax - \frac{3}{2}ax^2) = -6a(x - \frac{3}{2}x^2). \end{aligned}$$

In a similar manner if we substitute $\lambda = 2$ into (5.80) we arrive at $5c_1 + 6c_2 = 0$, from either equation. Setting $c_1 = b$, implies that $c_2 = -\frac{5}{6}b$, for nonzero constant b. Thus, using (5.73) with $f = 0$ we arrive at the infinitely many solutions

$$\begin{aligned} y(x) &= 0 + \lambda \sum_{i=1}^{2} c_i \alpha_i(x) = 2(c_1\alpha_1(x) + c_2\alpha_2(x)) \\ &= 2(bx - \frac{5}{6}bx^2) = 2b(x - \frac{5}{6}x^2). \end{aligned}$$

□

The next example illustrates the techniques for dealing with nonhomogeneous Fredholm integral equations.

Example 5.8 Consider the nonhomogeneous Fredholm equation

$$y(x) = f(x) + \lambda \int_0^1 (4x\xi - 5x^2\xi^2)y(\xi)d\xi. \tag{5.81}$$

Notice that the kernel and hence the matrix A and the values of λ are the same as in Example 5.7. We begin by addressing (i) of Theorem 5.3.

- If $f_i = \int_0^1 \beta_i(x)f(x)dx$, $i = 1,2$ are not all zero; that is

$$f_1 = \int_0^1 4xf(x)dx \neq 0, \quad \text{or} \quad f_2 = -\int_0^1 5x^2f(x)dx \neq 0,$$

and $\lambda \neq -6, 2$, then (5.81) has a unique solution

$$y(x) = f(x) + \lambda \sum_{i=1}^{2} c_i\alpha_i(x) = f(x) + \lambda(c_1\alpha_1(x) + c_2\alpha_2(x)),$$

where c_1 and c_2 is the unique solution of the system

$$(1 - \frac{4}{3}\lambda)c_1 - \lambda c_2 = \int_0^1 4xf(x)dx$$

$$\frac{5}{4}\lambda c_1 + (1+\lambda)c_2 = -\int_0^1 5x^2f(x)dx.$$

- If

$$f_1 = \int_0^1 4xf(x)dx \neq 0, \quad \text{or} \quad f_2 = -\int_0^1 5x^2f(x)dx \neq 0, \tag{5.82}$$

and $\lambda = -6$, then using (5.77) we arrive at the system

$$9c_1 + 6c_2 = \int_0^1 4xf(x)dx$$

$$\frac{15}{2}c_1 - 5c_2 = -\int_0^1 5x^2f(x)dx.$$

Multiplying the second equation by 2 and then simplifying the resulting system we arrive at the new system of equations

$$3c_1 + 2c_2 = \frac{1}{3}\int_0^1 4xf(x)dx$$

$$-3c_1 - 2c_2 = -2\int_0^1 x^2f(x)dx.$$

Adding both equations leads to

$$0 = \frac{1}{3}\int_0^1 4xf(x)dx - 2\int_0^1 x^2f(x)dx.$$

Thus, (5.81) has no solution if

$$\frac{1}{3}\int_0^1 4xf(x)dx \neq 2\int_0^1 x^2 f(x)dx.$$

On the other hand, (5.81) has infinitely many solutions when

$$\frac{1}{3}\int_0^1 4xf(x)dx = 2\int_0^1 x^2 f(x)dx.$$

In such case to determine the solutions we let $3c_1 = a$ and obtain $2c_2 = -a + 2\int_0^1 x^2 f(x)dx$. This gives the solutions

$$y(x) = f(x) - 6\sum_{i=1}^{2} c_i \alpha_i(x)$$

$$= f(x) - 6\left[\frac{a}{3}x + (\frac{-a}{2} + \int_0^1 x^2 f(x)dx)x^2\right]$$

As for $\lambda = 2$ and assuming (5.82), we have the system

$$-\frac{5}{3}c_1 - 2c_2 = \int_0^1 4xf(x)dx$$

$$\frac{5}{2}c_1 + 3c_2 = -\int_0^1 5x^2 f(x)dx.$$

By simplifying the above system we get

$$-5c_1 - 6c_2 = 3\int_0^1 4xf(x)dx$$

$$5c_1 + 6c_2 = -10\int_0^1 x^2 f(x)dx.$$

Adding both equations leads to

$$0 = 12\int_0^1 xf(x)dx - 10\int_0^1 x^2 f(x)dx.$$

Thus, (5.81) has no solution if

$$12\int_0^1 xf(x)dx \neq 10\int_0^1 x^2 f(x)dx.$$

On the other hand, (5.81) has infinitely many solutions when

$$12\int_0^1 4xf(x)dx = 10\int_0^1 x^2 f(x)dx.$$

In such case to determine the solutions we let $5c_1 = a$ and obtain $6c_2 = -a - 10\int_0^1 x^2 f(x)dx$. This gives the solutions

$$
\begin{aligned}
y(x) &= f(x) + 2\sum_{i=1}^{2} c_i \alpha_i(x) \\
&= f(x) + 2\left[\frac{a}{5}x + \left(\frac{-a}{6} - \frac{5}{3}\int_0^1 x^2 f(x)dx\right)x^2\right].
\end{aligned}
$$

Now we consider (ii) of Fredholm Theorem.

- Suppose

$$
\int_0^1 4xf(x)dx = -\int_0^1 5x^2 f(x)dx = 0. \tag{5.83}
$$

and $\lambda \neq -6, 2$, then (5.81) has the unique solution

$$
y(x) = f(x).
$$

- Assume (5.83) with $\lambda = -6$. Setting $\lambda = -6$ in (5.80) we arrive at $3c_1 + 2c_2 = 0$, from either equation. Setting $c_1 = a$, implies that $c_2 = -\frac{3}{2}a$, for nonzero constant a. Thus, using (5.73) we arrive at the infinitely many solutions

$$
\begin{aligned}
y(x) &= f(x) + \lambda \sum_{i=1}^{2} c_i \alpha_i(x) = -6\big(c_1 \alpha_1(x) + c_2 \alpha_2(x)\big) \\
&= f(x) - 6\big(ax - \frac{3}{2}ax^2\big) = f(x) - 6a\big(x - \frac{3}{2}x^2\big).
\end{aligned}
$$

In a similar manner if we substitute $\lambda = 2$ into (5.80) we arrive at $5c_1 + 6c_2 = 0$, from either equation. Setting $c_1 = b$, implies that $c_2 = -\frac{5}{6}b$, for nonzero constant b. Thus, using (5.73) we arrive at the infinitely many solutions

$$
\begin{aligned}
y(x) &= f(x) + \lambda \sum_{i=1}^{2} c_i \alpha_i(x) = 2\big(c_1 \alpha_1(x) + c_2 \alpha_2(x)\big) \\
&= f(x) + 2\big(bx - \frac{5}{6}bx^2\big) = f(x) + 2b\big(x - \frac{5}{6}x^2\big).
\end{aligned}
$$

$\square$

5.5.1   Exercises

Exercise 5.32 *Solve the Fredholm equation*

$$
y(x) = x^2 + \lambda \int_0^1 x\xi y(\xi)d\xi, \quad for \quad \lambda = -1.
$$

Exercise 5.33 *Consider the homogeneous Fredholm equation*

$$y(x) = \lambda \int_0^1 (x\xi^2 + x^2\xi)y(\xi)d\xi.$$

(a) Find the matrix A and show the roots of the equation

$$\det(I - \lambda A) = 0$$

are

$$\lambda = \frac{4\sqrt{15}}{\sqrt{15}-4}, \frac{4\sqrt{15}}{\sqrt{15}+4}.$$

(b) Use (a) and discuss the solutions of the Fredholm equation.

(c) Find the solution for nonhomogeneous Fredholm equation

$$y(x) = x + \lambda \int_0^1 (x\xi^2 + x^2\xi)y(\xi)d\xi.$$

Exercise 5.34 *Find all values of λ so that the nonhomogeneous Fredholm equation*

$$y(x) = e^x + \lambda \int_0^1 x\xi y(\xi)d\xi$$

has a solution and find it.

Exercise 5.35 *In light of Example 5.7 discuss the solutions of the homogeneous Fredholm equation*

$$y(x) = \lambda \int_{-1}^1 (x+\xi)y(\xi)d\xi.$$

Exercise 5.36 *(a) Show that the characteristic values of λ for the equation*

$$y(x) = \lambda \int_0^{2\pi} \sin(x+\xi) \, y(\xi) \, d\xi$$

are $\lambda_1 = 1/\pi$ and $\lambda_2 = -1/\pi$, with corresponding characteristic functions of the form $y_1(x) = \sin(x) + \cos(x)$ and $y_2(x) = \sin(x) - \cos(x)$.

(b) Obtain the most general solution of the equation

$$y(x) = \lambda \int_0^{2\pi} \sin(x+\xi) \, y(\xi) \, d\xi + F(x)$$

when $F(x) = x$ and when $F(x) = 1$, under the assumption that $\lambda \neq \pm 1/\pi$.

(c) Prove that the equation

$$y(x) = \frac{1}{\pi} \int_0^{2\pi} \sin(x+\xi) \, y(\xi) \, d\xi + F(x)$$

possesses no solution when $F(x) = x$, but that it possesses infinitely many solutions when $F(x) = 1$. Determine all such solutions.

(d) *Determine the most general form of the prescribed $F(x)$, for which the integral equation*

$$\int_0^{2\pi} \sin(x+\xi)\, y(\xi)\, d\xi = F(x),$$

of the first kind, possesses a solution.

Exercise 5.37 *In light of Example 5.8, discuss the solutions of the nonhomogeneous Fredholm equation*

$$y(x) = F(x) + \lambda \int_0^1 (1-3x\xi)y(\xi)d\xi.$$

Exercise 5.38 *In light of Example 5.8, discuss the solutions of the nonhomogeneous Fredholm equation*

$$y(x) = F(x) + \lambda \int_{-1}^1 (x+\xi)y(\xi)d\xi.$$

In addition, find an example of $F(x)$ that satisfies all the relevant condition(s) that you obtain in studying the solutions.

Exercise 5.39 *Solve*

$$y(x) = 1 + \lambda \int_0^1 (x+3x^2\xi)y(\xi)d\xi.$$

Exercise 5.40 *Solve*

$$y(x) = 1 + \lambda \int_0^1 (18x+4x^2\xi)y(\xi)d\xi.$$

Exercise 5.41 *Solve*

$$y(x) = 1 + \int_{-1}^1 (1+\xi+3x\xi)y(\xi)d\xi.$$

5.6   Symmetric Kernel

In Section 5.5 we looked at Fredholm integral equations with degenerate or separable kernels. Now, we consider the Fredholm integral equation

$$y(x) = f(x) + \lambda \int_a^b K(x,\xi)y(\xi)d\xi, \tag{5.84}$$

where all functions are continuous on their respective domains. We assume the kernel K in (5.84) is *symmetric*. That is

$$K(x,\xi) = K(\xi,x).$$

Throughout this section it is assumed that the kernel K is symmetric. As in the case of nonhomogeneous differential equations, first we learn how to find the solution of the homogeneous integral equation

$$y(x) = \lambda \int_a^b K(x,\xi)y(\xi)d\xi, \tag{5.85}$$

and then utilize it to find the general solution of (5.84).

Recall that, If λ and $y(x)$ satisfy (5.85) we say λ is an *eigenvalue* and $y(x)$ is the corresponding *eigenfunction*. It should cause no confusion between λ_n being all the eigenvalues of (5.85) and the value of λ for (5.84). In most cases, we will require $\lambda \neq \lambda_n$. We have the following theorem regarding eigenvalues and corresponding eigenfunctions of (5.85).

Theorem 5.4 *Assume the kernel of the homogeneous integral equation (5.85) is symmetric. Then the following statements hold.*

(i) *If λ_m and λ_n are two distinct eigenvalues, then the corresponding eigenfunctions $y_m(x)$ and $y_n(x)$ are orthogonal on the interval $[a,b]$. That is*

$$\int_a^b y_m(x)y_n(x)dx = 0, \quad \text{for} \quad m \neq n.$$

(ii) *The eigenvalues are real.*

Proof *Let λ_m and λ_n be two distinct eigenvalues with corresponding eigenfunctions $y_m(x)$ and $y_n(x)$, respectively. Then we have*

$$y_m(x) = \lambda_m \int_a^b K(x,\xi)y_m(\xi)d\xi,$$

and

$$y_n(x) = \lambda_n \int_a^b K(x,\xi)y_n(\xi)d\xi.$$

Multiplying $y_n(x) = \lambda_n \int_a^b K(x,\xi)y_n(\xi)d\xi$ by $y_m(x)$ and integrating the resulting equation from a to b we get

$$\int_a^b y_m(x)y_n(x)dx = \lambda_n \int_a^b y_m(x) \int_a^b K(x,\xi)y_n(\xi)d\xi dx$$

$$= \lambda_n \int_a^b \left(\int_a^b y_m(x)K(x,\xi)dx \right)y_n(\xi)d\xi$$

$$= \lambda_n \int_a^b \left(\int_a^b K(\xi,x)y_m(x)dx \right)y_n(\xi)d\xi \ (\text{since } K(x,\xi) = K(\xi,x))$$

$$= \lambda_n \int_a^b \left(\frac{1}{\lambda_m}y_m(\xi) \right)y_n(\xi)d\xi$$

$$= \frac{\lambda_n}{\lambda_m} \int_a^b y_m(\xi)y_n(\xi)d\xi.$$

This gives the relation

$$\left(1 - \frac{\lambda_n}{\lambda_m}\right) \int_a^b y_m(x) y_n(x) dx = 0. \tag{5.86}$$

Since $\lambda_n \neq \lambda_m$, then we must have from expression (5.86) that

$$\int_a^b y_m(x) y_n(x) dx = 0.$$

This completes the proof of (i).

Next we prove (ii). Assume one of the eigenvalues, say λ is complex. That is $\lambda = \alpha + i\beta$, where α and β are real numbers. Let $y(x)$ be the corresponding eigenfunction that might be complex. The conjugate of λ is $\bar{\lambda} = \alpha - i\beta$. If $y(x) = u(x) + iv(x)$, then $\bar{y}(x) = u(x) - iv(x)$ is the corresponding eigenfunction of $\bar{\lambda}$. Using a similar argument as in the proof of (i) and interchanging λ_m and λ_n with $\bar{\lambda}$ and λ, respectively we arrive at from (5.86) that

$$\left(1 - \frac{\bar{\lambda}}{\lambda}\right) \int_a^b y(x) \bar{y}(x) dx = 0.$$

Since

$$y(x) \bar{y}(x) = u^2(x) + v^2(x) > 0,$$

the above expression takes the form

$$\left(1 - \frac{\bar{\lambda}}{\lambda}\right) \int_a^b (u^2(x) + v^2(x)) dx > 0,$$

which is not identically zero. Thus, our assumption that $\beta \neq 0$, has led to a contradiction. Therefore, we must conclude that $\beta = 0$, and hence λ is real. This completes the proof.

Example 5.9 See Example 5.3. □

Next, we develop the solution for the nonhomogeneous integral equation (5.84).

We begin with Hilbert-Schmidt theorem.

Theorem 5.5 (Hilbert-Schmidt Theorem) *Assume that there is a continuous function g for which*

$$F(x) = \int_a^b K(x, \xi) g(\xi) d\xi.$$

Then $F(x)$ can be expressed as

$$F(x) = \sum_{n=1}^{\infty} c_n y_n(x),$$

where $y_n(x)$ are the normalized eigenfunctions of (5.85) and

$$c_n = \int_a^b F(x)y_n(x)dx. \tag{5.87}$$

As a result of Theorem 5.5, we may say the function F is *generated* by the continuous function g.

Theorem 5.6 *Let $y(x)$ be a solution to (5.84) where λ is not an eigenvalue of (5.85). Then*

$$y(x) = f(x) + \lambda \sum_{n=1}^{\infty} \frac{f_n}{\lambda_n - \lambda} y_n(x), \tag{5.88}$$

where

$$f_n = \int_a^b f(x)y_n(x)dx, \tag{5.89}$$

and the λ_n and y_n are the eigenvalues and normalized eigenfunctions of (5.85).

Proof *From (5.84), we have*

$$y(x) - f(x) = \int_a^b K(x,\xi)(\lambda y(\xi))d\xi$$

and hence $y - f$ is generated by the continuous function λy. Thus by Theorem 5.5, the function $y - f$ can be expressed by

$$y - f = \sum_{n=1}^{\infty} c_n y_n(x), \tag{5.90}$$

where $y_n(x)$ are the normalized eigenfunctions of (5.85) and

$$c_n = \int_a^b (y(x) - f(x))y_n(x)dx = \int_a^b y(x)y_n(x)dx - f_n, \tag{5.91}$$

with

$$f_n = \int_a^b f(x)y_n(x)dx.$$

Next we multiply (5.84) by $y_n(x)$ and integrate from a to b

$$\int_a^b y(x)y_n(x)dx = f_n + \lambda \int_a^b (\int_a^b K(x,\xi)y(\xi)d\xi)y_n(x)dx$$

$$= f_n + \lambda \int_a^b (\int_a^b K(\xi,x)y_n(x)dx)y(\xi)d\xi \ (since \ K(x,\xi) = K(\xi,x))$$

$$= f_n + \frac{\lambda}{\lambda_n} \int_a^b y_n(\xi)y(\xi)d\xi.$$

After replacing ξ with x in the right side and solving for $\int_a^b y(x)y_n(x)dx$ we arrive at

$$\int_a^b y(x)y_n(x)dx = \frac{f_n}{1 - \frac{\lambda}{\lambda_n}} = \frac{\lambda_n f_n}{\lambda_n - \lambda}.$$

Utilizing (5.91) we obtan

$$c_n = \frac{\lambda_n f_n}{\lambda_n - \lambda} - f_n = \frac{\lambda f_n}{\lambda_n - \lambda}.$$

Finally, using (5.90) we arrive at the solution

$$y(x) = f(x) + \lambda \sum_{n=1}^{\infty} \frac{f_n}{\lambda_n - \lambda} y_n(x).$$

This completes the proof.

Example 5.10 Consider the Fredholm integral equation

$$y(x) = x + \lambda \int_0^1 K(x,\xi)y(\xi)d\xi, \tag{5.92}$$

where $K(x,\xi)$ is defined by the relations

$$K(x,\xi) = \begin{cases} \xi(1-x) & \text{when } \xi \le x \le 1, \\ x(1-\xi) & \text{when } 0 \le x \le \xi. \end{cases}$$

The kernel K is symmetric and moreover, using Lemma 16 on the homogeneous integral equation

$$y(x) = \lambda \int_0^1 K(x,\xi)y(\xi)d\xi,$$

we arrive at the second-order boundary value problem

$$y''(x) + \lambda y(x) = 0, \quad 0 < x < 1,$$

$$y(0) = 0, \quad y(1) = 0.$$

See Example 5.3. This boundary value problem is a Sturm-Liouville problem. From Example 5.3 we have the eigenvalues $\lambda_n = n^2\pi^2, n = 1,2,\ldots$ with corresponding eigenfunctions

$$y_n(x) = \sin(n\pi x), \quad n = 1,2,\ldots$$

Then the normalized eigenfunctions are

$$y_n(x) = \sqrt{2}\sin(n\pi x), \quad n = 1,2,\ldots.$$

Moreover,

$$f_n = \int_0^1 f(x)y_n(x)dx = \int_0^1 x\sqrt{2}\sin(n\pi x)dx = \frac{(-1)^{n+1}\sqrt{2}}{n\pi}, \quad n = 1,2,\ldots.$$

Using (5.88), we have the solution to the nonhomogeneous integral equation

$$y(x) = x + \frac{2\lambda}{\pi} \sum_{n=1}^{\infty} \frac{(-1)^{n+1} \sin(n\pi x)}{n(n^2\pi^2 - \lambda)}, \quad \lambda \neq n^2\pi^2, \; n = 1,2,\ldots.$$

$\square$

Example 5.11 Consider the Fredholm integral equation

$$y(x) = (x+1)^2 + \lambda \int_{-1}^{1} (x\xi + x^2\xi^2) y(\xi) d\xi. \tag{5.93}$$

It is clear that the kernel $K(x,\xi) = x\xi + x^2\xi^2 = K(\xi,x)$. To apply Theorem 5.6, we first need to find the eigenvalues and corresponding normalized eigenfunctions of

$$y(x) = \lambda \int_{-1}^{1} (x\xi + x^2\xi^2) y(\xi) d\xi. \tag{5.94}$$

We write (5.94) in the following manner,

$$
\begin{aligned}
y(x) &= \lambda x \int_{-1}^{1} \xi y(\xi) d\xi + \lambda x^2 \int_{-1}^{1} \xi^2 y(\xi) d\xi \\
&= \lambda x C_1 + \lambda x^2 C_2,
\end{aligned} \tag{5.95}
$$

where

$$C_1 = \int_{-1}^{1} \xi y(\xi) d\xi, \; C_2 = \int_{-1}^{1} \xi^2 y(\xi) d\xi.$$

From (5.95), we see that

$$y(\xi) = \lambda \xi C_1 + \lambda \xi^2 C_2. \tag{5.96}$$

Substituting $y(\xi)$ given by (5.96) into C_1 and C_2 gives

$$C_1 = \int_{-1}^{1} \xi (\lambda \xi C_1 + \lambda \xi^2 C_2) d\xi = \frac{2}{3}\lambda C_1 + 0C_2,$$

and

$$C_2 = \int_{-1}^{1} \xi^2 (\lambda \xi C_1 + \lambda \xi^2 C_2) d\xi = 0C_1 + \frac{2}{5}\lambda C_2.$$

Thus, we have the system of equations

$$(1 - \frac{2}{3}\lambda)C_1 + 0C_2 = 0$$

$$0C_1 + (1 - \frac{2}{5}\lambda)C_2 = 0.$$

For nontrivial values of C_1 and C_2 we must have

$$\begin{vmatrix} 1 - \frac{2}{3}\lambda & 0 \\ 0 & 1 - \frac{2}{5}\lambda \end{vmatrix} = 0.$$

This gives the eigenvalues $\lambda_1 = \frac{3}{2}$, $\lambda_2 = \frac{5}{2}$. If $\lambda_1 = \frac{3}{2}$, then the above system is reduced to

$$0C_1 + 0C_2 = 0$$

$$0C_1 + \frac{2}{5}C_2 = 0.$$

Hence, $C_2 = 0$, and C_1 is arbitrary. Thus, our first eigenfunction is $y_1(x) = \lambda x C_1 + \lambda x^2 C_2 = \frac{3}{2}C_1 x = x$, by choosing $\frac{3}{2}C_1 = 1$. Similarly, when $\lambda_2 = \frac{5}{2}$, the above system reduces to

$$-\frac{2}{3}C_1 + 0C_2 = 0$$

$$0C_1 + 0C_2 = 0$$

and we get $C_1 = 0$ and C_2 is arbitrary. The corresponding eigenfunction is $y_2(x) = x^2$, by choosing $\frac{5}{2}C_2 = 1$. Next with normalize the eigenfunctions. Let

$$\phi_1(x) = \frac{y_1(x)}{\sqrt{\int_{-1}^{1} y_1^2(x)dx}}, \quad \phi_2(x) = \frac{y_2(x)}{\sqrt{\int_{-1}^{1} y_2^2(x)dx}}.$$

Then ϕ_1 and ϕ_2 are normalized and given by

$$\phi_1(x) = \frac{x}{\sqrt{\int_{-1}^{1} x^2 dx}} = \frac{x\sqrt{6}}{2},$$

and

$$\phi_2(x) = \frac{x^2}{\sqrt{\int_{-1}^{1} x^4 dx}} = \frac{x^2\sqrt{10}}{2}.$$

Moreover, f_1 and f_2, are found to be

$$f_1 = \int_{-1}^{1} f(x)\phi_1(x)dx = \int_{-1}^{1} (x+1)^2 \frac{x\sqrt{6}}{2} dx = \frac{2\sqrt{6}}{3},$$

and

$$f_2 = \int_{-1}^{1} f(x)\phi_2(x)dx = \int_{-1}^{1} (x+1)^2 \frac{x^2\sqrt{10}}{2} dx = \frac{8\sqrt{10}}{15}.$$

Thus, for $\lambda \neq \lambda_1 = \frac{3}{2}$ and $\lambda \neq \lambda_2 = \frac{5}{2}$, the solution is

$$
\begin{aligned}
y(x) &= (x+1)^2 + \lambda \sum_{n=1}^{2} \frac{f_n}{\lambda_n - \lambda} \phi_n(x) \\
&= (x+1)^2 + \lambda \left[\frac{f_1}{\lambda_1 - \lambda} \phi_1(x) + \frac{f_2}{\lambda_2 - \lambda} \phi_2(x) \right] \\
&= (x+1)^2 + \lambda \left[\frac{\frac{2\sqrt{6}}{3}}{\frac{3}{2} - \lambda} \frac{x\sqrt{6}}{2} + \frac{\frac{8\sqrt{10}}{15}}{\frac{5}{2} - \lambda} \frac{x^2\sqrt{10}}{2} \right].
\end{aligned}
$$

Notice for $\lambda = 1$, then the above solution reduces to

$$y(x) = \frac{25}{9}x^2 + 6x + 1,$$

which is the solution for

$$y(x) = (x+1)^2 + \int_{-1}^{1} (x\xi + x^2\xi^2)y(\xi)d\xi.$$

□

5.6.1 Exercises

Exercise 5.42 *(a) Find the eigenvalues and corresponding normalized eigenfunctions for the homogeneous integral equation*

$$y(x) = \lambda \int_0^1 K(x,\xi)y(\xi)d\xi$$

where $K(x,\xi)$ is defined by the relations

$$K(x,\xi) = \begin{cases} x(1-\xi) & \text{when } 0 \le x \le \xi \le 1, \\ \xi(1-x) & \text{when } 0 \le \xi \le x \le 1. \end{cases}$$

(b) Use part (a) to solve the Fredholm integral equation

$$y(x) = x + \lambda \int_0^1 K(x,\xi)y(\xi)d\xi.$$

(c) Does the Fredholm integral equation

$$y(x) = x + 4\pi^2 \int_0^1 K(x,\xi)y(\xi)d\xi$$

have a solution?

Exercise 5.43 *(a) Determine the eigenvalues and the corresponding normalized eigenfunctions for*

$$y(x) = \lambda \int_0^\pi \cos(x+\xi)\, y(\xi)\, d\xi.$$

(b) Solve

$$y(x) = F(x) + \lambda \int_0^\pi \cos(x+\xi)\, y(\xi)\, d\xi$$

when λ is not characteristic and $F(x) = 1$.

(c) *Obtain the general solution (when it exists) if* $F(x) = \sin(x)$, *considering all possible cases.*

Exercise 5.44 (a) *Determine the eigenvalues and the corresponding normalized eigenfunctions for*

$$y(x) = \lambda \int_0^1 K(x,\xi)y(\xi)\,d\xi$$

where K is the Green's function that was obtained in Example 5.5 for the boundary value problem

$$y''(x) = 0, \quad 0 < x < 1, \quad y(0) = 0, \quad y(1) - 3y'(1) = 0.$$

(b) *Find the solution of*

$$y(x) = 1 + \lambda \int_0^1 K(x,\xi)y(\xi)d\xi.$$

Exercise 5.45 (a) *Determine the eigenvalues and the corresponding normalized eigenfunctions for*

$$y(x) = \lambda \int_{-1}^1 (x\xi + 1)y(\xi)d\xi.$$

(b) *Solve*

$$y(x) = x + \lambda \int_{-1}^1 (x\xi + 1)y(\xi)d\xi.$$

when λ is not characteristic.

Exercise 5.46 (a) *Determine the eigenvalues and the corresponding normalized eigenfunctions for*

$$y(x) = \lambda \int_{-1}^1 (x^3\xi^3 + x^2\xi^2)y(\xi)d\xi.$$

(b) *Solve*

$$y(x) = x + \lambda \int_{-1}^1 (x^3\xi^3 + x^2\xi^2)y(\xi)d\xi.$$

when λ is not characteristic.

5.7 Iterative Methods and Neumann Series

At the beginning of this chapter, we briefly touched on an iteration method for solving Volterra integral equations of the second kind. In this section, we consider similar integral equations where the independent variable x is bounded. To be specific, we consider the Volterra integral equation

$$y(x) = f(x) + \lambda \int_a^x K(x,\xi)y(\xi)d\xi, \quad a \leq x \leq b \tag{5.97}$$

where f and K are continuous on their respective domains. We emphasize that the kernel K in (5.97) does not need to be symmetric or separable. We define a successive approximation by

$$
\begin{aligned}
y_0(x) &= f(x) \\
y_1(x) &= f(x) + \lambda \int_a^x K(x,\xi)y_0(\xi)d\xi \\
y_2(x) &= f(x) + \lambda \int_a^x K(x,\xi)y_1(\xi)d\xi \\
&\vdots \\
y_n(x) &= f(x) + \lambda \int_a^x K(x,\xi)y_{n-1}(\xi)d\xi, \quad n = 1,2,\ldots
\end{aligned}
\tag{5.98}
$$

Our aim here is to give a concise method on how to define a sequence of functions $\{y_n\}$ successively for (5.97) and obtain an infinite series that represents the solution. Let y_0 be an initial approximation. Then replacing y in the integrand by y_0 gives

$$y_1(x) = f(x) + \lambda \int_a^x K(x,\xi)y_0(\xi)d\xi.$$

Substituting this approximation for y in the integrand gives the approximation

$$
\begin{aligned}
y_2(x) &= f(x) + \lambda \int_a^x K(x,\xi)\left[f(\xi) + \lambda \int_a^\xi K(\xi,\xi_1)y_0(\xi_1)d\xi_1\right]d\xi \\
&= f(x) + \lambda \int_a^x K(x,\xi)f(\xi)d\xi \\
&\quad + \lambda^2 \int_a^x K(x,\xi)\int_a^\xi K(\xi,\xi_1)y_0(\xi_1)d\xi_1 d\xi.
\end{aligned}
$$

In a similar fashion, substituting y_2 into y_3 of (5.97) gives

$$y_3(x) = f(x) + \lambda \int_a^x K(x,\xi)f(\xi)d\xi$$

$$+ \quad \lambda^2 \int_a^x K(x,\xi) \int_a^\xi K(\xi,\xi_1) f(\xi_1) d\xi_1 d\xi$$

$$+ \quad \lambda^3 \int_a^x K(x,\xi) \int_a^\xi K(\xi,\xi_1) \int_a^{\xi_1} K(\xi_1,\xi_2) y_0(\xi_2) d\xi_2 d\xi_1 d\xi.$$

If we define the operator $\mathbf{L}$ by

$$\mathbf{L}y(x) = \int_a^x K(x,\xi) y(\xi) d\xi, \tag{5.99}$$

then we may write (5.97) in the form

$$y = f + \lambda \mathbf{L}y. \tag{5.100}$$

In addition, y_1, y_2, and y_3 may also be rewritten so that

$$y_1 = f + \lambda \mathbf{L}y_0, \quad y_2 = f + \lambda \mathbf{L}f + \lambda^2 \mathbf{L}^2 y_0$$

and

$$y_3 = f + \lambda \mathbf{L}f + \lambda^2 \mathbf{L}^2 f + \lambda^3 \mathbf{L}^3 y_0.$$

Continuing in this fashion we obtain the successive approximation

$$y_n(x) = f(x) + \sum_{i=1}^{n-1} \lambda^i \mathbf{L}^i f(x) + \lambda^n \mathbf{L}^n y_0(x), \tag{5.101}$$

where $\mathbf{L}^i = \underbrace{\mathbf{L}(\mathbf{L}(\cdots \mathbf{L}))}_{i \text{ times}}$, with $\mathbf{L}^0 f = f$. If we can show that $\mathbf{L}^n y_0(x) \to 0$ as $n \to \infty$, then from (5.101), the unique solution of (5.97) will be

$$y(x) = f(x) + \sum_{i=1}^{\infty} \lambda^i \mathbf{L}^i f(x). \tag{5.102}$$

Lemma 19 *Let $\mathbf{L}^n y_0(x)$ be define by (5.101). Then $\mathbf{L}^n y_0(x) \to 0$ and $n \to \infty$.*

Proof *Let*

$$M = \max_{a \le x, \xi \le b} |K(x,\xi)| \quad and \quad C = \max_{a \le x \le b} |y_0(x)|.$$

Then

$$\begin{aligned}
\left| \mathbf{L}y_0(x) \right| &= \left| \int_a^x K(x,\xi) y_0(\xi) d\xi \right| \\
&\le \int_a^x |K(x,\xi)| |y_0(\xi)| d\xi \\
&\le (x-a)MC, \quad a \le x \le b.
\end{aligned}$$

Similarly,

$$\left| \mathbf{L}^2 y_0(x) \right| = \left| \int_a^x K(x,\xi) \mathbf{L}y_0(\xi) d\xi \right|$$

$$\leq \int_a^x |K(x,\xi)||\mathbf{L}y_0(\xi)|d\xi$$

$$\leq \int_a^x M(\xi - a)MCd\xi$$

$$\leq \frac{(x-a)^2}{2}M^2C, \; a \leq x \leq b.$$

Continuing this way, we arrive at

$$|\mathbf{L}^n y_0(x)| \leq \frac{(x-a)^n}{n!}M^nC \leq \frac{(b-a)^n}{n!}M^nC. \tag{5.103}$$

To complete the induction argument, we assume (5.103) holds for n and show it holds for n + 1. Using (5.103), we arrive at

$$\begin{aligned}
|\mathbf{L}^{n+1}y_0(x)| &= \left|\int_a^x K(x,\xi)\mathbf{L}^n y_0(\xi)d\xi\right| \\
&\leq \int_a^x |K(x,\xi)||\mathbf{L}^n y_0(\xi)|d\xi \\
&\leq \int_a^x M\frac{(\xi-a)^n}{n!}M^nC\,d\xi \\
&\leq \frac{(x-a)^{n+1}}{(n+1)!}M^{n+1}C \\
&\leq \frac{(b-a)^{n+1}}{(n+1)!}M^{n+1}C.
\end{aligned}$$

This completes the induction argument. Now, it is clear from (5.103) that

$$\lim_{n\to\infty}|\lambda^n||\mathbf{L}^n y_0(x)| = \lim_{n\to\infty}|\lambda^n|\frac{(b-a)^n}{n!}M^nC = 0$$

uniformly for all $a \leq x \leq b$ and for all values of λ. This shows the infinite series (5.101) converges for any finite λ.

Lemma 20 *The solution of (5.97) is given by (5.102).*

Proof *By Lemma 19 we have the sequence $\{y_n(x)\}$ converges uniformly on $[a,b]$, say to a function $y(x)$. Consider the successive iterations*

$$y_n(x) = f(x) + \lambda \int_a^x K(x,\xi)y_{n-1}(\xi)d\xi, \; n = 1,2,\dots$$

Then,

$$\begin{aligned}
y(x) &= \lim_{n\to\infty} y_n(x) \\
&= f(x) + \lambda \lim_{n\to\infty}\int_a^x K(x,\xi)y_{n-1}(\xi)d\xi
\end{aligned}$$

$$= f(x) + \lambda \int_a^x K(x,\xi) \lim_{n \to \infty} y_{n-1}(\xi) d\xi$$

$$= f(x) + \lambda \int_a^x K(x,\xi) y(\xi) d\xi. \tag{5.104}$$

In the next lemma we show that if y satisfies (5.104), then it is unique. In addition, the author assume the reader is familiar with Banach spaces. For more on Banach spaces we refer to [19] of Chapter 4.

Lemma 21 *If y satisfies (5.104), then it is unique provided that*

$$\lambda < \frac{1}{M(b-a)}, \tag{5.105}$$

where $M = \max_{a \leq x, \xi \leq b} |K(x,\xi)|$.

Proof *Let $\mathbb{B} = \{g : g \in C([a,b], \mathbb{R})\}$. Then the space $\mathbb{B}$ endowed with the maximum norm $\|\cdot\|$ is a Banach space. Suppose y satisfies (5.104). By (5.105) there exists an $\alpha \in (0,1)$ such that $\lambda M(b-a) \leq \alpha$. For $y \in \mathbb{B}$, define the operator $\mathscr{P} : \mathbb{B} \to \mathbb{B}$ by*

$$\mathscr{P}(y)(x) = f(x) + \lambda \int_a^x K(x,\xi) y(\xi) d\xi.$$

Clearly, $\mathscr{P}(y)(x)$ is a continuous map. Let $w, z \in \mathbb{B}$. Then,

$$\|\mathscr{P}(w) - \mathscr{P}(z)\| \leq \max_{x \in [a,b]} |\mathscr{P}(w)(x) - \mathscr{P}(z)(x)|$$

$$\leq \lambda M \int_a^x |w(u) - z(u)| du$$

$$\leq \lambda M \int_a^b |w(u) - z(u)| du$$

$$\leq \lambda M(b-a)\|w - z\|$$

$$\leq \alpha \|w - z\|.$$

Thus, the operator $\mathscr{P}$ is a contraction and according to Banach fixed point theorem, it has a unique fixed point.

Finally, we use the above lemmas to state the following theorem.

Theorem 5.7 *Let f and K be continuous. Then the Volterra equation*

$$y(x) = f(x) + \lambda \int_a^x K(x,\xi) y(\xi) d\xi, \ a \leq x \leq b \tag{5.106}$$

has the solution

$$y(x) = f(x) + \sum_{i=1}^{\infty} \lambda^i \mathbf{L}^i f(x). \tag{5.107}$$

Moreover, if

$$|\lambda| < \frac{1}{M(b-a)}, \tag{5.108}$$

then the solution of (5.106) is unique and it is given by (5.107). The representation (5.107) is called the Neumann series.

Proof *The proof follows from Lemmas 19-21.*

The proof of the next corollary is a direct consequence of (5.107), since $f = 0$.

Corollary 9 *For any value of λ the Volterra equation*

$$y(x) = \lambda \int_a^x K(x,\xi)y(\xi)d\xi, \quad a \le x \le b, \tag{5.109}$$

has only the trivial solution and hence it has no eigenvalues.

Remark 22 *In our discussion we considered Volterra integral equation of the second kind. Similar results are easily derived for the Fredholm integral equation of the second kind. To see this, we consider Fredholm integral equation*

$$y(x) = f(x) + \lambda \int_a^b K(x,\xi)y(\xi)d\xi. \tag{5.110}$$

As before, we define the operator $\mathbf{L}$ by

$$\mathbf{L}y(x) = \int_a^b K(x,\xi)y(\xi)d\xi, \tag{5.111}$$

then we may write (5.110) in the form

$$y = f + \lambda\mathbf{L}y$$

Continuing in this fashion we obtain the successive approximation

$$y_n(x) = f(x) + \sum_{i=1}^{n-1} \lambda^i \mathbf{L}^i f(x) + \lambda^n \mathbf{L}^n y_0(x), \tag{5.112}$$

where

$$\left| \mathbf{L}^n y_0(x) \right| \le (b-a)^n M^n C. \tag{5.113}$$

Then,

$$|\lambda^n| \left| \mathbf{L}^n y_0(x) \right| \le |\lambda|^n (b-a)^n M^n C.$$

Furthermore,

$$\lim_{n\to\infty} |\lambda|^n \left| \mathbf{L}^n y_0(x) \right| \le \lim_{n\to\infty} |\lambda|^n (b-a)^n M^n C = 0,$$

provided that

$$|\lambda| < \frac{1}{M(b-a)}. \tag{5.114}$$

As a consequence, the infinite series (5.112) converges to the solution

$$y(x) = f(x) + \sum_{i=1}^{\infty} \lambda^i \mathbf{L}^i f(x). \tag{5.115}$$

Notice that condition (5.114) is a sufficient condition and hence without it the infinite series may or may not converge.

Observe that the convergence for the infinite series in the case of Volterra integral equation of the second kind was irrespective of the values of λ.

Example 5.12 Use both methods, iterations and Neumann series to solve the Volterra integral equation

$$y(x) = x + \lambda \int_0^x (x - \xi) y(\xi) d\xi.$$

We begin with the successive approximation or iterations

$$y_n(x) = x + \lambda \int_0^x (x - \xi) y_{n-1}(\xi) d\xi, \quad n = 1, 2, \dots$$

with $y_0(x) = x$. For $n = 1$ we have

$$y_1(x) = x + \lambda \int_0^x (x - \xi) \xi d\xi = x + \lambda \frac{x^3}{3!}.$$

For $n = 2$, with $y_1(\xi) = \xi + \lambda \frac{\xi^3}{3!}$, we have that

$$y_2(x) = x + \lambda \int_0^x (x - \xi)\left(\xi + \lambda \frac{\xi^3}{3!}\right) d\xi = x + \lambda \frac{x^3}{3!} + \lambda^2 \frac{x^5}{5!}.$$

Similarly, for $n = 3$ with $y_2(\xi) = \xi + \lambda \frac{\xi^3}{3!} + \lambda^2 \frac{\xi^5}{5!}$, we have

$$y_3(x) = x + \lambda \int_0^x (x - \xi)\left(\xi + \lambda \frac{\xi^3}{3!} + \lambda^2 \frac{\xi^5}{5!}\right) d\xi = x + \lambda \frac{x^3}{3!} + \lambda^2 \frac{x^5}{5!} + \frac{x^7}{7!}.$$

A continuation of this process leads to the sequence of functions

$$y_n(x) = x + \lambda \frac{x^3}{3!} + \lambda^2 \frac{x^5}{5!} + \lambda^3 \frac{x^7}{7!} + \cdots = x + \sum_{n=1}^{\infty} \lambda^n \frac{x^{2n+1}}{(2n+1)!},$$

which converges for all values of λ and x. Next we compute the Neumann series. Let

$$\mathbf{L}y(x) = \int_0^x K(x, \xi) y(\xi) d\xi,$$

then

$$\mathbf{L}^1 f(x) = \mathbf{L}^1 x = \int_0^x (x - \xi) \xi d\xi = \frac{x^3}{3!}.$$

Using the value of $\mathbf{L}^1 x$, we obtain

$$\mathbf{L}^2 x = \int_0^x (x - \xi) \frac{\xi^3}{3!} d\xi = \frac{x^5}{5!}.$$

In a similar approach, we arrive at

$$\mathbf{L}^3 x = \int_0^x (x - \xi) \frac{\xi^5}{5!} d\xi = \frac{x^7}{7!},$$

and so on. The Neumann series takes the form

$$\begin{aligned} y(x) &= x + \sum_{i=1}^{\infty} \lambda^i \mathbf{L}^i x \\ &= x + \lambda \frac{x^3}{3!} + \lambda^2 \frac{x^5}{5!} + \lambda^3 \frac{x^7}{7!} + \dots \\ &= x + \sum_{n=1}^{\infty} \lambda^n \frac{x^{2n+1}}{(2n+1)!}. \end{aligned}$$

$\square$

In what to follow, we twist the Neumann series and define the *resolvent* for the Volterra integral equation given by (5.97). As before, define the operator

$$(\mathbf{L}f)(x) = \int_a^x K(x, \xi) f(\xi) d\xi.$$

Then,

$$\begin{aligned} (\mathbf{L}^2 f)(x) &= \mathbf{L}(\mathbf{L}f)(x) \\ &= \int_a^x K(x, \xi) \int_a^\xi K(\xi, \xi_1) f(\xi_1) d\xi_1 d\xi \\ &= \int_a^x \left(\int_{\xi_1}^x K(x, \xi) K(\xi, \xi_1) d\xi \right) f(\xi_1) d\xi_1, \end{aligned}$$

where we changed the order of integrations. If we let

$$K_2(x, \xi_1) = \int_{\xi_1}^x K(x, \xi) K(\xi, \xi_1) d\xi,$$

then we have

$$(\mathbf{L}^2 f)(x) = \int_a^x K_2(x, \xi_1) f(\xi_1) d\xi_1.$$

Following in the same steps we arrive at

$$(\mathbf{L}^3 f)(x) = \int_a^x K_3(x, \xi_1) f(\xi_1) d\xi_1,$$

where

$$K_3(x,\xi_1) = \int_{\xi_1}^x K(x,\xi)K_2(\xi,\xi_1)d\xi,$$

and in general,

$$(\mathbf{L}^n f)(x) = \int_a^x K_n(x,\xi_1)f(\xi_1)d\xi_1,$$

where

$$K_n(x,\xi_1) = \int_{\xi_1}^x K(x,\xi)K_{n-1}(\xi,\xi_1)d\xi.$$

The kernels $K_1 = K$, K_2, $K_3,\ldots$ are called the *iterated kernels*. Consequently, the Neumann series (5.107) can be written as

$$\begin{aligned}
y(x) &= f(x) + \lambda \sum_{i=1}^{\infty} \lambda^{i-1} \int_a^x K_i(x,\xi)f(\xi)d\xi \\
&= f(x) + \lambda \int_a^x \left(\sum_{i=1}^{\infty} \lambda^{i-1} K_i(x,\xi) \right) f(\xi)d\xi \\
&= f(x) + \lambda \int_a^x \Gamma(x,\xi;\lambda)f(\xi)d\xi, \quad (5.116)
\end{aligned}$$

where

$$\Gamma(x,\xi;\lambda) = \sum_{i=1}^{\infty} \lambda^{i-1} K_i(x,\xi)d\xi, \quad (5.117)$$

is the *resolvent kernel*. We arrived at the following theorem.

Theorem 5.8 *Let f and K be continuous. Then the Volterra equation*

$$y(x) = f(x) + \lambda \int_a^x K(x,\xi)y(\xi)d\xi, \quad a \le x \le b \quad (5.118)$$

has the solution

$$y(x) = f(x) + \lambda \int_a^x \Gamma(x,\xi;\lambda)f(\xi)d\xi, \quad (5.119)$$

where the resolvent kernel $\Gamma(x,\xi;\lambda)$ is given by (5.117).

Example 5.13 Find the resolvent kernel and the solution for the Volterra integral equation

$$y(x) = e^{x^2} + \int_0^x e^{x^2 - \xi^2} y(\xi)d\xi. \quad (5.120)$$

Here we have $f(x) = e^{x^2}$, $K(x,\xi) = e^{x^2 - \xi^2}$, and $\lambda = 1$. Set

$$K_1(x,\xi) = K(x,\xi) = e^{x^2 - \xi^2}.$$

Then

$$K_2(x,\xi_1) = \int_{\xi_1}^x K(x,\xi)K(\xi,\xi_1)d\xi$$

$$= \int_{\xi_1}^{x} e^{x^2-\xi^2} e^{\xi^2-\xi_1^2} d\xi$$

$$= e^{x^2-\xi_1^2} \int_{\xi_1}^{x} d\xi = e^{x^2-\xi_1^2}(x-\xi_1).$$

Similarly

$$K_3(x,\xi_1) = \int_{\xi_1}^{x} K(x,\xi)K_2(\xi,\xi_1)d\xi$$

$$= \int_{\xi_1}^{x} e^{x^2-\xi^2} e^{\xi^2-\xi_1^2}(\xi-\xi_1)d\xi$$

$$= e^{x^2-\xi_1^2} \int_{\xi_1}^{x} (\xi-\xi_1)d\xi = e^{x^2-\xi_1^2}\frac{(x-\xi_1)^2}{2}.$$

Additionally

$$K_4(x,\xi_1) = \int_{\xi_1}^{x} K(x,\xi)K_3(\xi,\xi_1)d\xi$$

$$= \int_{\xi_1}^{x} e^{x^2-\xi^2} e^{\xi^2-\xi_1^2}\frac{(\xi-\xi_1)^2}{2}d\xi$$

$$= e^{x^2-\xi_1^2}\frac{(x-\xi_1)^3}{3!}.$$

Inductively, we arrive at the formula

$$K_n(x,\xi_1) = e^{x^2-\xi_1^2}\frac{(x-\xi_1)^{n-1}}{(n-1)!}, \quad n=1,2,\dots.$$

Thus, the resolvent kernel is

$$\Gamma(x,\xi;\lambda) = \sum_{n=1}^{\infty} \lambda^{n-1} K_n(x,\xi)d\xi$$

$$= \sum_{n=1}^{\infty} e^{x^2-\xi_1^2}\frac{(x-\xi_1)^{n-1}}{(n-1)!}$$

$$= e^{x^2-\xi_1^2} \sum_{n=1}^{\infty} \frac{(x-\xi_1)^{n-1}}{(n-1)!}$$

$$= e^{x^2-\xi_1^2} e^{x-\xi_1}.$$

Using (5.119), we arrive at the solution

$$y(x) = e^{x^2} + \int_0^x e^{x^2-\xi_1^2} e^{x-\xi_1} e^{\xi_1^2} d\xi_1$$

$$= e^{x^2} + e^{x^2+x} \int_0^x e^{-\xi_1} d\xi_1$$

$$= e^{x^2} + e^{x^2+x}(1 - e^{-x})$$

$$= e^{x^2+x}.$$

□

The next example displays another approach of finding the resolvent kernel.

Example 5.14 Consider the Volterra equation

$$\varphi(x) = f(x) + \lambda \int_a^x \xi \varphi(\xi) d\xi,$$

where f is continuously differentiable. Using Lemma 16, we arrive at the first-order differential equation

$$\varphi'(x) - \lambda x \varphi(x) = f'(x).$$

Multiplying the equation by the integrating factor $e^{-\lambda \frac{x^2}{2}}$ we arrive at

$$\frac{d}{dx}\left(e^{-\lambda \frac{x^2}{2}} \varphi(x)\right) = e^{-\lambda \frac{x^2}{2}} f'(x).$$

Integrating both sides from a to x leads to

$$e^{-\lambda \frac{x^2}{2}} \varphi(x) - e^{-\lambda \frac{a^2}{2}} \varphi(a) = \int_a^x e^{-\lambda \frac{\xi^2}{2}} f'(\xi) d\xi.$$

The term on the right hand side can be integrated by parts to obtain

$$e^{-\lambda \frac{x^2}{2}} \varphi(x) - e^{-\lambda \frac{a^2}{2}} \varphi(a) = -e^{-\lambda \frac{a^2}{2}} f(a) + f(x)e^{-\lambda \frac{x^2}{2}} + \lambda \int_a^x \xi e^{-\lambda \frac{\xi^2}{2}} f\xi) d\xi.$$

Since $f(a) = \varphi(a)$, the above expression reduces to

$$e^{-\lambda \frac{x^2}{2}} \varphi(x) = f(x)e^{-\lambda \frac{x^2}{2}} + \lambda \int_a^x \xi e^{-\lambda \frac{\xi^2}{2}} f\xi) d\xi.$$

Multiplying both sides with $e^{\lambda \frac{x^2}{2}}$ yields

$$\varphi(x) = f(x) + \lambda \int_a^x \xi e^{\frac{\lambda}{2}(x^2 - \xi^2)} f\xi) d\xi. \qquad (5.121)$$

An comparison between (5.119) and (5.121) indicates that

$$\Gamma(x, \xi; \lambda) = \xi e^{\frac{\lambda}{2}(x^2 - \xi^2)}.$$

If $f(x)$ is given then we can use (5.121) to compute the solution.   □

We end this section by considering *nonlinear* Volterra integral equations. Recall from Definition 5.4 that an expression of the solution of the nonlinear ordinary differential equation $x'(t) = f(t,x(t))$ for $t \in I$, and $x(t_0) = x_0$ is given by the nonlinear integral equation

$$x(t) = x_0 + \int_{t_0}^{t} f(\xi, x(\xi))d\xi.$$

To be consistent with our notations, we consider the nonlinear Volterra integral equation

$$y(x) = h(x) + \int_{t_0}^{x} f(\xi, y(\xi))d\xi.$$

where the functions h and f are continuous on their respective domains. As we have done before, we define a successive approximation or Picard's iteration by

$$
\begin{aligned}
y_0(x) &= h(x) \\
y_1(x) &= h(x) + \int_0^x f(\xi, y_0(\xi))d\xi \\
y_2(x) &= h(x) + \int_0^x f(\xi, y_1(\xi))d\xi \\
&\vdots \\
y_n(x) &= h(x) + \int_0^x f(\xi, y_{n-1}(\xi))d\xi, \quad n = 1, 2, \ldots
\end{aligned}
\tag{5.122}
$$

Example 5.15 Consider the nonlinear integral equation

$$y(x) = \int_0^x (\xi + y^2(\xi))d\xi.$$

We use the successive approximation or iterations

$$y_n(x) = \int_0^x (\xi + y_{n-1}^2(\xi))d\xi, \quad n = 1, 2, \ldots$$

with $y_0(x) = 0$. For $n = 1$ we have

$$y_1(x) = \int_0^x \xi d\xi = \frac{x^2}{2}.$$

For $n = 2$, with $y_1(\xi) = \frac{\xi^2}{2}$, we have that

$$y_2(x) = \int_0^x \left(\xi + \frac{\xi^4}{4}\right)d\xi = \frac{x^2}{2} + \frac{x^5}{20}.$$

Similarly, for $n = 3$ with $y_2(\xi) = \frac{\xi^2}{2} + \frac{\xi^5}{20}$, we have

$$y_3(x) = \int_0^x \left[\xi + \left(\frac{\xi^2}{2} + \frac{\xi^5}{20}\right)^2\right]d\xi = x + \frac{x^5}{20} + \frac{x^8}{16} + \frac{x^{11}}{4400}.$$

A continuation of this process leads to higher approximation. $\qquad \square$

5.7.1   Exercises

Exercise 5.47 *Provide all the detail for* (5.113).

Exercise 5.48 *Find the Neumann series for the Volterra integral equation*

$$y(x) = 1 - \int_0^x (x - \xi)y(\xi)d\xi,$$

and then find its solution.

Answer: $y(x) = \cos(x)$.

Exercise 5.49 *Find the solution of the Fredholm integral equation*

$$y(x) = e^x + \frac{1}{e} \int_0^1 y(\xi)d\xi,$$

using:

(a) iterations,

(b) Neumann series. Answer: $y(x) = e^x + 1$.

Exercise 5.50 *Consider the Fredholm integral equation*

$$y(x) = 1 + \lambda \int_0^1 (x - \xi)y(\xi)d\xi.$$

Find:

(a) $y_1(x)$, $y_2(x)$ *and* $y_3(x)$,

(b) the first three terms of the Neumann series.

Exercise 5.51 *Find the Neumann series for the following integral equations.*

(a)

$$y(x) = 1 - 2\int_0^x \xi y(\xi)d\xi.$$

(b)

$$y(x) = x + \frac{1}{2}\int_{-1}^1 (x + \xi)y(\xi)d\xi.$$

Exercise 5.52 *Solve the Volterra integral equation using the Neumann series*

(a)

$$y(x) = 1 + x^2 - 2\int_0^x (x - \xi)y(\xi)d\xi.$$

Answer: $y(x) = 1$.

(b)

$$y(x) = x\cos(x) + \int_0^x \xi y(\xi)d\xi.$$

Very hard to simplify.

Answer: $y(x) = \sin(x)$.

Exercise 5.53 *Consider the integral equation*

$$y(x) = \frac{23}{6}x + \frac{1}{8}\int_0^1 x\xi y(\xi)d\xi. \tag{5.123}$$

(a) *Show that the iterative kernel* $K_n(x,\xi) = \frac{x\xi}{3^{n-1}}$.

(b) *Show that the resolvent kernel simplifies to* $\Gamma(x,\xi;\lambda) = x\xi\frac{1}{1-1/24} = \frac{24}{23}x\xi$.

(c) *Find the solution of* (5.123).

Exercise 5.54 *Consider the integral equation*

$$y(x) = x + \lambda \int_0^x y(\xi)d\xi. \tag{5.124}$$

(a) *Show that the iterative kernel* $K_n(x,\xi) = \frac{(x-\xi)^{n-1}}{(n-1)!}$.

(b) *Show that the resolvent kernel simplifies to* $\Gamma(x,\xi;\lambda) = e^{\lambda(x-\xi)}$.

(c) *Find the solution of* (5.124).

Exercise 5.55 *Use the idea of Example 5.14 to compute the resolvent kernel for the Volterra equation*

$$y(x) = f(x) + \lambda \int_a^x y(\xi)d\xi.$$

Find the solution when

(a) $f(x) = 1$,

(b) $f(x) = x$.

Exercise 5.56 *Use the idea of Example 5.14 and redo Example 5.13.*

Exercise 5.57 *Use any method that you wish to solve the Volterra equation*

$$y(x) = \cos(x) - x - 2 + \int_0^x (\xi - x)y(\xi)d\xi.$$

Exercise 5.58 *Consider the integral equation*

$$y(x) = x + \int_0^{\frac{1}{2}} y(\xi)d\xi. \tag{5.125}$$

(a) *Show that the iterative kernel* $K_n(x,\xi) = \frac{1}{2^{n-1}}$.

(b) *Show that the resolvent kernel simplifies to* $\Gamma(x,\xi;\lambda) = 2$.

(c) *Find the solution of* (5.125).

Exercise 5.59 *Apply Picard iteration to the (IVPs)*

1. $x'(t) = tx, \quad x(0) = 1$

2. $x'(t) = 2t(1+x), \quad x(0) = 2$

and show the obtained $\{x_n(t)\}$ *of each of the of iterate converges to the true solution of each of the (IVP) (True solution is the solution found by solving the (IVP)).*

Exercise 5.60 *Consider the coupled system of differential equations*

$$\frac{dy}{dx} = z(x), \quad \frac{dz}{dx} = x^3(y(x)+z(x)); \quad y(0) = 1 \ \text{and} \ z(0) = \frac{1}{2}.$$

Convert the system to integral equations and find the iterates

$$\{y_1(x), y_2(x), y_3(x), z_1(x), z_2(x), z_3(x)\}.$$

5.8 Approximating Non-Degenerate Kernels

In some cases, it is useful to approximate a non-degenerate kernel by a finite terms using Maclaurin expansion so that it is degenerate. This is accomplished by approximating a given kernel $K(x,\xi)$, through a sum of a finite number of products of functions of x alone by functions of ξ alone. To better explain the concept, we consider the Fredholm integral equation of second kind

$$y(x) = f(x) + \lambda \int_a^b K(x,\xi)y(\xi)d\xi. \tag{5.126}$$

Let $D(x,\xi)$ be the approximate and degenerta kernel of K. Then, the approximate Fredholm integral equation of the second kind of (5.126) may take the form

$$e(x) = f(x) + \lambda \int_a^b D(x,\xi)e(\xi)d\xi, \tag{5.127}$$

where the kernel D is degenerate. We may use Section 5.5 to obtain the solution $e(x)$ of (5.127), which is the approximate solution of (5.126). Such approximation will involve an error which we denote by

$$\varepsilon = |y(x) - e(x)|$$

for small and positive ε. For illustrational purpose we propose the following example.

Example 5.16 Consider the Fredholm equation

$$y(x) = \cos(x) + \int_0^1 x\sin(x\xi)y(\xi)d\xi. \tag{5.128}$$

Then, $K(x,\xi) = x\sin(x\xi)$ is non-degenerate. However, a finite terms of its Maclaurin series

$$x\sin(x\xi) = x\left(x\xi - \frac{(x\xi)^3}{3!} + \frac{(x\xi)^5}{5!} + \cdots\right)$$

is degenerate. We only consider the first two terms of its Maclaurin series and set

$$D(x,\xi) = x^2\xi - \frac{x^4\xi^3}{3!},$$

which is degenerate. The approximate Fredholm integral equation is then

$$e(x) = \cos(x) + \lambda \int_0^1 \left(x^2\xi - \frac{x^4\xi^3}{3!}\right)e(\xi)d\xi. \tag{5.129}$$

Our task is to find the solution $e(x)$ of (5.129). We begin by rewriting

$$D(x,\xi) = x^2\xi - \frac{x^4\xi^3}{3!} = \sum_{i=1}^2 \alpha_i(x)\beta_i(\xi),$$

with

$$\alpha_1(x) = x^2, \quad \alpha_2(x) = -x^4, \quad \beta_1(\xi) = \xi, \quad \beta_2(\xi) = \frac{\xi^3}{3!}.$$

Next we use

$$a_{ij} = \int_0^1 \beta_i(x)\alpha_j(x)dx, \quad i,j = 1,2$$

to compute the matrix $A = (a_{ij})$.

$$a_{11} = \int_0^1 \beta_1(x)\alpha_1(x)dx = \int_0^1 x^3 dx = \frac{1}{4},$$

$$a_{12} = \int_0^1 \beta_1(x)\alpha_2(x)dx = -\int_0^1 x^5 dx = -\frac{1}{6},$$

$$a_{21} = \int_0^1 \beta_2(x)\alpha_1(x)dx = \int_0^1 \frac{x^5}{6}dx = \frac{1}{36},$$

$$a_{22} = \int_0^1 \beta_2(x)\alpha_2(x)dx = -\int_0^1 \frac{x^7}{6}dx = -\frac{1}{48}.$$

So we have

$$A = \begin{pmatrix} \frac{1}{4} & -\frac{1}{6} \\ \frac{1}{36} & -\frac{1}{48} \end{pmatrix}.$$

Since $\lambda = 1$, we have that

$$\det(I - \lambda A) = \det(I - A) = \begin{vmatrix} \frac{3}{4} & \frac{1}{6} \\ -\frac{1}{36} & -\frac{49}{48} \end{vmatrix} \neq 0.$$

Thus, $\lambda = 1$ is not an eigenvalue. We make use of $f_i = \int_0^1 \beta_i(x)f(x)dx$, $i = 1,2$ to compute f_1 and f_2.

$$f_1 = \int_0^1 x\cos(x)dx \approx 0.38177,$$

and

$$\begin{aligned}
f_2 &= \frac{1}{6}\int_0^1 x^3 \cos(x)dx \\
&= \frac{1}{6}\left[x^3 \sin(x) + 3x^2 \cos(x) - 6x\sin(x) - 6\cos(x)\right]\Big|_0^1 \\
&\approx 0.02862.
\end{aligned}$$

Left to find c_1, c_2 as the the unique solution of the system

$$\frac{3}{4}c_1 + \frac{1}{6}c_2 = 0.38177$$

$$-\frac{1}{36}c_1 + \frac{49}{48}c_2 = 0.02862.$$

After some calculations, we obtained

$$c_1 \approx 0.49977, \quad c_2 \approx 0.041635.$$

Thus, the approximate solution $e(x)$ of (5.129) is given by

$$e(x) = \cos(x) + \sum_{i=1}^{2} c_i\alpha_i(x) = \cos(x) + 0.49977x^2 - 0.041635x^4.$$

The values of $e(x)$ are compared to the values of the actual solution of (5.128) which can be easily proved to be $y(x) = 1$, at various values of $x \in [0,1]$, in the table below.
□

x	0	0.25	0.5	0.75	1
$y(x)$	1	1	1	1	1
$e(x)$	1	0.99998	0.99926	0.99963	0.998437

5.8.1 Exercises

Exercise 5.61 *Verify that $y(x) = 1$ is a solution of the Fredholm integral equation given by (5.128).*

Exercise 5.62 *Redo Example 5.16 by taking*

$$D(x,\xi) = x^2\xi - \frac{x^4\xi^3}{3!} + \frac{x^6\xi^5}{5!}.$$

Exercise 5.63 *Consider the Fredholm equation*

$$y(x) = 1 - \cos(x) + \int_0^1 (1 - x\sin(x\xi))y(\xi)d\xi. \tag{5.130}$$

(a) Verify $y(x) = 1$ is a solution of (5.130).

(b) Use the first two terms of the Maclaurin series of the kernel and compute the approximate solution $e(x)$.

(c) Compare both solutions at $x = 0, 0.25, 0.5, 0.75, 1$.

Exercise 5.64 *Repeat Exercise 5.63 for the Fredholm integral equation*

$$y(x) = e^x - x - \int_0^1 x(e^{x\xi} - 1)y(\xi)d\xi$$

by considering the first three terms of the Maclaurin series of the kernel.

5.9 Laplace Transform and Integral Equations

The Laplace transform is a powerful tool for solving differential equations and integral equations of convolution types. In this introductory section, we first define the Laplace transform and develop some of its basic properties. We begin with the following definition:

Definition 5.8 *Let $f(t)$ be define on $0 \le t < \infty$. The Laplace transform of f is denoted by $F(s)$*

$$Ł[f(t)] = F(s) = \int_0^\infty e^{-st} f(t) dt, \quad \text{for } s > 0. \tag{5.131}$$

The Laplace transform of f is said to exist if the improper integral (5.131) converges.

Note that the right side of (5.131) is a function of s and hence the notation $F(s)$. We shall use small letters of functions of t such as f, g or h and shall denote their Laplace transforms by $Ł[f(t)] = F(s), Ł[g(t)] = G(s), Ł[h(t)] = H(s)$. Laplace transform is linear. That is for functions f and g and constants a and b it follows that

$$Ł[af + bg] = aŁ[f] + bŁ[g].$$

Example 5.17 In this example we develop the Laplace transform of basic functions. We do so by considering different values of $f(t)$.

(a) For $f(t) = 1$, then

$$Ł[1] = \int_0^\infty e^{-st} dt = \left[-\frac{1}{s} e^{-st} \right]_0^\infty = \frac{1}{s}, \ s > 0.$$

(b) For $f(t) = t$, then

$$\text{L}[t] = \int_0^\infty e^{-st} t \, dt = \left[-\frac{1}{s} e^{-st} t \right]_0^\infty + \int_0^\infty \frac{1}{s} e^{-st} = \frac{1}{s^2}, \, s > 0.$$

(c) For $f(t) = \frac{dy}{dt}$, then by performing an integration by parts we get

$$\text{L}\left[\frac{dy}{dt}\right] = \int_0^\infty e^{-st} \frac{dy}{dt} \, dt = \left[e^{-st} y\right]_0^\infty + \int_0^\infty s e^{-st} y \, dt = -y(0) + sY(s).$$

(d) For $f(t) = e^{at}$, a constant, then

$$\text{L}[e^{at}] = \int_0^\infty e^{-st} e^{at} \, dt = \int_0^\infty e^{-(s-a)t} \, dt = \left[-\frac{1}{s-a} e^{-(s-a)t} \right]_0^\infty = \frac{1}{s-a}, \, s > a.$$

(e)

$$\text{L}[\cos at + i \sin at] = \text{L}[e^{iat}] \text{ by DeMoivre.}$$

$$\text{L}[e^{iat}] = \frac{1}{s - ia} = \frac{s}{s^2 + a^2} + \frac{ia}{s^2 + a^2}.$$

Hence, equating real and imaginary parts and using linearity

$$\text{L}[\cos at] = \frac{s}{s^2 + a^2}$$

$$\text{L}[\sin at] = \frac{a}{s^2 + a^2}.$$

We can apply the convolution property from the table to find

$$\text{L}^{-1}\left[\frac{f(s)}{s}\right].$$

$$\text{L}^{-1}[f(s)] = f(t), \text{ and } \text{L}^{-1}[\frac{1}{s}] = 1 = g(t),$$

so

$$\text{L}^{-1}\left[\frac{f(s)}{s}\right] = \int_0^t f(\theta) \, d\theta.$$

(f) For $f(t) = t^n, n = 0, 1, 2, \ldots$, then

$$\text{L}[t^n] = \frac{n!}{s^{n+1}}, \, n = 0, 1, 2, \ldots.$$

□

You can access the Laplace Transforms of all the functions you are likely to meet online thanks to computer algebra tools like Mathematica, Matlab, and Maple. The packages also provide an inversion technique to find a function f from a given $F(s)$. For example

$$Ł[t^{1/2}] = \frac{1}{2}\left(\frac{\pi}{s^3}\right)^{1/2},$$

and

$$Ł[t^{-1/2}] = \left(\frac{\pi}{s}\right)^{1/2}.$$

Theorem 5.9 (Shift Theorem) *If $F(s) = Ł[f(t)]$, then*

$$Ł[e^{at} f(t)] = F(s-a).$$

Proof *Using Definition 5.8 we have*

$$Ł[e^{at} f(t)] = \int_0^\infty e^{-st} e^{at} f(t) dt = \int_0^\infty e^{-(s-a)t} dt = F(s-a).$$

This completes the proof.

Next we define the Laplace inverse.

Definition 5.9 *Let $F(s)$ be the Laplace transform of given a function $f(t)$. We denote the Laplace inverse of f by $Ł^{-1}[F(s)]$ such that*

$$Ł^{-1}[F(s)] = f.$$

Example 5.18 We use Laplace transform to solve the initial value problem

$$2\frac{dy}{dt} - y = \sin t, \quad y(0) = 1.$$

We begin by taking the Laplace transform on both sides and obtain

$$2(sY(s) - 1) - Y(s) = \frac{1}{s^2 + 1}.$$

Solving for $Y(s)$ gives

$$Y(s) = \frac{2s^2 + 3}{(2s - 1)(s^2 + 1)}.$$

Taking the Laplace inverse we arrive at

$$y(t) = Ł^{-1}\left[\frac{2s^2 + 3}{(2s - 1)(s^2 + 1)}\right].$$

Next we use partial fractions. That is

$$\frac{2s^2 + 3}{(2s - 1)(s^2 + 1)} = \frac{A}{2s - 1} + \frac{Bs + C}{s^2 + 1},$$

and after some calculations we obtain $A = \frac{7}{5}$, $B = \frac{-2}{5}$, $C = \frac{-1}{5}$. Thus, the solution $y(t)$ is

$$y(t) = \frac{7}{5}e^{t/2} - \frac{2}{5}\cos t - \frac{1}{5}\sin t.$$

$\square$

Next we address the convolution between two functions.

Definition 5.10 *Let $f(t)$ and $g(t)$ be define on $0 \le t < \infty$. The function*

$$h(t) = \int_0^t f(t - \tau)g(\tau)\,d\tau, \tag{5.132}$$

is called the convolution of f and g and is written

$$h = f * g.$$

Theorem 5.10

$$f * g = g * f.$$

Proof *Let $u = t - \tau$ in (5.132). Then*

$$
\begin{aligned}
h(t) &= \int_t^0 f(u)g(t - u)(-du) = \int_0^t f(u)g(t - u)\,du \\
&= \int_0^t g(t - u)f(u)\,du = (g * f)(t).
\end{aligned}
$$

This completes the proof.

Let $F(s)$ and $G(s)$ be the Laplace transform of the functions f, and g, respectively. We are interested in computing $Ł^{-1}[F(s)G(s)]$. We have the following theorem

Theorem 5.11 (Convolution Theorem) *Let $F(s)$ and $G(s)$ be the Laplace transform of the functions f, and g, respectively. Then*

$$Ł[f * g] = Ł\left[\int_0^t f(t - \tau)g(\tau)\,d\tau\right] = F(s)G(s).$$

Proof *For $h = Ł^{-1}[H(s)]$, with $H(s) = F(s)G(s)$, let*

$$H(s) = \int_0^\infty e^{-st}h(t)\,dt = F(s)G(s). \tag{5.133}$$

Then

$$H(s) = \left(\int_0^\infty e^{-st}f(t)\,dt\right)\left(\int_0^\infty e^{-s\tau}g(\tau)\,d\tau\right)$$

$$= \int_0^\infty \left(\int_0^\infty e^{-s(t+\tau)} f(t) dt \right) g(\tau) d\tau.$$

Make the change of variables $u = t + \tau$ for the inside integral. Then

$$H(s) = \int_0^\infty \left(\int_\tau^\infty e^{-su} f(u-\tau) du \right) g(\tau) d\tau.$$

By changing the order of integrations we obtain

$$H(s) = \int_0^\infty \left(\int_0^u f(u-\tau) g(\tau) d\tau \right) e^{-su} du.$$

Replacing the dummy variable of integration u with t and then compare the result with (5.133), we clearly see that

$$\int_0^\infty e^{-st} h(t) dt = \int_0^t f(t-\tau) g(\tau) d\tau$$
$$= \int_0^\infty \left(\int_0^t f(t-\tau) g(\tau) d\tau \right) e^{-st} dt.$$

This completes the proof.

Example 5.19 Express h in the form $f * g$, when

$$H(s) = \frac{1}{(s^2+4)^2}.$$

Let $F(s) = G(s) = \frac{1}{s^2+4}$. Then $H(s) = F(s) G(s)$. Moreover,

$$\mathcal{L}^{-1}[F(s)] = \mathcal{L}^{-1}[G(s)] = \mathcal{L}^{-1}\left[\frac{1}{s^2+4}\right] = \frac{1}{2} \sin(2t).$$

Thus,

$$h(t) = \frac{1}{2} \sin(2t) * \frac{1}{2} \sin(2t) = \frac{1}{4} \int_0^t \sin 2(t-\tau) \sin(2\tau) d\tau.$$

$\square$

Before we consider the next example, we define the *error function*.

Definition 5.11 *The error function is the following improper integral considered as a real function $erf : \mathbb{R} \to \mathbb{R}$, such that*

$$erf(x) = \frac{2}{\sqrt{\pi}} \int_0^x e^{-z^2} dz,$$

where exponential is the real exponential function. In addition the complementary error function,

$$erfc(x) = 1 - erf(x) = \frac{2}{\sqrt{\pi}} \int_x^\infty e^{-z^2} dz.$$

We can easily verify that

$$erfc(-\infty) = 2, \quad erfc(0) = 1, \quad erfc(\infty) = 0.$$

Next we state the *gamma function*, which is needed in future work. We denote the Gamma function by Γ and it is defined by

$$\Gamma(x) = \int_0^\infty u^{x-1}e^{-u}du, \quad x > 0.$$

Note that, it can be easily shown

$$\Gamma(x+1) = x\Gamma(x),$$

which can be used to show that, for every positive integer n

$$\Gamma(n) = (n-1)!.$$

We will also need the following formula. For positive integer n, we have

$$\Gamma(\frac{1}{2}n) = \frac{(n-2)!!\sqrt{\pi}}{2^{(n-1)/2}}, \tag{5.134}$$

where $n!!$ is a *double factorial*. For example,

$$\Gamma(1/2) = \sqrt{\pi}, \quad \Gamma(3/2) = \sqrt{\pi}/2, \quad \Gamma(5/2) = (3\sqrt{\pi})/4, \text{ etc,}$$

$$\Gamma(\frac{1}{2}+n) = \frac{(2n-1)!!}{2^n}\sqrt{\pi},$$

and

$$\Gamma(\frac{1}{2}-n) = \frac{(-1)^n 2^n}{(2n-1)!!}\sqrt{\pi}.$$

The next example deals with integral equations of convolution type.

Example 5.20 We are interested in finding the solution of the integral equation of convolution type

$$r(t) = 1 - \int_0^t (t-\tau)^{-1/2}r(\tau)\,d\tau. \tag{5.135}$$

In convolution form, equation (5.135) becomes

$$r(t) = 1 - t^{-1/2} * r(t). \tag{5.136}$$

Let $Ł[r(t)] = R(s)$ and take Laplace transform on both sides of (5.136).

$$Ł[r(t)] = Ł[1] - Ł[t^{1/2}]Ł[r(t)].$$

This gives

$$R(s) = \frac{1}{s} - \frac{\Gamma(1-1/2)}{s^{1-1/2}}R(s) = \frac{1}{s} - \frac{\sqrt{\pi}}{s^{1/2}}R(s).$$

Solving for $R(s)$ in the above equation gives

$$R(s) = \frac{1}{s^{1/2}(s^{1/2} + \sqrt{\pi})}.$$

Using partial fractions, we write

$$R(s) = \frac{A}{s^{1/2}} + \frac{B}{s^{1/2} + \sqrt{\pi}}. \tag{5.137}$$

Next we compute the coefficient A and B. Taking a common denominator and equating both sides give

$$1 = A(s^{1/2} + \sqrt{\pi}) + Bs^{1/2}.$$

For $s = 0$, we have $A = 1/\sqrt{\pi}$, and for $s^{1/2} = 1$, we obtain $B = -1/\sqrt{\pi}$. Hence, (5.137) becomes

$$R(s) = \frac{1}{\sqrt{\pi}(s^{1/2})} - \frac{1}{\sqrt{\pi}(s^{1/2} + \sqrt{\pi})}.$$

Then, by taking the inverse Laplace of $R(s)$, we get

$$r(t) = \mathbf{L}^{-1}[R(s)] = \mathbf{L}^{-1}\left[\frac{1}{\sqrt{\pi}(s^{1/2})}\right] - \mathbf{L}^{-1}\left[\frac{1}{\sqrt{\pi}(s^{1/2} + \sqrt{\pi})}\right].$$

Or,

$$r(t) = \frac{1}{\sqrt{\pi}}\mathbf{L}^{-1}\left[\frac{1}{s^{1/2}}\right] - \frac{1}{\sqrt{\pi}}\mathbf{L}^{-1}\left[\frac{1}{s^{1/2} + \sqrt{\pi}}\right]. \tag{5.138}$$

By our provided table, we see that $\mathbf{L}^{-1}\left[\frac{1}{\sqrt{s+a}}\right] = \frac{1}{\sqrt{\pi}\sqrt{t}} - a\,e^{a^2 t}\,\text{erf}(a\sqrt{t})$, then

$$r(t) = \frac{1}{\sqrt{\pi}}\frac{1}{\sqrt{\pi}\sqrt{t}} - \frac{1}{\sqrt{\pi}}\left[\frac{1}{\sqrt{\pi}\sqrt{t}} - \sqrt{\pi}\,e^{\pi t}\,\text{erf}(\sqrt{\pi}\sqrt{t})\right].$$

This simplifies to

$$r(t) = \frac{1}{\pi\sqrt{t}} - \frac{1}{\pi\sqrt{t}} + e^{\pi t}\,\text{erf}(\sqrt{\pi}\sqrt{t}).$$

Finally,

$$r(t) = e^{\pi t}\,\text{erf}(\sqrt{\pi}\sqrt{t}).$$

□

5.9.1 Frequently used Laplace transforms

Function $f(t)$	**Transform** $F(s) = \int_0^\infty e^{-st} f(t)\,dt$
1	$1/s$
t^n, for $n = 0, 1, 2, \ldots$	$n!/s^{n+1}$
$t^{1/2}$	$\frac{1}{2}(\pi/s^3)^{1/2}$
$t^{-1/2}$	$\left(\frac{\pi}{s}\right)^{1/2}$

e^{at}	$1/(s-a)$
$\sin \omega t$	$\omega/(s^2+\omega^2)$
$\cos \omega t$	$s/(s^2+\omega^2)$
$t \sin \omega t$	$2\omega s/(s^2+\omega^2)^2$
$t \cos \omega t$	$(s^2-\omega^2)/(s^2+\omega^2)^2$
$e^{at} t^n$	$n!/(s-a)^{n+1}$
$e^{at} \sin \omega t$	$\omega/\left((s-a)^2+\omega^2\right)$
$e^{at} \cos \omega t$	$(s-a)/\left((s-a)^2+\omega^2\right)$
$\sinh \omega t$	$\omega/(s^2-\omega^2)$
$\cosh \omega t$	$s/(s^2-\omega^2)$
Shift of g: $e^{at} g(t)$	$G(s-a)$
Convolution: $f(t) * g(t) = \int_0^t f(t-\tau)g(\tau)\,d\tau$	$G(s)F(s)$
Integration: $1 * g(t) = \int_0^t g(\tau)\,d\tau$	$\frac{1}{s}G(s)$
Derivative: y'	$sY(s)-y(0)$
y''	$s^2 Y(s)-sy(0)-y'(0)$
$(1+2at)/\sqrt{\pi t}$	$(s+a)/s\sqrt{s}$
$e^{-at}/\sqrt{\pi t}$	$1/\sqrt{s+a}$
$(e^{bt}-e^{-at})/2t\sqrt{\pi t}$	$\sqrt{s-a}-\sqrt{s-b}$
$(e^{-bt}-e^{-at})/2t\sqrt{\pi t}$	$\sqrt{s+a}+\sqrt{s+b}$
$erf(\sqrt{at})/\sqrt{a}$	$1/(s\sqrt{s+a})$
$e^{at} erf(\sqrt{at})/\sqrt{a}$	$1/(\sqrt{s}\sqrt{s-a})$
$\frac{1}{\sqrt{\pi t}}-be^{b^2 t}erf(b\sqrt{t})]$	$1/(\sqrt{s}+b)$
$f(ct)$	$1/(cF(1/c)),c>0$
$f^{(n)}(t)$	$s^n F(s)-s^{n-1}f(0)-\ldots-f^{(n-1)}(0)$
$(-t)^n f(t)$	$F^{(n)}(s)$
$u(t-a)f(t-a)$	$e^{-as}F(s)$
$u(t-a)$	e^{-as}/s
$t^v,\ (v>-1)$	$\frac{\Gamma(v+1)}{s^{v+1}}.$

5.9.2   Exercises

Exercise 5.65 *Solve the initial value problem using Laplace transform*

(a) $y''+9y=u(t-3),\ y(0)=1,\ y'(0)=2.$

(b) $2\frac{dy}{dt}-y=\sin(t),\ y(0)=1.$

Exercise 5.66 *Express h in the form $f * g$, when*

(a) $H(s)=\frac{1}{s^3-3s}.$

(b) $H(s)=\frac{1}{(s^2+4)(s^2+9)}.$

(c) $H(s)=\frac{1}{s^{\frac{3}{2}}(s^2+4)}.$

Exercise 5.67 *Use Laplace transform and write down the solution of the integral equation*

$$y(t) = f(t) + \lambda \int_0^t e^{(t-\tau)} y(\tau) \, d\tau.$$

Answer: $y(t) = f(t) + \lambda \int_0^t e^{(\lambda+1)(t-\tau)} f(\tau) \, d\tau.$

Exercise 5.68 *Use Exercise 5.67 to solve the integral equation*

$$y(t) = \cos(t) - \int_0^t e^{(t-\tau)} y(\tau) \, d\tau.$$

Answer: $y(t) = \cos(t) - \sin(t).$

Exercise 5.69 *Solve the the system of ODEs using Laplace transform*

$$\frac{dy}{dt} - \frac{dx}{dt} + y + 2x = e^t,$$

$$\frac{dy}{dt} + \frac{dx}{dt} - x = e^{2t},$$

$$x(0), y(0) = 1.$$

Exercise 5.70 *Show that*

$$(\Gamma(\tfrac{1}{2}))^2 = \pi.$$

Exercise 5.71 *Use Laplace transform to solve the following integral equations.*

(a) $f(t) = t + \int_0^t (t - \tau) \, d\tau,$

(b) $f(t) + 2 \int_0^t f(\tau) \cos(t - \tau) \, d\tau = 4e^{-t} + \sin(t),$

(c) $y'(t) = 1 - \sin(t) - \int_0^t y(\tau) \, d\tau, \ y(0) = 0.$

Exercise 5.72 *Show that if*

$$r(t) = -a(t) + \int_0^t a(t - s) r(s) \, ds$$

and

$$x(t) = f(t) + \int_0^t a(t - s) x(s) \, ds,$$

then

$$x(t) = f(t) - \int_0^t r(t - s) f(s) \, ds.$$

Exercise 5.73 *Solve the Abel equation*

$$\int_0^t \frac{1}{\sqrt{t - \tau}} y(\tau) \, d\tau = f(t),$$

where $f(t)$ is a given function with $f(0) = 0$ and f' admits a Laplace transform.

Find the solution when

(a) $f(t) = 1 + t + t^2$,

(b) $f(t) = t^3$.

Exercise 5.74 *Use Laplace transform to solve*

$$\int_0^t (t-\tau)^{1/3} y(\tau) d\tau = t^{3/2}.$$

Answer: $y(t) = (3\sqrt{\pi}\, t^{1/6})/(4\Gamma(4/3)\Gamma(7/6))$.

Exercise 5.75 *Use Laplace transform to solve*

$$\int_0^t e^{-2(t-\tau)} (t-\tau)^{-1/2} y(\tau) d\tau = 1.$$

Answer: $y(t) = (t^{-1/2}/\pi)e^{-2t} + \sqrt{2/\pi}\, erf(\sqrt{2t})$.

5.10   Odd Behavior

In this brief section, we skim over some integral equations that display odd behavior, either in the sense of solutions existing only over finite time or the existence of more than one solution. In Chapter 1, Section 1.1, Example 1.4, we considered a first-order initial value problem and showed that it had more than one solution. Since initial value problems and integral equations have a direct relationship, one might expect such strange behavior to apply to integral equations as well. Thus, strange behavior requires qualitative analysis of solutions using different means. For this particular section, the author assume the reader is familiar with complete metric spaces and Banach spaces and we refer to [19] of Chapter 4.

In the next example we show an integral equation has its solution exists over a finite time.

Example 5.21 Consider the integral equation

$$y(x) = y_0 + \int_0^x y^2(\xi) d\xi, \quad y_0 > 0$$

which is equivalent to the initial value problem

$$y'(x) = y^2(x), \quad y(0) = y_0 > 0.$$

Separating the variables, the initial value problem has the solution

$$y(x) = \frac{y_0}{1 - y_0 x}.$$

This solution exists only on the interval $x \in [0, \frac{1}{y_0})$.   ☐

The next example is concerned with the existence of multiple solutions on an integral equation.

Example 5.22 Consider the integral equation

$$y(x) = \int_0^x \frac{y(\xi)}{\sqrt{x^2 - \xi^2}} d\xi.$$

It is clear that $y(x) = 0$, is a solution. Additionally, $y(x) = x$ is another solution since

$$\int_0^x \frac{y(\xi)}{\sqrt{x^2 - \xi^2}} d\xi = \int_0^x \frac{\xi}{\sqrt{x^2 - \xi^2}} d\xi$$

$$= \frac{1}{2} \int_0^{x^2} \frac{1}{\sqrt{u}} d\xi = x,$$

where we have used the transformation $u = x^2 - \xi^2$. Note that the kernel $K(x, \xi) = \frac{1}{\sqrt{x^2 - \xi^2}}$, is well behaved under integration. That is for any $T > 0$ we see that

$$\int_0^T \int_0^x |K(x, \xi)| d\xi dx < \infty,$$

and moreover, the function $g(x) = x$ is certainly Lipschitz continuous. However, the kernel is *singular*, in the sense that

$$K(x, \xi) \to \infty, \text{ as } x \to \xi.$$

□

The next theorem provide necessary conditions for the existence of unique solutions of integral equations of the form

$$x(t) = f(t) + \int_0^t g(t, s, x(s)) ds \tag{5.139}$$

in which x is an n vector, $f : [0, \infty) \to \mathbb{R}^n$, and $g : \pi \times \mathbb{R}^n \to \mathbb{R}^n$ is continuous in all of its arguments where $\pi = \{(t, s) \in [0, \infty) \times [0, \infty) : 0 \le s \le t < \infty\}$. We will use the contraction mapping principle and show the existence of solutions of (5.139) over a short interval, say $[0, T]$.

Theorem 5.12 *Suppose there are positive constants a, b, and $\alpha \in (0, 1)$. Suppose*

(a) f is continuous on $[0, a]$,

(b) g is continuous on

$$U = \{(t, s, x) : (t, s) \in [0, \infty) \times [0, \infty) : 0 \le s \le t < \infty \text{ and } |x - f(t)| \le b\},$$

(c) g satisfies a Lipschitz condition with respect to x on U

$$|g(t, s, x) - g(t, s, y)| \le L|x - y|$$

for $(t,s,x),\ (t,s,y) \in U$.

If $M = \max_U |g(t,s,x)|$, *then there is a unique solution of* (5.139) *on* $[0,T]$, *where* $c = \alpha/L$ *for fixed* α *and* $T = min\{a, b/M, c\}$.

Proof *Let* X *denote the space of continuous functions* $\phi : [0,T] \to \mathbb{R}^n$, *such that*

$$||\phi - f|| = \max_{t \in [0,T]} \{|\phi(t) - f(t)|\} \leq b,$$

where for $\Psi \in X$ *the norm* $|| \cdot ||$ *is taken to be* $||\Psi|| = \max_{t \in [0,T]} \{|\Psi_i(t)|\}$. *Let* $\phi \in X$ *and define an operator* $D : X \to X$, *by*

$$D(\phi)(t) = f(t) + \int_0^t g(t,s,\phi(s))ds.$$

Since ϕ *is continuous we have that* $D(\phi)$ *is continuous, and*

$$
\begin{aligned}
||D(\phi) - f|| &= \max_{t \in [0,T]} \left| \int_0^t g(t,s,\phi(s))ds \right| \\
&\leq MT \leq b.
\end{aligned}
$$

This shows that D *maps* X *into itself. For the contraction part, we let* $\phi, \psi \in X$. *Then*

$$
\begin{aligned}
||D(\phi) - D(\psi)|| &= \max_{t \in [0,T]} \left| \int_0^t g(t,s,\phi(s))ds - \int_0^t g(t,s,\psi(s))ds \right| \\
&\leq \max_{t \in [0,T]} \int_0^t \left| g(t,s,\phi(s)) - g(t,s,\psi(s)) \right| ds \\
&\leq \max_{t \in [0,T]} L \int_0^t \left| \phi(s) - \psi(s) \right| ds \\
&\leq T \max_{t \in [0,T]} L \left| \phi(s) - \psi(s) \right| \\
&= TL||\phi - \psi|| \leq cL||\phi - \psi|| \\
&= \alpha||\phi - \psi||.
\end{aligned}
$$

Thus, by the contraction mapping principle, there is a unique function $x \in X$ *with*

$$D(x)(t) = x(t) = f(t) + \int_0^t g(t,s,x(s))ds.$$

Next we state and prove Gronwall's inequality, which plays an important role in the next results.

Theorem 5.13 (*Gronwall's inequality*) *Let* C *be a nonnegative constant and let* u, v *be nonnegative continuous functions on* $[a,b]$ *such that*

$$v(t) \leq C + \int_a^t v(s)u(s)ds, \quad a \leq t \leq b, \tag{5.140}$$

then

$$v(t) \leq Ce^{\int_a^t u(s)ds}, \quad a \leq t \leq b. \tag{5.141}$$

In particular, if $C = 0$, then $v = 0$.

Proof *Assume $C > 0$ and let $h(t) = C + \int_a^t v(s)u(s)ds$. Then*

$$h'(t) = v(t)u(t) \leq h(t)u(t).$$

So we have the differential equation

$$h'(t) - h(t)u(t) \leq 0.$$

Multiply both sides of the above expression by the integrating factor $e^{-\int_a^t u(s)ds}$, to get

$$\left(h(t)e^{-\int_a^t u(s)ds} \right)' \leq 0.$$

Integrating both sides from a to t gives

$$h(t)e^{-\int_a^t u(s)ds} - h(a) \leq 0, \quad or \quad h(t) \leq h(a)e^{\int_a^t u(s)ds}.$$

Finally,

$$v(t) \leq h(t) \leq Ce^{\int_a^t u(s)ds}, \quad C = h(a).$$

If $C = 0$ then form (5.140) it follows that

$$v(t) \leq \int_a^t v(s)u(s)ds \leq \frac{1}{m} + \int_a^t v(s)u(s)ds \quad a \leq t \leq b,$$

for any $m \geq 1$. Then from what we have just proved we arrive at

$$v(t) \leq \frac{1}{m}e^{\int_a^t u(s)ds}, \quad a \leq t \leq b.$$

Thus for any fixed $t \in [a,b]$, we can let $m \to \infty$ to conclude that $v(t) \leq 0$ and it follows that $v(t) = 0$, for all $t \in [a,b]$. This completes the proof.

Theorem 5.14 *Consider the integral equation given by (5.139). Suppose there are positive constants α, β, λ, and A such that*

$$|f(t)| \leq Ae^{-\alpha t} \quad and \quad |g(t,s,x(s))| \leq \lambda e^{-\alpha(t-s)}|x|.$$

If $\alpha - \lambda = \beta > 0$ and if $x(t)$ is any solution of (5.139) then

$$|x(t)| \leq Ae^{-\beta t}.$$

Proof *From (5.139) we see that*

$$
\begin{aligned}
|x(t)| &= \left| f(t) + \int_0^t g(t,s,x(s))ds \right| \\
&\leq Ae^{-\alpha t} + \int_0^t \lambda e^{-\alpha(t-s)}|x(s)|ds.
\end{aligned}
$$

Multiplying both side of the above expression by $e^{\alpha t}$ we arrive at

$$
e^{\alpha t}|x(t)| \leq A + \int_0^t \lambda e^{\alpha s}|x(s)|ds.
$$

Applying Gronwall's inequality, we obtain the estimate

$$
e^{\alpha t}|x(t)| \leq Ae^{\int_0^t \lambda ds}.
$$

Or,

$$
|x(t)| \leq Ae^{(\lambda - \alpha)t} = Ae^{-\beta t}.
$$

This completes the proof.

The next theorem shows that if the signs of the function g are right, then the growth of g has nothing to do with continuation of solutions. Before we embark on the details, the following is needed. Let $x : \mathbb{R} \to \mathbb{R}$ be continuous. Observing that

$$
|x| = \sqrt{x^2} = (x^2)^{\frac{1}{2}},
$$

and by using the chain rule we arrive at

$$
\begin{aligned}
\frac{d}{dt}|x(t)| &= \frac{1}{2}(x^2(t))^{-\frac{1}{2}}(2x(t)x'(t)) \\
&= \frac{x(t)}{(x^2(t))^{\frac{1}{2}}}x'(t) \\
&= \frac{x(t)}{|x(t)|}x'(t).
\end{aligned}
$$

Theorem 5.15 *Consider the scalar nonlinear integral equation*

$$
y(t) = f(t) + \int_0^t K(t,s)g(y(s))ds, \quad t \geq 0. \tag{5.142}
$$

Suppose f and f' are continuous. In addition, we assume $\frac{\partial K(t,s)}{\partial s}$ and $K(t,s)$ are continuous for $0 \leq s \leq t < \infty$. If for $y \neq 0$, $yg(y) > 0$ and for each $T > 0$ we have

$$
K(t,t) + \int_t^T \left| \frac{\partial K(u,t)}{\partial u} \right| du \leq 0,
$$

then each solution $y(t)$ of (5.142) can be continued for all future times.

Proof *Let $\eta > 0$ and set $K_u(u,t) = \frac{\partial K(u,t)}{\partial u}$. Then $|f'(t)| \leq M$, on $[0,\eta)$ for positive constant M. It suffices to show that if a solution $y(t)$ of (5.142) is defined on $[0,\eta)$, then it is bounded. Let*

$$H(t, y(\cdot)) = e^{-Mt}\left[1 + |y(t)| + \int_0^t \int_t^\eta |K_u(u,s)| du |g(y(s))| ds\right].$$

Then along the solutions of (5.142) we have

$$\begin{aligned}
H'(t, y(\cdot)) &= -Me^{-Mt}\left[1 + |y(t)| + \int_0^t \int_t^\eta |K_u(u,s)| du |g(y(s))| ds\right] \\
&+ e^{-Mt}\left[\frac{y(t)}{|y(t)|} y'(t) + \int_t^\eta |K_u(u,t)| du |g(y(t))|\right. \\
&- \left.\int_0^t |K_t(t,s)||g(y(s))| ds\right] \\
&\leq e^{-Mt}\left[-M - M|y(t)| + \frac{y(t)}{|y(t)|} y'(t) + \int_t^\eta |K_u(u,t)| du |g(y(t))|\right. \\
&- \left.\int_0^t |K_t(t,s)||g(y(s))| ds\right].
\end{aligned} \tag{5.143}$$

Notice that, the condition $yg(y) > 0$ implies

$$\frac{y(t)g(y(t))}{|y(t)|} = \frac{|y(t)||g(y(t))|}{|y(t)|} = |g(y(t))|.$$

Hence, by differentiating (5.142) we have that

$$\begin{aligned}
\frac{y(t)}{|y(t)|} y'(t) &= \frac{y(t)}{|y(t)|}\left(f'(t) + K(t,t)g(y(t)) + \int_0^t K_t(t,s)g(y(s)) ds\right) \\
&\leq |f'(t)| + K(t,t)\frac{|y(t)||g(y(t))|}{|y(t)|} + \int_0^t |K_t(t,s)||g(y(s))| ds \\
&= |f'(t)| + K(t,t)|g(y(t))| + \int_0^t |K_t(t,s)||g(y(s))| ds.
\end{aligned}$$

Substituting into (5.143) we obtain

$$\begin{aligned}
H'(t, y(\cdot)) &\leq e^{-Mt}\left[-M|y(t)| + \left(K(t,t) + \int_t^\eta |K_u(u,t)| du\right)|g(y)|\right] \\
&\leq 0.
\end{aligned}$$

Since $H > 0$ and H is decreasing along the solutions, we see that H is bounded by some constant, and hence $|y(t)|$ is bounded on $[0,\eta)$. As a matter of fact, we have from the definition of H that

$$e^{-Mt}|y(t)| \leq H(t, y(\cdot)) \leq D, \text{ for some } D.$$

This yields

$$|y(t)| \le De^{Mt} \le De^{M\alpha}.$$

This completes the proof.

We furnish the following example.

Example 5.23   Consider the integral equation

$$y(t) = e^t - \int_0^t \frac{y^5(s)}{\sqrt{t-s+1}} ds.$$

Then, for any $\eta > 0$ we have $|f'(t)| = e^t \le e^\eta := M$. It readily follows that $yg(y) = y^6 > 0$ when $y \ne 0$. Let $K(t,s) = -(t-s+1)^{-1/2}$. Then $K_u(u,t) = \frac{1}{2}(u-t+1)^{-3/2}$. Moreover, for any $T > 0$ we have

$$\begin{aligned}
K(t,t) + \int_t^T |K_u(u,t)| du &= -1 + \int_t^T \frac{1}{2}(u-t+1)^{-3/2} du \\
&= -1 - (T-t+1)^{-1/2} + 1 \le 0.
\end{aligned}$$

Hence, by Theorem 5.15 solutions can be continued, or continuable, for all future times. $\qquad\qquad\square$

5.10.1   Exercises

Exercise 5.76   *Construct an example that satisfies the hypothesis of Theorem 5.14.*

Exercise 5.77   *Use Theorem 5.15 to show that solutions of the integral equation*

$$y(t) = e^t - \int_0^t \frac{y^3(s)}{(t-s+1)^2} ds$$

can be continued for all future times.

Exercise 5.78   *Consider the scalar nonlinear integral equation*

$$y(t) = f(t) + \int_0^t g(t,s,y(s)) ds, \quad t \ge 0.$$

Suppose f is continuous. In addition, assume g is continuous for $0 \le s \le t < \infty$, and for each $T > 0$ there is a continuous function $M(s,T)$ with

$$|g(t,s,y)| \le M(s,T)(1+|y|), \quad for \ 0 \le s \le t \le T.$$

Show that if $y(t)$ is a solution of the above integral equation on some interval $[0,\alpha)$, then it is bounded, and, hence, it can be continued for all future times.

Hint: Convince yourself of the fact that $|f(t)| + \int_0^\alpha M(s,\alpha) ds \le Q$, for some positive constant Q, and then apply Gronawall's inequality.

Appendices

A

Fourier Series

This appendix covers the basic main topics of Fourier series. We briefly discuss Fourier series expansion, including sine and cosine. We provide applications to the heat problem in a finite slab by utilizing the concept of separation of variables. We end this appendix by studying the Laplacian equation in circular domains.

A.1   Preliminaries

We start with some basic definitions.

Definition A.1 *A function $f(x)$ is said to be periodic with period p if $f(x+p) = f(x)$ for all x in the domain of f. This means that the function will repeat itself every p units. The main period is the smallest positive period of a function.*

For example, the trig functions $\sin x$ and $\cos x$ are periodic with period 2π, as well as with period $4\pi, 6\pi, 8\pi$, etc. The function $\sin nx$ is periodic, with main period $\frac{2\pi}{n}$, though it also has period 2π. If two functions are period with the same period, then any linear combination of those functions is periodic with the same period. This is important fact since the infinite sum

$$\frac{a_0}{2} + \sum_{n=1}^{\infty}(a_n \cos nx + b_n \sin nx), \tag{A.1}$$

has period 2π. Expression (A.1) is known as the *Fourier series*, where a_n, b_n are called Fourier coefficients. Given a function $f(x)$ that is periodic with period 2π, then we write

$$f(x) = \frac{a_0}{2} + \sum_{n=1}^{\infty}(a_n \cos nx + b_n \sin nx), \tag{A.2}$$

where the Fourier coefficients of $f(x)$ are given by the Euler formulas

$$a_0 = \frac{1}{\pi}\int_{-\pi}^{\pi} f(x)dx, \tag{A.3}$$

$$a_n = \frac{1}{\pi}\int_{-\pi}^{\pi} f(x)\cos(nx)dx, \quad n = 1,2\ldots \tag{A.4}$$

DOI: 10.1201/9781003449881-A

and

$$b_n = \frac{1}{\pi} \int_{-\pi}^{\pi} f(x) \sin(nx)dx, \quad n = 1, 2\ldots \tag{A.5}$$

This is an alternative way of expressing a function in an infinite series in terms of sine and cosine. The above extension of f can be easily extended to periodic function with period $2L$. In such a case the above formulae takes the form

$$\frac{a_0}{2} + \sum_{n=1}^{\infty} \left(a_n \cos\left(\frac{n\pi x}{L}\right) + b_n \sin\left(\frac{n\pi x}{L}\right) \right), \tag{A.6}$$

has period $2L$. Given a function $f(x)$ that is periodic with period $2L$, then we write

$$f(x) = \frac{a_0}{2} + \sum_{n=1}^{\infty} \left(a_n \cos\left(\frac{n\pi x}{L}\right) + b_n \sin\left(\frac{n\pi x}{L}\right) \right), \tag{A.7}$$

where the Fourier coefficients of $f(x)$ are given by the Euler formulas

$$a_0 = \frac{1}{L} \int_{-L}^{L} f(x)dx, \tag{A.8}$$

$$a_n = \frac{1}{L} \int_{-L}^{L} f(x) \cos\left(\frac{n\pi x}{L}\right)dx, \quad n = 1, 2\ldots \tag{A.9}$$

and

$$b_n = \frac{1}{L} \int_{-L}^{L} f(x) \sin\left(\frac{n\pi x}{L}\right)dx, \quad n = 1, 2\ldots \tag{A.10}$$

We have the following definition.

Definition A.2 *Let x_0 be a point in the domain of a function f. Then,*

(a) the right-hand limit of f at x_0, denoted by $f(x_0^+)$ is defined by

$$\lim_{\substack{x \to x_0 \\ x > x_0}} f(x) = f(x_0^+),$$

(b) the left-hand limit of f at x_0, denoted by $f(x_0^-)$ is defined by

$$\lim_{\substack{x \to x_0 \\ x < x_0}} f(x) = f(x_0^-),$$

(c) the right-hand derivative of f at x_0, denoted by $f'(x_0^+)$ is defined by

$$f'(x_0^+) = \lim_{\substack{x \to x_0 \\ x > x_0}} \frac{f(x) - f(x_0^+)}{x - x_0},$$

(d) the left-hand derivative of f at x_0, denoted by $f'(x_0^-)$ is defined by

$$f'(x_0^-) = \lim_{\substack{x \to x_0 \\ x < x_0}} \frac{f(x) - f(x_0^-)}{x - x_0}.$$

Remark 23 *If (a) and (b) of Definition A.2 are satisfied for ever $x \in (a_*, b_*)$, then we say f is piecewise continuous on (a_*, b_*), and we write $f \in C_p(a_*, b_*)$. In addition to (a) and (b), if (c) and (d) of Definition A.2 are satisfied for ever $x \in (a_*, b_*)$, then we say f is piecewise smooth on (a_*, b_*), and we write $f \in C_p'(a_*, b_*)$.*

We furnish the following example.

Example A.1 Consider

$$f(x) = \begin{cases} -x, & x < 0 \\ x + 1, & x > 0 \end{cases}$$

Then

$$\lim_{\substack{x \to 0 \\ x > 0}} f(x) = f(0^+) = 1, \quad \text{and} \quad \lim_{\substack{x \to 0 \\ x < 0}} f(x) = f(0^-) = 0.$$

Moreover,

$$f'(0^+) = \lim_{\substack{x \to 0 \\ x > 0}} \frac{f(x) - 1}{x} = \lim_{\substack{x \to 0 \\ x > 0}} \frac{(x + 1) - 1}{x} = 1,$$

and

$$f'(0^-) = \lim_{\substack{x \to 0 \\ x < 0}} \frac{f(x) - 1}{x} = \lim_{\substack{x \to 0 \\ x < 0}} \frac{-x}{x} = -1.$$

We see that $f \in C_p'(-\infty, \infty)$. $\qquad\qquad\square$

The next theorem is known as the *Fourier convergence theorem*.

Theorem A.1 *[Fourier convergence theorem] Suppose $f \in C_p'(-L, L)$. Then the Fourier series given by (A.6) with Fourier coefficients given by (A.7), (A.8), and (A.10) will converge to the function $f(x)$ at the point x at which f is continuous and it will converge to*

$$\frac{f(x_0^+) + f(x_0^-)}{2}$$

at which f is discontinuous at x_0.

A.2 Finding the Fourier Coefficients

Finding the Fourier coefficient depends on the concept of orthogonality of the sine and cosine functions. The concept of orthogonality was discussed and used in Chapters 3, 4 and 5. Recall the following trigonometric identities $\cos(nx)\cos(mx) = \frac{1}{2}\big(\cos(n + m) + \cos(n - m)\big)$, $\sin(nx)\sin(mx) = \frac{1}{2}\big(\cos(n - m)x - \cos(n + m)x\big)$, and $\sin(nx)\cos(mx) = \frac{1}{2}\big(\sin(n + m)x + \sin(n - m)x\big)$. Thus, utilizing those trigonometric identities we may easily show now that for any integers $m \neq n$, we have $\int_{-\pi}^{\pi} \cos(nx)\cos(mx)dx = 0$, $\int_{-\pi}^{\pi} \sin(nx)\sin(mx)dx = 0$, and $\int_{-\pi}^{\pi} \sin(nx)\cos(mx)dx = 0$. In addition, if $m = n$ then $\int_{-\pi}^{\pi} \sin(nx)\cos(nx)dx = 0$. Because these integrals are zero, we say that $\sin nx$, $\cos mx$ forms an orthogonal system of functions. This is proved using the fact that $n \pm m \neq 0$ is an integer, and so $\int_{-\pi}^{\pi} \cos(n \pm m)dx = 0$ and $\int_{-\pi}^{\pi} \sin(n \pm m)dx = 0$. If $n = m$, then

$$\cos(nx)\cos(nx) = \frac{1}{2}(\cos(2nx) + 1) \text{ and } \sin(nx)\sin(nx) = \frac{1}{2}(1 - \cos(2nx)),$$

so we can compute

$$
\begin{aligned}
\int_{-\pi}^{\pi} \cos(nx)\cos(nx)dx &= \frac{1}{2}\int_{-\pi}^{\pi}(\cos(2nx) + 1)dx \\
&= \frac{1}{2}(x + \frac{\sin(2nx)}{2n+x})|_{-\pi}^{\pi}dx \\
&= \pi,
\end{aligned}
$$

$$
\begin{aligned}
\int_{-\pi}^{\pi} \sin nx \sin nx dx &= \frac{1}{2}\int_{-\pi}^{\pi}(1 - \cos(2nx))dx \\
&= \frac{1}{2}(x - \frac{\sin(2nx)}{2n})|_{-\pi}^{\pi}dx \\
&= \pi.
\end{aligned}
$$

Now, if we multiply both sides of

$$f(x) = \frac{a_0}{2} + \sum_{n=1}^{\infty}\left(a_n \cos(nx) + b_n \sin(nx)\right)$$

by $\cos(mx)$, and then integrate term by term, we have by using the orthogonality concept that

$$
\begin{aligned}
\int_{-\pi}^{\pi} f(x)\cos(mx)dx &= \int_{-\pi}^{\pi} \frac{a_0}{2}\cos(mx)dx \\
&+ \sum_{n=1}^{\infty}(a_n \int_{-\pi}^{\pi}\cos(nx)\cos(mx)dx + b_n \int_{-\pi}^{\pi}\sin(nx)\cos(mx)dx) \\
&= 0 + a_m \pi.
\end{aligned}
$$

Hence

$$a_m = \frac{1}{\pi}\int_{-\pi}^{\pi} f(x)\cos(mx)dx.$$

The other coefficients are derived similarly. Now we work out some examples.

Example A.2 Let

$$f(x) = \begin{cases} 2, & 0 < x < \pi \\ -1, & -\pi < x < 0 \end{cases}.$$

The function $f(x)$ has period 2π. First we compute

$$
\begin{aligned}
a_0 &= \frac{1}{\pi} \int_{-\pi}^{\pi} f(x) dx \\
&= \frac{1}{\pi} \left(\int_{-\pi}^{0} -1 dx + \int_{0}^{\pi} 2 dx \right) \\
&= \frac{1}{\pi} (-\pi + 2\pi) \\
&= 1,
\end{aligned}
$$

$$
\begin{aligned}
a_n &= \frac{1}{\pi} \int_{-\pi}^{\pi} f(x) \cos(nx) dx \\
&= \frac{1}{\pi} \left(\int_{-\pi}^{0} -\cos(nx) dx + \int_{0}^{\pi} 2\cos(nx) dx \right) \\
&= \frac{1}{\pi} \left(-\frac{\sin nx}{n} \Big|_{-\pi}^{0} + 2\frac{\sin nx}{n} \Big|_{0}^{\pi} \right) \\
&= 0, \ n = 1, 2, \ldots.
\end{aligned}
$$

Finally,

$$
\begin{aligned}
b_n &= \frac{1}{\pi} \int_{-\pi}^{\pi} f(x) \sin(nx) dx \\
&= \frac{1}{\pi} \left(\int_{-\pi}^{0} -\sin(nx) dx + \int_{0}^{\pi} 2\sin(nx) dx \right) \\
&= \frac{1}{\pi} \left(\frac{\cos nx}{n} \Big|_{-\pi}^{0} - 2\frac{\cos nx}{n} \Big|_{0}^{\pi} \right) \\
&= \frac{1}{\pi} \left(\frac{1}{n} - \frac{\cos n\pi}{n} - 2\frac{\cos n\pi}{n} + \frac{2}{n} \right) \\
&= \frac{3}{n\pi} (1 - \cos n\pi).
\end{aligned}
$$

Note that when n is even then $\cos n\pi = 1$ and when n is odd $\cos n\pi = -1$. Hence $b_n = \frac{6}{n\pi}$ if n is odd and $b_n = 0$ if n is even. This means that can replace n by $2n - 1$ in the sum and obtain

$$
f(x) = \begin{cases} 2, & 0 < x < \pi \\ -1, & -\pi < x < 0 \end{cases} = \frac{1}{2} + \sum_{n=1}^{\infty} \frac{6}{(2n-1)\pi} \sin((2n-1)x).
$$

According to Theorem A.1, the infinite sum given by the above expression converges to the function

$$
g(x) = \begin{cases} \frac{1}{2} & \text{at } x = 0, \pm\pi \\ f(x) & \text{otherwise} \end{cases}
$$

□

We now consider a function with period 8.

Example A.3 Consider the function

$$f(x) = \begin{cases} 0, & -4 < x < -2 \\ 1, & -2 < x < 2 \\ 0, & 2 < x < 4 \end{cases}.$$

This function is 1 for $-2 < x < 2$, $6 < x < 10$, etc. It is a regular pulse which is on for 4 units of time, and then off for four units of time. Since the period is not 2π, but instead $2L = 8$, we have $L = 4$. The Fourier coefficients are

$$a_0 = \frac{1}{4} \int_{-4}^{4} f(x)dx = \frac{1}{4} \int_{-2}^{2} 1\,dx = 1,$$

$$\begin{aligned} a_n &= \frac{1}{L} \int_{-4}^{4} f(x)\cos\frac{n\pi x}{4}dx = \frac{1}{4} \int_{-2}^{2} \cos\frac{n\pi x}{4}dx \\ &= \frac{1}{4}\frac{4}{n\pi}\sin\frac{n\pi x}{4}\Big|_{-2}^{2} = \frac{1}{n\pi}\sin\frac{n\pi x}{4}\Big|_{-2}^{2} \\ &= \frac{1}{n\pi}\left(\sin\frac{2n\pi}{4} - \sin\frac{-2n\pi}{4}\right) = \frac{1}{n\pi}(2\sin\frac{n\pi}{2}), \end{aligned}$$

$$\begin{aligned} b_n &= \frac{1}{L} \int_{-4}^{4} f(x)\sin\frac{n\pi x}{4}dx = \frac{1}{4} \int_{-2}^{2} \sin\frac{n\pi x}{4}dx \\ &= -\frac{1}{4}\frac{4}{n\pi}\cos\frac{n\pi x}{4}\Big|_{-2}^{2} = -\frac{1}{n\pi}\cos\frac{n\pi x}{4}\Big|_{-2}^{2} \\ &= -\frac{1}{n\pi}\left(\cos\frac{n\pi}{2} - \cos\frac{-n\pi}{2}\right) = \frac{1}{n\pi}(0) = 0. \end{aligned}$$

If n is even, then $a_n = 0$ as sine is 0 at integer values. Thus, a_n contribute nonzero values for odd n, and so we may replace n by $2n - 1$. With this in mind, the Fourier series can be written as

$$f(x) = \begin{cases} 0 & -4 < x < -2 \\ 1 & -2 < x < 2 \\ 0 & 2 < x < 4 \end{cases} = \frac{1}{2} + \frac{2}{\pi} \sum_{n=1}^{\infty} \frac{\sin(\frac{(2n-1)\pi}{2})}{2n-1}\cos(\frac{(2n-1)\pi x}{4}).$$

□

A.3 Even and Odd Extensions

Any function $f(x)$ which satisfies $f(-x) = f(x)$ is called an *even function*. Similarly, any function that satisfies $f(-x) = -f(x)$ is called an *odd function*. Even functions are symmetric about the *y*-axis and odd functions are symmetric about the origin. For example, $f(x) = \cos(x)$ is an even function since $\cos(-x) = \cos(x)$ and $f(x) = \sin(x)$ is an odd function since $\sin(-x) = -\sin(x)$. Using the concept of odd and even functions, one can easily show, using (A.9) and (A.10) that the Fourier coefficients of an even function are simply

$$a_n = \frac{2}{L} \int_0^L f(x) \cos \frac{n\pi x}{L} dx, \quad n = 0, 1, \ldots$$

and

$$b_n = 0, \quad n = 1, 2, \ldots$$

and the corresponding Fourier series is called a Fourier cosine series. Similarly, for an odd function the coefficients are

$$a_n = 0, \quad n = 0, 1, \ldots$$

and

$$b_n = \frac{2}{L} \int_0^L f(x) \sin \frac{n\pi x}{L} dx, \quad n = 1, 2, \ldots$$

and the corresponding Fourier series is called a Fourier sine series. This comes because the product of two even functions is even, the product of two odd functions is even, and the product of an even and an odd function is odd. In addition, integration from $-L$ to L of an odd function is zero, while integration from $-L$ to L of an even function is twice the integral of 0 to L.

Consider the sawtooth wave, which is given by the function $f(x) = x + \pi$ for $-\pi < x < \pi$, and $f(x + 2\pi) = f(x)$. It can be written as the sum of an even function $f_1(x) = \pi$ and an odd function $f_2(x) = x$. The corresponding Fourier cosine and sine series are $f_1 = \pi$ and $f_2 = 2\left(\sin x - \frac{1}{2}\sin 2x + \frac{1}{3}\sin 3x - \frac{1}{4}\sin 4x + \cdots\right)$. Addition of series gives $f(x) = \pi + 2\left(\sin x - \frac{1}{2}\sin 2x + \frac{1}{3}\sin 3x - \frac{1}{4}\sin 4x + \cdots\right)$. (The coefficients b_n are obtained using integration by parts and $b_n = -\frac{2}{n}\cos n\pi$.)

If a function is defined on the interval $[0, L]$, then it is possible to expand the function periodically onto the interval $[-L, 0]$ by either using an even expansion (reflection about the *y* axis), or an odd expansion (reflection about the origin). Both expansions are called half-range expansions. The Fourier series of an even half-range expansion is the Fourier cosine series, and the Fourier series of an odd half-range expansion is the Fourier sine series. We have the following. Let f_e and f_o denote the even and odd extensions of $2l$-periodic function f. Then,

$$f_e(x) = \begin{cases} f(x), & 0 < x < l, \\ f(-x), & -l < x < 0, \end{cases} \qquad f_o(x) = \begin{cases} f(x), & 0 < x < l, \\ -f(-x), & -l < x < 0 \end{cases}$$

Thus, the *periodic odd extension* of a function f that is piecewise continuous on the interval $(0, L)$, is the *Fourier sine series* of f given by

$$f(x) = \sum_{n=1}^{\infty} b_n \sin\left(\frac{n\pi x}{L}\right), \quad 0 < x < L, \tag{A.11}$$

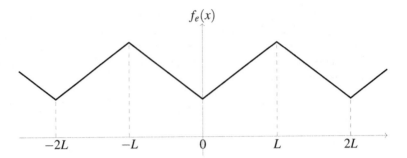

FIGURE A.1
Periodic even extension.

where

$$b_n = \frac{2}{L} \int_0^L f(x) \sin(\frac{n\pi x}{L}) dx, \ n = 1, 2, \ldots . \tag{A.12}$$

Similarly, the *periodic even extension* of a function f that is piecewise continuous on the interval $(0, L)$, is the *Fourier cosine series* of f given by

$$f(x) = \frac{a_0}{2} + \sum_{n=1}^{\infty} a_n \cos(\frac{n\pi x}{L}), \quad 0 < x < L, \tag{A.13}$$

where

$$a_0 = \frac{2}{L} \int_0^L f(x) dx,$$

and

$$a_n = \frac{2}{L} \int_0^L f(x) \cos(\frac{n\pi x}{L}) dx, \ n = 1, 2, \ldots .$$

In Fig. A.1, we display the cosine Fourier series of

$$f(x) = \frac{x}{L} + 1, \quad 0 < x < L.$$

Every term in the Fourier cosine series is $2L$-periodic. Note that the periodic even extension does not introduce new jumps. Similarly, if we consider

$$f(x) = \frac{x}{L} + 1, \quad 0 < x < L,$$

Every term in the Fourier sine series is $2L$-periodic. Note that the periodic odd extension does not introduce new jumps if and only if $f(0) = f(L) = 0$. The two figures below illustrate both cases.

We provide the following examples.

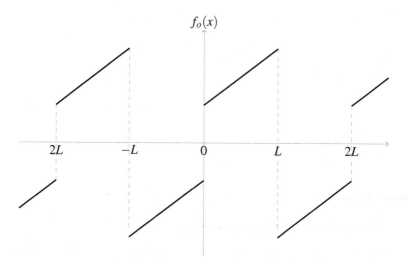

FIGURE A.2
Discontinuous periodic odd extension.

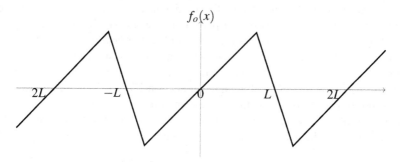

FIGURE A.3
Continuous periodic odd extension when we require $f(0) = f(L) = 0$.

Example A.4 Let us find the Fourier cosine series of $f(x) = x$, $0 < x < \pi$. It is easy to see that

$$a_0 = \frac{2}{\pi} \int_0^\pi x\,dx = \pi.$$

Using integration by parts, we find that

$$a_n = \frac{2}{\pi} \int_0^\pi x\cos(nx)\,dx = \frac{2}{\pi} \frac{(-1)^n - 1}{n^2}, \quad n = 1, 2, \ldots.$$

By noticing that $a_n = 0$ for n is even and $a_n = -2$ for n is odd, we may replace $(-1)^n - 1$ with -2 and use $2n - 1$ for n in the summation. Thus,

$$x = \frac{\pi}{2} - \frac{4}{\pi} \sum_{n=1}^\infty \frac{\cos(2n-1)x}{(2n-1)^2}, \quad 0 < x < \pi,$$

which is the periodic even extension of $f(x) = x$. Note that

$$f_e(x) = |x|, \quad -\pi \le x \le \pi.$$

In conclusion we may safely write

$$|x| = \frac{\pi}{2} - \frac{4}{\pi} \sum_{n=1}^{\infty} \frac{\cos(2n-1)x}{(2n-1)^2}, \quad -\pi \le x \le \pi.$$

$\square$

A.4   Applications of Fourier Series

Fourier series are particularly useful in the telecommunications and graphics industries in areas such as cell phones, internet, land lines, and radio communication, etc. Radio transmitters send out radio waves, which are essentially periodic vibrations of space. Antennae detect these vibrations as they are transmitted in all directions. In order to determine the received signal's coefficients, your radio receiver computes Fourier integrals. Fourier series also play a significant part in heat transfer modeling. Engineers use the Fourier series to simulate the heat transfer in spacecraft, automobiles, jet engines, and any other system that can malfunction due to overheating. In this section we will use Fourier series to solve the heat equation with Neumann condition on bounded domain. In Section 2.7.3 we studied the heat equation on semi-infinite domain with a Neumann condition. In our analysis, we will utilize the concept of *separation of variables* after we recast the PDE into an ODE. Thus, we consider the heat problem on a finite slab

$$\begin{cases} u_t - k u_{xx} = 0, & 0 < x < c, t > 0 \\ u(x,0) = f(x), & 0 < x < c \\ u_x(0,t) = 0, \ u_x(c,t) = 0, & 0 < x < c. \end{cases} \tag{A.14}$$

We are interested in finding a non trivial solution $u(x,t)$ of (A.14) that satisfies the boundary and the initial conditions. We seek separated solutions of functions of the

$$u(x,y) = X(x)T(t), \tag{A.15}$$

where X is a function of x alone and T is a function of t alone. Note, too, that X and T must be nontrivial. That is $X \ne 0$, and $T \ne 0$. By differentiating (A.15) with respect to t and x and substituting into (A.14) we obtain the relation

$$X(x)T'(t) = kX''(x)T(t).$$

Since $X(x) \ne 0$, and $T(t) \ne 0$, we may divide by the term $X(x)T(t)$ to separate the variables. That is,

$$\frac{X''(x)}{X(x)} = \frac{T'(t)}{kT(t)}.$$

Since the left-hand side is a function of x alone, and the right-hand side is a function of t alone, the two sides must have a common constant value $-\lambda$. That is,

$$\frac{X''(x)}{X(x)} = \frac{T'(t)}{kT(t)} = -\lambda.$$

Now we check the Neumann boundary conditions. $0 = u_x(0,t) = X'(0)T(t)$, implies that $X'(0) = 0$. Similarly, $0 = u_x(c,t) = X'(c)T(t)$, implies that $X'(c) = 0$. Thus, we arrive at the Sturm-Liouville problem

$$X''(x) + \lambda X(x) = 0, \quad X'(0) = 0, \, X'(a) = 0, \tag{A.16}$$

and at the first-order ordinary differential equation

$$T'(t) + \lambda T(t) = 0, \quad t > 0. \tag{A.17}$$

One can easily argue as in Section 4.14, and determine that (A.16) has the trivial solution for $\lambda < 0$. For $\lambda = 0$, we have from (A.16) that $X''(x) = 0$, which has the solution $X(x) = Ax + B$. Applying the boundary conditions we get $B = 0$ and A is arbitrary, and so we set it equal to one. Thus, for $\lambda_0 = 0$, the corresponding eigenfunction is $X_0(x) = 1$. Now for $\lambda > 0$, we assume $\lambda = \alpha^2$ for positive α. Then the general solution of (A.16) is

$$X(x) = A\cos(\alpha x) + B\sin(\alpha x),$$

and hence

$$X'(x) = -A\alpha\sin(\alpha x) + B\alpha\cos(\alpha x).$$

Applying $X'(0) = 0$, we automatically get $B = 0$. Applying $X'(c) = 0$, with $B = 0$ already, we arrive at

$$-A\alpha\sin(\alpha x) = 0.$$

To obtain a nontrivial solution we set $\sin(\alpha c) = 0$. This gives $\alpha c = n\pi$, $n = 1, 2, \ldots$ and obtain $\alpha = \frac{n\pi}{c}$. Thus, for $\lambda_n = \left(\frac{n\pi}{c}\right)^2$, the corresponding eigenfunctions are given by

$$X_n(x) = \cos\left(\frac{n\pi x}{c}\right), \quad n = 1, 2, \ldots,$$

where we set $A = 1$. Turning to (A.17), we need to solve it based on the already determined eigenvalues λ_0 and λ_n, $n = 1, 2, \ldots$. For $\lambda_0 = 0$, equation (A.17) has the solution constant multiple of $T_0(t) = 1$. Similarly, for λ_n we have the corresponding eigenfunctions

$$T_n(t) = e^{-\frac{n^2\pi^2 k}{c^2}t}, \quad n = 1, 2, \ldots.$$

Thus, we may write the solution as

$$u(x,t) = u_0(x,t) + u_n(x,t) = 1 + e^{-\frac{n^2\pi^2 k}{c^2}t}\cos\left(\frac{n\pi x}{c}\right).$$

Note that u satisfies both of Neumann conditions. Now by the *superposition principle* the general solution of (A.14) maybe written as

$$u(x,t) = \frac{a_0}{2} + \sum_{n=1}^{\infty} a_n e^{-\frac{n^2\pi^2 k}{c^2}t} \cos\left(\frac{n\pi x}{c}\right). \tag{A.18}$$

By applying the initial condition $u(x,0) = f(x)$ to (A.18) we obtain the Fourier cosine series

$$f(x) = \frac{a_0}{2} + \sum_{n=1}^{\infty} a_n \cos\left(\frac{n\pi x}{c}\right),$$

where

$$a_0 = \frac{2}{c} \int_0^c f(x)dx,$$

and

$$a_n = \frac{2}{c} \int_0^c f(x) \cos\left(\frac{n\pi x}{c}\right)dx, \quad n = 1, 2, \ldots.$$

A.5   Laplacian in Polar, Cylindrical and Spherical Coordinates

In this section, we will discuss the Laplacian $\nabla^2 u$, in *polar, cylindrical, and spherical coordinates*. When solving boundary value problems in more than one dimension, it is often necessary to use other coordinate systems than the cartesian. It is then important to be able to express the Laplacian operator in these coordinate systems.

In two dimension the Laplacian can be written

$$\nabla^2 u = \frac{\partial^2 u(x,y)}{\partial x^2} + \frac{\partial^2 u(x,y)}{\partial y^2}$$

while in three dimensions we write

$$\nabla^2 u = \frac{\partial^2 u(x,y,z)}{\partial x^2} + \frac{\partial^2 u(x,y,z)}{\partial y^2} + \frac{\partial^2 u(x,y,z)}{\partial z^2}.$$

Either equation may be written

$$\nabla^2 u = 0$$

and we will have to learn how to express the **Laplacian**

$$\nabla^2 = \frac{\partial^2}{\partial x^2} + \frac{\partial^2}{\partial y^2} + \frac{\partial^2}{\partial z^2}$$

in different ways such as in polar, cylindrical or spherical coordinates.

We begin by considering the Lapacian in polar coordinates. Thus, we make use of the transformation

$$x = r\cos\theta$$

$$y = r\sin\theta$$

Laplace's equation in this coordinate system can be shown to be,

$$\nabla^2 u(r,\theta) = \frac{\partial^2 u(r,\theta)}{\partial r^2} + \frac{1}{r}\frac{\partial u(r,\theta)}{\partial r} + \frac{1}{r^2}\frac{\partial^2 u(r,\theta)}{\partial\theta^2}. \tag{A.19}$$

In *cylindrical coordinates,* we set

$$x = \rho\cos\phi$$

$$y = \rho\sin\phi$$

$$z = z$$

and it can also be shown that Laplace's equation in cylindrical coordinates takes the form

$$\nabla^2 u(\rho,\phi,z) = \frac{\partial^2 u(\rho,\phi,z)}{\partial\rho^2} + \frac{1}{\rho}\frac{\partial u(\rho,\phi,z)}{\partial\rho} + \frac{1}{\rho^2}\frac{\partial^2 u(\rho,\phi,z)}{\partial\phi^2} + \frac{\partial^2 u(\rho,\phi,z)}{\partial z^2}. \tag{A.20}$$

Note that expression (A.19) is a special case of (A.20) by simply holding z constant. Finally, in *spherical coordinates*

$$x = \rho\sin\theta\cos\phi$$

$$y = \rho\sin\theta\sin\phi$$

$$z = \rho\cos\theta$$

we have

$$\nabla^2 u(r,\theta,z) = \frac{1}{r^2}[\frac{\partial}{\partial r}(r^2\frac{\partial u(r\theta,\phi)}{\partial r})$$

$$+ \frac{1}{\sin\theta}\frac{\partial}{\partial\theta}(\sin\theta\frac{\partial u(r,\theta,\phi)}{\partial\theta}) + \frac{1}{\sin^2\theta}\frac{\partial^2 u(r,\theta,\phi)}{\partial\phi^2}]. \tag{A.21}$$

Next, we give a brief derivation of (A.20). We already know that

$$\rho = \sqrt{x^2+y^2}, \quad\text{and}\quad \phi = \arctan\left(\frac{y}{x}\right).$$

Using the chain rule we have

$$\frac{\partial u}{\partial x} = \frac{\partial u}{\partial\rho}\frac{\partial\rho}{\partial x} + \frac{\partial u}{\partial\phi}\frac{\partial\phi}{\partial x} + \frac{\partial u}{\partial z}\frac{\partial z}{\partial x}.$$

However,

$$\frac{\partial\rho}{\partial x} = \frac{x}{\sqrt{x^2+y^2}} = \frac{x}{\rho} = \frac{\rho\cos\phi}{\rho} = \cos\phi.$$

In similar fashions, we arrive at

$$\frac{\partial \rho}{\partial y} = \sin \phi, \quad \frac{\partial \phi}{\partial x} = -\frac{\sin \phi}{\rho}, \quad \text{and} \quad \frac{\partial \phi}{\partial y} = \frac{\cos \phi}{\rho}.$$

Substituting into $\dfrac{\partial u}{\partial x}$ we have

$$\frac{\partial u}{\partial x} = \cos \phi \frac{\partial u}{\partial \rho} - \frac{\sin \phi}{\rho} \frac{\partial u}{\partial \phi}, \tag{A.22}$$

since $\dfrac{\partial z}{\partial x} = 0$. To obtain $\dfrac{\partial^2 u}{\partial x^2}$, we replace the function u in (A.22) by $\dfrac{\partial u}{\partial x}$. That is,

$$
\begin{aligned}
\frac{\partial^2 u}{\partial x^2} &= \cos \phi \frac{\partial}{\partial \rho}\left(\frac{\partial u}{\partial x}\right) - \frac{\sin \phi}{\rho} \frac{\partial}{\partial \phi}\left(\frac{\partial u}{\partial x}\right) \\
&= \cos \phi \frac{\partial}{\partial \rho}\left(\cos \phi \frac{\partial u}{\partial \rho} - \frac{\sin \phi}{\rho} \frac{\partial u}{\partial \phi}\right) - \frac{\sin \phi}{\rho} \frac{\partial}{\partial \phi}\left(\cos \phi \frac{\partial u}{\partial \rho} - \frac{\sin \phi}{\rho} \frac{\partial u}{\partial \phi}\right) \\
&= \cos \phi \left(\cos \phi \frac{\partial^2 u}{\partial \rho^2} + \frac{\sin \phi}{\rho^2} \frac{\partial u}{\partial \phi} - \frac{\sin \phi}{\rho} \frac{\partial^2 u}{\partial \rho \partial \phi}\right) \\
&\quad - \frac{\sin \phi}{\rho}\left(-\sin \phi \frac{\partial u}{\partial \rho} + \cos \phi \frac{\partial^2 u}{\partial \phi \partial \rho} - \frac{\cos \phi}{\rho} \frac{\partial u}{\partial \phi} - \frac{\sin \phi}{\rho} \frac{\partial^2 u}{\partial \phi^2}\right).
\end{aligned}
$$

Using the fact that

$$\frac{\partial^2 u}{\partial \rho \partial \phi} = \frac{\partial^2 u}{\partial \phi \partial \rho},$$

the above expression simplifies to

$$
\begin{aligned}
\frac{\partial^2 u}{\partial x^2} &= \cos^2 \phi \frac{\partial^2 u}{\partial \rho^2} - \frac{2 \sin \phi \cos \phi}{\rho} \frac{\partial^2 u}{\partial \phi \partial \rho} + \frac{\sin^2 \phi}{\rho^2} \frac{\partial^2 u}{\partial \phi^2} \\
&\quad + \frac{\sin^2 \phi}{\rho} \frac{\partial u}{\partial \rho} + \frac{2 \sin \phi \cos \phi}{\rho^2} \frac{\partial u}{\partial \phi}
\end{aligned} \tag{A.23}
$$

By similar steps, we see that

$$
\begin{aligned}
\frac{\partial u}{\partial y} &= \frac{\partial u}{\partial \rho} \frac{\partial \rho}{\partial y} + \frac{\partial u}{\partial \phi} \frac{\partial \phi}{\partial y} + \frac{\partial u}{\partial z} \frac{\partial z}{\partial x} \\
&= \sin \phi \frac{\partial u}{\partial \rho} + \frac{\cos \phi}{\rho} \frac{\partial u}{\partial \phi}.
\end{aligned}
$$

Moreover,

$$
\begin{aligned}
\frac{\partial^2 u}{\partial y^2} &= \sin^2 \phi \frac{\partial^2 u}{\partial \rho^2} + \frac{2 \sin \phi \cos \phi}{\rho} \frac{\partial^2 u}{\partial \phi \partial \rho} + \frac{\cos^2 \phi}{\rho^2} \frac{\partial^2 u}{\partial \phi^2} \\
&\quad + \frac{\cos^2 \phi}{\rho} \frac{\partial u}{\partial \rho} - 2 \frac{\sin \phi \cos \phi}{\rho^2} \frac{\partial u}{\partial \phi}.
\end{aligned} \tag{A.24}
$$

A substitution of (A.23) and (A.24) into

$$\nabla^2 u = \frac{\partial^2 u(x,y,z)}{\partial x^2} + \frac{\partial^2 u(x,y,z)}{\partial y^2} + \frac{\partial^2 u(x,y,z)}{\partial z^2} = 0,$$

leads to (A.20). For an application, we consider the Laplacian given by equation (A.19) on an *annulus*. Using simplified notations we write

$$u_{\rho\rho} + \frac{1}{\rho} u_\rho + \frac{1}{\rho^2} u_{\phi\phi} = 0, \quad 1 < \rho < 2, \ 0 < \phi < \pi, \tag{A.25}$$

along with boundary conditions

$$u(\rho,0) = 0, \quad u(\rho,\pi) = u_0, \tag{A.26}$$

$$u(1,\phi) = 0, \quad u(2,\phi) = 0, \tag{A.27}$$

as depicted in Fig. A.4. As in Section A.4, we seek a nontrivial solution of the form

$$u(\rho,\phi) = R(\rho)\Theta(\phi).$$

Differentiate and then substitute into (A.25), and then divide the resulting equation with $R(\rho)\Theta(\phi)$. Finally separating the variables and setting both sides equal to $-\lambda$, we arrive at the two second-order differential equations

$$\Theta'' - \lambda\Theta = 0,$$

$$\Theta(0) = 0$$

and

$$R''(\rho) + \frac{1}{\rho} R'(\rho) + \frac{\lambda}{\rho} R(\rho) = 0, \quad \rho > 0,$$

$$R(1) = R(2) = 0.$$

Using the method of Section 1.11, we obtain

$$R''(t) + \lambda R(t) = 0, \quad \text{where } \rho = e^t.$$

It is easy to verify that $R(t) = 0$ for $\lambda \le 0$. For $\lambda > 0$, we assume $\lambda = \alpha^2$, $\alpha > 0$. Then

$$R''(t) + \alpha^2 R(t) = 0,$$

has the general solution

$$R(t) = A\cos(\alpha t) + B\sin(\alpha t),$$

and in term of ρ, the solution takes the form

$$R(\rho) = A\cos(\alpha \ln(\rho)) + B\sin(\alpha \ln(\rho)).$$

Applying $0 = R(1)$, we obtain $A = 0$. Applying the second boundary condition and keeping in mind that $A = 0$, we arrive at $B\sin(\alpha\ln(2)) = 0$, from which we obtain $\alpha\ln(2) = n\pi$, $n = 1, 2, \ldots$. This yields

$$\alpha_n = \frac{n\pi}{\ln(2)}, \quad n = 1, 2, \ldots.$$

Thus, the corresponding eigenfunctions are given by

$$R_n(\rho) = \sin(\alpha_n\ln(\rho)), \quad n = 1, 2, \ldots.$$

Next we normalize the eigenfunctions with respect the weight function $p = \frac{1}{\rho}$, by setting

$$\zeta_n = \frac{R_n(\rho)}{\|R_n\|}, \quad n = 1, 2, \ldots,$$

$$\|R_n\|^2 = \int_1^2 \frac{\sin^2(\alpha\ln(\rho))}{\rho}d\rho.$$

For learning purposes, we show the necessary steps for evaluating the integral.

$$
\begin{aligned}
\int_1^2 \frac{\sin^2(n\pi\ln(\rho))}{\rho}d\rho &= \int_1^2 \frac{1 - \cos(2\alpha\ln(\rho))}{2\rho}d\rho \\
&= \frac{1}{2}\ln(\rho)\Big|_1^2 - \int_1^2 \frac{\cos(2\alpha\ln(\rho))}{2\rho}d\rho \\
&= \frac{\ln(2)}{2} - \int_1^2 \frac{\cos(2\alpha\ln(\rho))}{2\rho}d\rho.
\end{aligned}
$$

To evaluate the integral, we make the substitution $u = 2\alpha\ln(\rho)$. Then, $\frac{du}{2\alpha} = \frac{d\rho}{\rho}$ and

$$
\begin{aligned}
\int_1^2 \frac{\cos(2\alpha\ln(\rho))}{2\rho}d\rho &= \int_0^{2\alpha\ln(2)} \frac{\cos(u)}{4\alpha}du \\
&= \frac{\sin(2\alpha\ln(2))}{4\alpha} - 0 \\
&= \frac{\sin(2\frac{n\pi}{\ln(2)}\ln(2))}{4\alpha} = 0.
\end{aligned}
$$

Thus,

$$\|R_n\| = \sqrt{\frac{\ln(2)}{2}}.$$

Hence, the normalized eigenfunctions are given by

$$\zeta_n(\rho) = \sqrt{\frac{2}{\ln(2)}}\sin(\alpha_n\ln(\rho)), \quad n = 1, 2, \ldots.$$

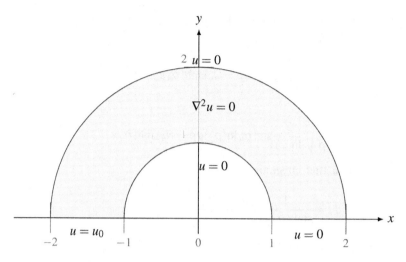

FIGURE A.4
Laplacian on an annulus.

Similarly, for the same eigenvalues $\lambda_n = \alpha_n^2$, the system

$$\Theta'' - \lambda\Theta = 0, \quad \Theta(0) = 0,$$

has the general solution

$$\Theta_n(\phi) = c_1 e^{\alpha_n \phi} + c_2 e^{-\alpha_n \phi}.$$

An application of the boundary condition, leads to $c_2 = -c_1$. By setting $c_1 = \frac{1}{2}$, we then have $c_2 = -\frac{1}{2}$, and the solution takes the form

$$\Theta_n(\phi) = \sinh(\alpha_n \phi), \quad n = 1, 2, \ldots.$$

So, the solution of (A.25) can be written as

$$u(\rho, \phi) = \sum_{n=1}^{\infty} d_n \sinh(\alpha_n \phi) \zeta_n(\rho).$$

Using $u(\rho, \pi) = u_0$, we get

$$u_0 = \sum_{n=1}^{\infty} d_n \sinh(\alpha_n \pi) \zeta_n(\rho). \tag{A.28}$$

Recall that the set of functions given by ζ_n are normalized with respect to the weight function $p = \frac{1}{\rho}$ and therefore,

$$\int_1^2 \zeta_n(\rho) \zeta_m(\rho) d\rho = \begin{cases} 1, & m = n \\ 0, & m \neq n. \end{cases}$$

Hence, by multiplying both sided of (A.28) with $\zeta_m(\rho)$ and then integrate with respect to ρ from 1 to 2 we arrive at

$$u_0 \int_1^2 \zeta_n(\rho)d\rho = d_n \sinh(\alpha_n \pi),$$

or,

$$u_0 \int_1^2 \frac{1}{\rho} \sqrt{\frac{2}{\ln(2)}} \sin(\alpha_n \ln(\rho))d\rho = d_n \sinh(\alpha_n \pi).$$

From which we obtain after integrating,

$$d_n = \frac{1}{\sinh(\alpha_n \pi)} \left(\frac{u_0 \sqrt{2\ln(2)}}{\pi} \frac{1-(-1)^n}{n} \right).$$

Accordingly, the solution takes the form

$$u(\rho,\phi) = \frac{2u_0}{\pi} \sum_{n=1}^{\infty} \frac{1-(-1)^n}{n} \frac{\sinh(\alpha_n \phi)}{\sinh(\alpha_n \pi)} \sin(\alpha_n \ln(\rho)),$$

where

$$\alpha_n = \frac{n\pi}{\ln(2)}, \quad n = 1,2,\ldots.$$

Bibliography

Chapter 1

1. Bellman, R., *Stability Theory of Differential Equations,* McGraw-Hill Book Company, New York, London, 1953.

2. Berezansky, L., and Braverman, E., *Exponential stability of difference equations with several delays: Recursive approach,* Adv. Difference. Equ. Vol. 2009, Article ID 104310, 13.

3. Driver, R. D., *Introduction to Ordinary Differential Equations,* Harper & Row, Publishers, New York, 1978.

4. Hartman, P., *Ordinary Differential Equations,* John Wiley & Sons, Inc., New York, 1964.

5. Kelley, W., and Peterson, A., *The Theory of Differential Equations, Classical and Qualitative,* Pearson Prentice Hall, 2004.

6. Miller, R. K., *Nonlinear Volterra Integral Equations*, Benjamin, New York, 1971.

7. Miller, R. K., *Introduction to Differential Equations*, Prentice Hall 1987.

8. Raffoul, Y. N., *Class Notes on Ordinary Differential Equations,* University of Dayton, 2022.

9. Raffoul, Y. N., *Advanced Differential Equations,* Elsevier/Academic Press, N Y, 2022.

Chapter 2

1. Brown, J. W., *Fourier Series and Boundary Value Problems,* 8th edition, McGrawhill, 2012.

2. Jeffrey, A., *Applied Partial Differential Equations: An Introduction,* Academic Press, 2003.

3. Logan, D. J., *Applied Partial Differential Equations,* Springer, 1998.

4. Myint-U, T. and L. Debnath, *Linear Partial Differential Equations for Scientists and Engineers,* 4th edition, Birkhauser, 2006.

5. Olver, P. J., *Introduction to Partial Differential Equations,* Springer, 2014.

6. Prasad, P. and R. Ravindran, *Linear Partial Differential Equations,* Wiley Eastern, 1985.

7. Pinsky, M. A., *Partial Differential Equations and Boundary Value Problems with Applications,* 3rd edition, Waveland Press Inc., Prospect Heights, Illinois, 2003.

8. Raffoul, Y. N., *Class Notes on Partial Differential Equations,* University of Dayton, 2022.

9. Rauch, J., *Partial Differential Equations,* Springer-Verlag, 1991.

10. Ioannis P. Stavroulakis, I. P., and Tersian, S. A., *Partial differential equations,* World Scientific Publishing Co. Inc., River Edge, NJ, 2nd edition, 2004.

11. Strauss, W. A., *Partial Differential Equations,* 2nd edition, John Wiley & Sons Ltd., Chichester, 2008.

Chapter 3

1. Axler, S., *Linear Algebra Done Right,* 3rd revised edition, Springer, 2015.

2. Barker, G. P., and Schneider. H., *Matrices and Linear Algebra,* 2nd revised edition, Dover, 1989.

3. Boyd, S, and Vandenberghe, L., *Introduction to Applied Linear Algebra: Vectors, Matrices, and Least Square,* Cambridge University Press, 2018.

4. Cohen, M. X., *Linear Algebra: Theory, Intuition, Code,* sincXpress, 2021.

5. Cullen, C. G., *Matrices and Linear Transformations,* 2nd edition, Dover, 1990.

6. Friedberg, S. H., Insel, A. J, and Spence, L. E., *Linear Algebra,* 2nd edition, Prentice Hall, Englwood Cliffs, New Jersey, 1989.

7. Hildebrand, F. B., *Methods of Applied Mathematics,* Prentice-Hall, 1965.

8. Nering, E. D., *Linear Algebra and Matrix Theory,* 2nd edition, John Wiley & Sons, 1970.

9. O'Nan, M., *Linear Algebra,* 3rd edition, Hardcourt Brace Jovamovich, publishers and its subsidiary, Academic Press, 1990.

10. Raffoul, Y. N., *Class Notes on Linear Algebra and Matrices,* University of Dayton, 2022.

11. Shilov, G. E., *Linear Algebra,* Dover, 1977.

Chapter 4

1. Anderson, I. and Thompson, G., *The Inverse Problem of the Calculus of Variations for Ordinary Differential Equations,* Memoirs of the Amer. Math. Soc., vol. 98, No. 473, 1992.

2. Arfken, G., *Mathematical Methods for Physicists,* 2nd edition, Academic Press, 1970.

3. Arnold, V. I., *Mathematical Methods of Classical Mechanics,* Springer-Verlag, 1978.

4. Bliss, G. A., *Lectures on the Calculus of Variations,* University of Chicago Press, 1946.

5. Bolza, O., *Lectures on the Calculus of Variations,* G.E. Stechert and Co., 1931.

6. Brechtken-Manderscheid, U., *Introduction to the Calculus of Variations,* Chapman & Hall, 1991.

7. Carathéodory, C., *Calculus of Variations and Partial Differential Equations of the First Order,* Chelsea, 1982.

8. Ewing, G. M., *Calculus of Variations with Applications,* Dover, 1985.

9. Forsyth, A. R., *Calculus of Variations, Cambridge University Press,* 1927.

10. Fox, C., *An Introduction to the Calculus of Variations,* Dover, 1987.

11. Fulks, W., *Advanced Calculus,* 3rd edition, John Wiley, 1978.

12. Gelfand, I. M. and Fomin, S. V., *Calculus of Variations,* Prentice-Hall, 1963.

13. Giaquinta, M. and Hildebrandt, S., *Calculus of Variations I*: The Lagrangian Formalism, Springer-Verlag, 1996.

14. Giaquinta, M. and Hildebrandt, S., *Calculus of Variations II*: The Hamiltonian Formalism, Springer-Verlag, 1996.

15. Hildebrand, F. B., *Methods of Applied Mathematics,* Prentice-Hall, 1965.

16. Morse, M., *The Calculus of Variations in the Large,* American Math. Soc. Colloquium Pub., Vol. 18, 1932.

17. Pars, L. A., *A Treatise on Analytical Dynamics,* Heinemann, 1965.

18. Postnikov, M. M., *The Variational Theory of Geodesics,* Dover, 1983.

19. Raffoul, Y. N., *Advanced Differential Equations,* Elsevier/Academic Press, N Y, 2022.

20. Raffoul, Y. N., *Class Notes on Calculus of Variations,* University of Dayton, 2022.

21. Sagan, H., *Introduction to the Calculus of Variations,* Dover, 1992.

22. Wan, F. W., *Introduction to the Calculus of Variations and its Applications,* Chapman & Hall, 1995.

Chapter 5

1. Constanda., C., *Integral methods in science and engineering,* CRC, Press, 2000.

2. Hackbusch, W., *Integral Equations: Theory and Numerical Treatment,* Birkhäuser, 1995.

3. Hochstadt, H., *Integral Equations,* Wiley, 1973.

4. Colton, D., and Kress, R., *Integral equation methods in scattering theory: Classics In Applied mathematics,* SIAM, 2013.

5. Lovitt, W. V., *Linear Integral Equations,* Dover Publications Inc.: New York, 1950.

6. Mikhlin, S. G., *Linear Integral Equations,* Dover Publications, 2020.

7. Porter, D, Stirling, D. G., and et al. *Integral Equations: A Practical Treatment, from Spectral Theory to Applications,* Cambridge University Press, 1991.

8. Raffoul, Y. N., *Class Notes on Integral Equations,* University of Dayton, 2022.

9. Rahman, M., *Mathematical Methods with Applications,* WIT Press: Southampton, 2000.

10. Sharma, D. C., and Goyal, M. C., *Integral equations,* PHI Learning, Delhi, 2017.

11. Tricomi, F. G., *Integral Equations,* Dover, 1985.

12. Wazwaz, A. M., *A First Course in Integral Equations,* World Scientific: Singapore, 2015.

13. Yosida, K., *Lectures on differential and integral equations,* Dover Publications, 1991.

14. Zabreyko, P. P., *Integral equations: A Reference Text,* Springer 1976.

Index

Printed in the United States
by Baker & Taylor Publisher Services